Serono Symposia, USA
Norwell, Massachusetts

PROCEEDINGS IN THE SERONO SYMPOSIA, USA SERIES

GROWTH HORMONE II: Basic and Clinical Aspects
Edited by Barry B. Bercu and Richard F. Walker

TROPHOBLAST CELLS: Pathways for Maternal-Embryonic Communication
Edited by Michael J. Soares, Stuart Handwerger, and Frank Talamantes

IN VITRO FERTILIZATION AND EMBRYO TRANSFER IN PRIMATES
Edited by Don P. Wolf, Richard L. Stouffer, and Robert M. Brenner

OVARIAN CELL INTERACTIONS: Genes to Physiology
Edited by Aaron J.W. Hsueh and David W. Schomberg

CELL BIOLOGY AND BIOTECHNOLOGY: Novel Approaches to Increased Cellular Productivity
Edited by Melvin S. Oka and Randall G. Rupp

PREIMPLANTATION EMBRYO DEVELOPMENT
Edited by Barry D. Bavister

MOLECULAR BASIS OF REPRODUCTIVE ENDOCRINOLOGY
Edited by Peter C.K. Leung, Aaron J.W. Hsueh, and Henry G. Friesen

MODES OF ACTION OF GnRH AND GnRH ANALOGS
Edited by William F. Crowley, Jr., and P. Michael Conn

FOLLICLE STIMULATING HORMONE: Regulation of Secretion and Molecular Mechanisms of Action
Edited by Mary Hunzicker-Dunn and Neena B. Schwartz

SIGNALING MECHANISMS AND GENE EXPRESSION IN THE OVARY
Edited by Geula Gibori

GROWTH FACTORS IN REPRODUCTION
Edited by David W. Schomberg

UTERINE CONTRACTILITY: Mechanisms of Control
Edited by Robert E. Garfield

NEUROENDOCRINE REGULATION OF REPRODUCTION
Edited by Samuel S.C. Yen and Wylie W. Vale

FERTILIZATION IN MAMMALS
Edited by Barry D. Bavister, Jim Cummins, and Eduardo R.S. Roldan

GAMETE PHYSIOLOGY
Edited by Ricardo H. Asch, Jose P. Balmaceda, and Ian Johnston

GLYCOPROTEIN HORMONES: Structure, Synthesis, and Biologic Function
Edited by William W. Chin and Irving Boime

THE MENOPAUSE: Biological and Clinical Consequences of Ovarian Failure: Evaluation and Management
Edited by Stanley G. Korenman

Serono Symposia, USA
Norwell, Massachusetts

Barry B. Bercu Richard F. Walker
Editors

Growth Hormone II

Basic and Clinical Aspects

With 101 Figures

Springer-Verlag
New York Berlin Heidelberg London Paris
Tokyo Hong Kong Barcelona Budapest

Barry B. Bercu, M.D.
Richard F. Walker, Ph.D.
Pediatric Endocrinology
All Children's Hospital
St. Petersburg, FL 33701
USA

Proceedings of the Symposium on Growth Hormone II: Basic and Clinical Aspects, sponsored by Serono Symposia, USA, held December 3 to 6, 1992, in Tarpon Springs, Florida.

For information on previous volumes, please contact Serono Symposia, USA.

Library of Congress Cataloging-in-Publication Data
Growth hormone II: basic and clinical aspects/edited by Barry B. Bercu and Richard F. Walker; Serono Symposia, USA.
p. cm.
Proceedings of the Second International Symposium on Growth Hormone, held Dec. 3–6, 1992, in Tarpon Springs, Fla.
Includes bibliographical references and index.
ISBN-13: 978-1-4613-8374-1 e-ISBN-13: 978-1-4613-8372-7
DOI: 10.1007/978-1-4613-8372-7
1. Somatotropin—Physiological effect—Congresses.
2. Somatotropin—Therapeutic use—Congresses. I. Bercu, Barry B.
II. Walker, Richard F., 1939- . III. Serono Symposia, USA.
IV. International Symposium on Growth Hormone (Serono Symposia, USA)
(2nd: 1992: Tarpon Springs, Fla.) V. Title: Growth hormone 2.
[DNLM: 1. Somatotropin—physiology—congresses. 2. Somatostatin—physiology—congresses. WK 515 G8843 1992]
QP572.S6G756 1993
612.6—dc20
DNLM/DLC 93-28882

Printed on acid-free paper.

Softcover reprint of the hardcover 1st edition 1994

Production coordinated by Marilyn Morrison and managed by Francine McNeill; manufacturing supervised by Vincent Scelta.
Typeset by Best-set Typesetter Ltd., Hong Kong.

9 8 7 6 5 4 3 2 1

SYMPOSIUM ON GROWTH HORMONE II: BASIC AND CLINICAL ASPECTS

Scientific Committee

Barry B. Bercu, M.D., Chairman
St. Petersburg, Florida

Richard F. Walker, Ph.D., Chairman
St. Petersburg, Florida

Cyril Y. Bowers, M.D.
New Orleans, Louisiana

Kazuo Chihara, M.D.
Kobe, Japan

Jacob Kraicer, M.D., Ph.D.
London, Ontario, Canada

Michael A. Preece, M.D.
London, United Kingdom

Ron G. Rosenfeld, M.D.
Stanford, California

Organizing Secretary

Bruce K. Burnett, Ph.D.
Serono Symposia, USA
100 Longwater Circle
Norwell, Massachusetts

Preface

This book is a compendium of proceedings from the Symposium on Growth Hormone II: Basic and Clinical Aspects. The intent of the symposium was to review current aspects of the rapidly expanding science of growth hormone (GH) neuroendocrinology, especially those developments that occurred since our last meeting in 1985. The meeting was timely since there have been many new and exciting developments in basic research on the GH axis, and several new clinical applications for GH other than growth have been identified. For example, significant strides have been made into use of the bioengineered hormone as an anabolic in debilitated states and as an immunorestorative agent. Both these areas have received most attention as they relate to geriatric populations. Furthermore, recent focus on a novel family of GH secretagogues that complement the action of GHRH has shown that in animal models, these compounds have the potential to restore GH secretion in deficient or insufficient states, including growth retardation, obesity, and aging. Accordingly, these xenobiotics may have significant potential as drug products to stimulate physiological GH secretion and, thus, resist or reverse some of the maladaptive changes in form and function resulting from GH insufficiency. The importance of these agents as drug candidates is underscored by the diversity of molecules that have been studied by various pharmaceutical companies and the recent discovery that non-peptide analogs with high oral bioavailability can be synthesized.

The format of the book follows that of the symposium, which was designed to proceed from a background of neuroendocrine regulatory processes incorporating the roles of GHRH and somatostatin (SRIF) as major controllers of GH secretion. Such information is the basis for a subsequent analysis of the cellular and molecular mechanisms by which these compounds regulate GH secretion, also incorporating information on how structural modification can enhance GH secretagogue potency and how the binding characteristics of SRIF vary in the brain and pituitary. Presumably, these differences subserve different functions ranging from modulation of GH secretion to control of behavioral events. The book then proceeds with chapters on non-growth-related functions of GH. A major focus is placed on GH and immune function, progressing from an analysis of how GH affects immunity to identification of GH and GHRH within cellular elements of the immune system. Also analyzed are the potential uses of GH for immunoreconstitution as an approach for treating certain diseases or preventing disease in individuals with immune dysfunction. From there, the focus is on generalized effects of GH in the aged, ranging from its anabolic effects in the frail elderly to treatment of osteoporosis and hypercholesterolemia. Chapters are also devoted to the subject of GH insufficiency, especially involving loss of pituitary responsiveness to GHRH resulting from receptor or second messenger defects and how these factors contribute to the process of senescence. Incorporated into the theme of the chapters on immune function and aging is a section devoted to structural and functional analyses of the family of novel GH secretagogues that, in combination with GHRH, constitute a potential therapy for various aspects of GH insufficiency. The last section deals with current clinical studies involving GH and IGF-I ranging from acromegaly to dwarfism.

The book, therefore, covers a broad spectrum of topics in GH neuroendocrinology from the theoretical to the practical with contributions from an outstanding group of scientists and physicians involved in state-of-the-art studies. It is our hope that the proceedings from this meeting provide a view of current issues that are of interest to basic as well as clinical scientists. We thank Serono Symposia, USA for making this valuable project possible.

Barry B. Bercu
Richard F. Walker

Contents

Preface vii
Contributors xiii

Part I. Somatostatin/GRF Regulation

1. Pharmacology of GRF and Somatostatin: Positive and Negative Control of GH Release 3
DAVID H. COY, WILLIAM A. MURPHY, SIMON J. HOCART, AND JOHN TAYLOR

2. Ionic Mechanisms Governing the Control of Growth Hormone Secretion by Somatostatin 17
JACOB KRAICER AND STEPHEN M. SIMS

3. Biochemical and Functional Properties of Somatostatin Receptors 33
TERRY REISINE, KAREN RAYNOR, HAEYOUNG KONG, SUSAN F. LAW, MAGALI THEVENIAU, MELANIE TALLENT, STEPHANIE RENS-DOMIANO, JOHN HINES, KAZUKI YASUDA, AND GRAEME I. BELL

4. Regulation of Growth Hormone Output: The GRF Signal ... 47
I.C.A.F. ROBINSON

5. Regulation of Growth Hormone Secretion in Man 66
MARY LEE VANCE

6. Mechanisms for Insulin-Like Growth Factor I Regulation of Growth Hormone Secretion 73
DIANE PRAGER AND SHLOMO MELMED

Part II. The Immune System and Growth Hormone Interaction

7. Role of Growth Hormone and Insulin-Like Growth Factor I in Immunoregulation 87
KEITH W. KELLEY, SEAN ARKINS, AND YONG MING LI

8. Production and Function of Growth Hormone in the Immune System ... 104
Douglas A. Weigent and J. Edwin Blalock

9. Immunomodulating Role of Growth Hormone ... 116
Nicola Fabris and Eugenio Mocchegiani

Part III. Novel Aspects of Growth Hormone Research I

10. Further Observations on the Effects of Human Growth Hormone in Elderly Hyposomatomedinemic Men ... 133
Lester Cohn, Axel G. Feller, Inge W. Rudman, and Daniel Rudman

11. Effects of Growth Hormone in Older People ... 150
Robert Marcus, Leah Holloway, and Gail E. Butterfield

12. Alteration of Growth Hormone Secretion in Aging: Peripheral Effects ... 155
Thierry Abribat, Nathalie Deslauriers, Isabelle Reeves, Dominique R. Garrel, and Paul Brazeau

13. Effects of Growth Hormone Releasing Peptides on Stimulated Growth Hormone Secretion in Old Rats ... 167
Richard F. Walker, Sei-Won Yang, Ryuji Masuda, Cheng-Shih Hu, and Barry B. Bercu

14. Changes in Growth Hormone Secretory Dynamics in Chronic Renal Failure and in Adaptation to Moderately High Altitude Living ... 193
German Ramirez

Part IV. Novel Aspects of Growth Hormone Research II

15. Atypical Growth Hormone Releasing Peptides ... 203
C.Y. Bowers, K. Veeraragavan, and K. Sethumadhavan

16. Growth Hormone Releasing Hexapeptide: Basic Research and Clinical Application ... 223
Kazuo Chihara, Hidesuke Kaji, Satoko Hayashi, Hiroji Yagi, Yasuhiro Takeshima, Maki Mitani, Shin-ichiro Ohashi, and Hiromi Abe

17. Molecular Endocrinology of Human Growth Hormone ... 231
James Wells, Brian Cunningham, Germaine Fuh, Henry Lowman, Mark Ultsch, and Bart de Vos

18. Growth Hormone Binding Protein: A Modulator of Growth Hormone Activity 242
Ross Clark

Part V. Clinical Application of Growth Hormone Treatment

19. Growth Hormone Treatment, Acromegaly, and Relationship to Cancer and Leukemia 259
Louis E. Underwood

20. Clinical Applications of Growth Hormone Treatment in Childhood 269
M.A. Preece

21. Medical and Scientific Aspects of Growth Hormone Receptor Deficiency (Laron Syndrome) in Ecuador 279
Pinchas Cohen, Jaime Guevara-Aguirre, Paul J. Fielder, Kristin Wilson, Sharron E. Gargosky, Mary A. Vaccarello, Frank B. Diamond, Jr., Arlan L. Rosenbloom, and Ron G. Rosenfeld

22. Spontaneous Growth Hormone Secretion in Children With Normal and Abnormal Growth 286
E. Verena Jorgensen, Dorothy I. Shulman, Frank B. Diamond, Jr., Allen W. Root, and Barry B. Bercu

Part VI. Poster Presentation Manuscripts

23. Expression of the Short Isoform of the Growth Hormone Receptor in Adipocytes 301
H. Maurice Goodman, G. Peter Frick, Tova Bick, Lih-Ruey Tai, and Jack L. Leonard

24. Effects of Porcine Somatotropin on Thymic Weight, Thymosin α1, and Thymosin β4 in Gilts and Barrows 317
T. Wise, J. Klindt, F.C. Buonomo, and J.T. Yen

25. Effects of an Acetylcholine Precursor on GH Secretion in Elderly Subjects 328
Gian Paolo Ceda, Gian Piero Marzani, Valeria Tontodonati, Emanuela Piovani, Alberto Banchini, Maria Teresa Baffoni, Giorgio Valenti, and Andrew R. Hoffman

26. Even Short-Term hGH Treatment Induces Endocrine Changes in the Elderly ... 338
F. Strollo, A. Semprini, G. Strollo, M. Morè, L. Bollanti, A. Ciarmatori, D. Pizzabiocca, N. Mangrossa, and G. Riondino

27. Dichotomic Role of Glucocorticoids in the Regulation of Growth Hormone Secretion In Vivo ... 347
Andrea Giustina, Anna Rosa Bussi, and William B. Wehrenberg

Author Index ... 359
Subject Index ... 361

Contributors

HIROMI ABE, Third Division, Department of Medicine, Kobe University School of Medicine, Kobe, Japan.

THIERRY ABRIBAT, Neuroendocrinology Laboratory, Notre-Dame Hospital Research Center, Montreal, Quebec, Canada.

SEAN ARKINS, Department of Animal Sciences, Laboratory of Immunophysiology, University of Illinois, Urbana, Illinois, USA.

MARIA TERESA BAFFONI, Department of Geriatrics, University of Parma, Parma, Italy.

ALBERTO BANCHINI, Department of Geriatrics, University of Parma, Parma, Italy.

GRAEME I. BELL, Departments of Biochemistry and Molecular Biology and Medicine, Howard Hughes Medical Institute, University of Chicago, Chicago, Illinois, USA.

BARRY B. BERCU, Department of Pediatrics, University of South Florida, Tampa, Florida, and All Children's Hospital, St. Petersburg, Florida, USA.

TOVA BICK, Department of Physiology, University of Massachusetts Medical School, Worcester, Massachusetts, USA.

J. EDWIN BLALOCK, Department of Physiology and Biophysics, University of Alabama, Birmingham, Alabama, USA.

LUCILLA BOLLANTI, Endocrinology Department, Endocrine Unit, Italian National Research Centers on Aging (INRCA), Rome, Italy.

C.Y. BOWERS, Department of Medicine, Tulane University School of Medicine, New Orleans, Louisiana, USA.

PAUL BRAZEAU, Neuroendocrinology Laboratory, Notre-Dame Hospital Research Center, Montreal, Quebec, Canada.

F.C. BUONOMO, Monsanto Agriculture Company, St. Louis, Missouri, USA.

ANNA ROSA BUSSI, Medical Clinic, University of Brescia, Brescia, Italy.

GAIL E. BUTTERFIELD, Geriatrics Research, Education and Clinical Center, VA Medical Center, Palo Alto, California, USA.

GIAN PAOLO CEDA, Department of Geriatrics, University of Parma, Parma, Italy.

KAZUO CHIHARA, Third Division, Department of Medicine, Kobe University School of Medicine, Kobe, Japan.

ANNA CIARMATORI, Endocrinology Department, Endocrine Unit, Italian National Research Centers on Aging (INRCA), Rome, Italy.

ROSS CLARK, Department of Endocrine Research, Genentech, Inc., South San Francisco, California, USA.

PINCHAS COHEN, Department of Pediatrics, University of Pennsylvania School of Medicine, Philadelphia, Pennsylvania, USA.

LESTER COHN, Department of Medicine, Chicago Medical School, and Medical Service, Veterans Affairs Medical Center, North Chicago, Illinois, USA.

DAVID H. COY, Peptide Research Laboratories, Tulane University School of Medicine, New Orleans, Louisiana, USA.

BRIAN CUNNINGHAM, Department of Protein Engineering, Genentech, Inc., South San Francisco, California, USA.

NATHALIE DESLAURIERS, Wyeth Ayerst Research, St-Laurent, Quebec, Canada.

BART DE VOS, Department of Protein Engineering, Genentech, Inc., South San Francisco, California, USA.

FRANK B. DIAMOND, JR., Department of Pediatrics, University of South Florida, Tampa, Florida, and All Children's Hospital, St. Petersburg, Florida, USA.

NICOLA FABRIS, Immunology Center, Gerontological Research Department, Italian National Research Centers on Aging (INRCA), Ancona, and Institute of Pharmacology II, University of Pavia, Pavia, Italy.

AXEL G. FELLER, Department of Medicine, Chicago Medical School, and Medical Service, Veterans Affairs Medical Center, North Chicago, Illinois, USA.

PAUL J. FIELDER, Department of Pediatrics, Stanford University School of Medicine, Stanford, California, USA.

G. PETER FRICK, Department of Physiology, University of Massachusetts Medical School, Worcester, Massachusetts, USA.

GERMAINE FUH, Department of Protein Engineering, Genentech, Inc., South San Francisco, California, USA.

SHARRON E. GARGOSKY, Department of Pediatrics, Stanford University School of Medicine, Stanford, California, USA.

DOMINIQUE R. GARREL, Burn Center, Hotel Dieu Hospital, Montreal, Quebec, Canada.

ANDREA GIUSTINA, Medical Clinic, University of Brescia, Brescia, Italy.

H. MAURICE GOODMAN, Department of Physiology, University of Massachusetts Medical School, Worcester, Massachusetts, USA.

JAIME GUEVARA-AGUIRRE, Institute of Endocrinology, Metabolism and Reproduction, Quito, Ecuador.

SATOKO HAYASHI, Division of Metabolism and Nutrition, International Center for Exchanging Medical Research, Kobe University School of Medicine, Kobe, Japan.

JOHN HINES, Department of Pharmacology, University of Pennsylvania School of Medicine, Philadelphia, Pennsylvania, USA.

SIMON J. HOCART, Peptide Research Laboratories, Tulane University School of Medicine, New Orleans, Louisiana, USA.

ANDREW R. HOFFMAN, Medical Service, VA Medical Center, Palo Alto, California, and Department of Medicine, Stanford University Medical Center, Stanford, California, USA.

Leah Holloway, Geriatrics Research, Education and Clinical Center, VA Medical Center, Palo Alto, California, USA.

Cheng-Shih Hu, Department of Pediatrics, University of South Florida, Tampa, Florida, and All Children's Hospital, St. Petersburg, Florida, USA.

E. Verena Jorgensen, Department of Pediatrics, University of South Florida, Tampa, Florida, and All Children's Hospital, St. Petersburg, Florida, USA.

Hidesuke Kaji, Division of Metabolism and Nutrition, International Center for Exchanging Medical Research, Kobe University School of Medicine, Kobe, Japan.

Keith W. Kelley, Department of Animal Sciences, Laboratory of Immunophysiology, University of Illinois, Urbana, Illinois, USA.

J. Klindt, USDA-ARS, Roman L. Hruska U.S. Meat Animal Research Center, Clay Center, Nebraska, USA.

Haeyoung Kong, Department of Pharmacology, University of Pennsylvania School of Medicine, Philadelphia, Pennsylvania, USA.

Jacob Kraicer, Department of Physiology, The University of Western Ontario, London, Ontario, Canada.

Susan F. Law, Graduate Group in Cell Biology, University of Pennsylvania, Philadelphia, Pennsylvania, USA.

Jack L. Leonard, Departments of Nuclear Medicine and Physiology, University of Massachusetts Medical School, Worcester, Massachusetts, USA.

Yong Ming Li, Department of Animal Sciences, Laboratory of Immunophysiology, University of Illinois, Urbana, Illinois, USA.

Henry Lowman, Department of Protein Engineering, Genentech, Inc., South San Francisco, California, USA.

Novella Mangrossa, Hormone Laboratory, Endocrine Unit, Italian National Research Centers on Aging (INRCA), Rome, Italy.

Robert Marcus, Stanford University, Stanford, California, and Geriatrics Research, Education and Clinical Center, VA Medical Center, Palo Alto, California, USA.

Gian Piero Marzani, Department of Geriatrics, University of Parma, Parma, Italy.

Ryuji Masuda, Department of Pediatrics, University of South Florida, Tampa, Florida, and All Children's Hospital, St. Petersburg, Florida, USA.

Shlomo Melmed, Division of Endocrinology and Metabolism, Cedars-Sinai Medical Center, Los Angeles, California, USA.

Maki Mitani, Division of Metabolism and Nutrition, International Center for Exchanging Medical Research, Kobe University School of Medicine, Kobe, Japan.

Eugenio Mocchegiani, Immunology Center, Gerontological Research Department, Italian National Research Centers on Aging (INRCA), Ancona, Italy.

Massimo Morè, Endocrinology Department, Endocrine Unit, Italian National Research Centers on Aging (INCRA), Rome, Italy.

William A. Murphy, Peptide Research Laboratories, Tulane University School of Medicine, New Orleans, Louisiana, USA.

Shin-ichiro Ohashi, Third Division, Department of Medicine, Kobe University School of Medicine, Kobe, Japan.

Emanuela Piovani, Department of Geriatrics, University of Parma, Parma, Italy.

Donatella Pizzabiocca, Clinical Biochemistry Laboratory, Italian National Research Centers on Aging (INRCA), Rome, Italy.

Diane Prager, Division of Endocrinology and Metabolism, Cedars-Sinai Medical Center, Los Angeles, California, USA.

M.A. Preece, Clinical Trials Unit, Division of Public Health, Institute of Child Health, University of London, London, UK.

German Ramirez, Division of Nephrology and Hypertension, College of Medicine, University of South Florida, and the James A. Haley Veterans Administration Hospital, Tampa, Florida, USA.

Karen Raynor, Department of Pharmacology, University of Pennsylvania School of Medicine, Philadelphia, Pennsylvania, USA.

Isabelle Reeves, Neuroendocrinology Laboratory, Notre-Dame Hospital Research Center, Montreal, Quebec, Canada.

Terry Reisine, Department of Pharmacology, University of Pennsylvania School of Medicine, Philadelphia, Pennsylvania, USA.

Stephanie Rens-Domiano, Department of Pharmacology, University of Pennsylvania School of Medicine, Philadelphia, Pennsylvania, USA.

Giuseppe Riondino, Endocrine Unit, Italian National Research Centers on Aging (INRCA), Rome, Italy.

I.C.A.F. Robinson, Division of Neurophysiology and Neuropharmacology, National Institute for Medical Research, London, UK.

Allen W. Root, Department of Pediatrics, University of South Florida, Tampa, Florida, and All Children's Hospital, St. Petersburg, Florida, USA.

Arlan L. Rosenbloom, Department of Pediatrics, University of Florida, Gainesville, Florida, USA.

Ron G. Rosenfeld, Department of Pediatrics, Stanford University School of Medicine, Stanford, California, USA.

Daniel Rudman, Department of Medicine, Medical College of Wisconsin, and Medical Service, Veterans Affairs Medical Center, Milwaukee, Wisconsin, USA.

Inge W. Rudman, Department of Medicine, Medical College of Wisconsin, and Medical Service, Veterans Affairs Medical Center, Milwaukee, Wisconsin, USA.

Alberto Semprini, Clinical Biochemistry Laboratory, Italian National Research Centers on Aging (INRCA), Rome, Italy.

K. Sethumadhavan, Department of Medicine, Tulane University School of Medicine, New Orleans, Louisiana, USA.

Dorothy I. Shulman, Department of Pediatrics, University of South Florida, Tampa, Florida, and All Children's Hospital, St. Petersburg, Florida, USA.

Stephen M. Sims, Department of Physiology, The University of Western Ontario, London, Ontario, Canada.

FELICE STROLLO, Hormone Laboratory, Endocrine Unit, Italian National Research Centers on Aging (INRCA), Rome, Italy.

GIOVANNA STROLLO, Diabetology Department, Endocrine Unit, Italian National Research Centers on Aging (INRCA), Rome, Italy.

LIH-RUEY TAI, Department of Physiology, University of Massachusetts Medical School, Worcester, Massachusetts, USA.

YASUHIRO TAKESHIMA, Division of Metabolism and Nutrition, International Center for Exchanging Medical Research, Kobe University School of Medicine, Kobe, Japan.

MELANIE TALLENT, Institute of Neurological Sciences, University of Pennsylvania School of Medicine, Philadelphia, Pennsylvania, USA.

JOHN TAYLOR, Biomeasure, Inc., Hopkinton, Massachusetts, USA.

MAGALI THEVENIAU, Department of Pharmacology, University of Pennsylvania School of Medicine, Philadelphia, Pennsylvania, USA.

VALERIA TONTODONATI, Department of Geriatrics, University of Parma, Parma, Italy.

MARK ULTSCH, Department of Protein Engineering, Genentech, Inc., South San Francisco, California, USA.

LOUIS E. UNDERWOOD, Department of Pediatrics, University of North Carolina at Chapel Hill, Chapel Hill, North Carolina, USA.

MARY A. VACCARELLO, Department of Pediatrics, University of Florida, Gainesville, Florida, USA.

GIORGIO VALENTI, Department of Geriatrics, University of Parma, Parma, Italy.

MARY LEE VANCE, Department of Internal Medicine, University of Virginia Health Sciences Center, Charlottesville, Virginia, USA.

K. VEERARAGAVAN, Department of Medicine, Tulane University School of Medicine, New Orleans, Louisiana, USA.

RICHARD F. WALKER, Department of Pediatrics, University of South Florida, Tampa, Florida, and All Children's Hospital, St. Petersburg, Florida, USA.

William B. Wehrenberg, Department of Health Sciences, University of Wisconsin, Milwaukee, Wisconsin, USA.

Douglas A. Weigent, Department of Physiology and Biophysics, University of Alabama, Birmingham, Alabama, USA.

James Wells, Department of Protein Engineering, Genentech, Inc., South San Francisco, California, USA.

Kristin Wilson, Department of Pediatrics, Stanford University School of Medicine, Stanford, California, USA.

T. Wise, USDA-ARS, Roman L. Hruska U.S. Meat Animal Research Center, Clay Center, Nebraska, USA.

Hiroji Yagi, Division of Metabolism and Nutrition, International Center for Exchanging Medical Research, Kobe University School of Medicine, Kobe, Japan.

Sei-Won Yang, Department of Pediatrics, University of South Florida, Tampa, Florida, and All Children's Hospital, St. Petersburg, Florida, USA.

Kazuki Yasuda, Departments of Biochemistry and Molecular Biology and Medicine, Howard Hughes Medical Institute, University of Chicago, Chicago, Illinois, USA.

J.T. Yen, USDA-ARS, Roman L. Hruska U.S. Meat Animal Research Center, Clay Center, Nebraska, USA.

Part I

Somatostatin/GRF Regulation

1

Pharmacology of GRF and Somatostatin: Positive and Negative Control of GH Release

DAVID H. COY, WILLIAM A. MURPHY, SIMON J. HOCART, AND JOHN TAYLOR

The physiological mechanism(s) controlling the release of *growth hormone* (GH) from the adenohypophysis is unique in having two hormones, *growth hormone releasing factor* (GRF) (1, 2) and *somatostatin* (SRIF) (3), that cause either the release or inhibition of release of GH, respectively. In addition to these two primary control substances, however, many other events also impinge on the process, some of which are summarized in Figure 1.1. These events include negative feedback loops involving GRF, SRIF, GH, and IGF, as well as the involvement of additional peptide factors such as bombesin/gastrin releasing peptide, galanin, and the opiate peptides (in which group can be classified the synthetic *growth hormone releasing peptides* [GHRPs], which are also discussed in another chapter). Additionally, it is recently apparent that SRIF analogs can act at the hypothalamic level to inhibit the release of GRF directly (4).

In view of the great physiological importance of SRIF and GRF, both have been the subject of extensive structure-activity relationship investigations utilizing synthetic analogs. These have been aimed at elucidating receptor binding/activation mechanisms and the development of more therapeutically interesting compounds, such as those possessing improved potency, specificity, or pharmacokinetic properties.

Somatostatin

The SRIF cyclic tetradecapeptide was the first of the hypothalamic peptides associated with GH control to be isolated and subsequently became the first of the now ubiquitous neuro-GI family of peptides when

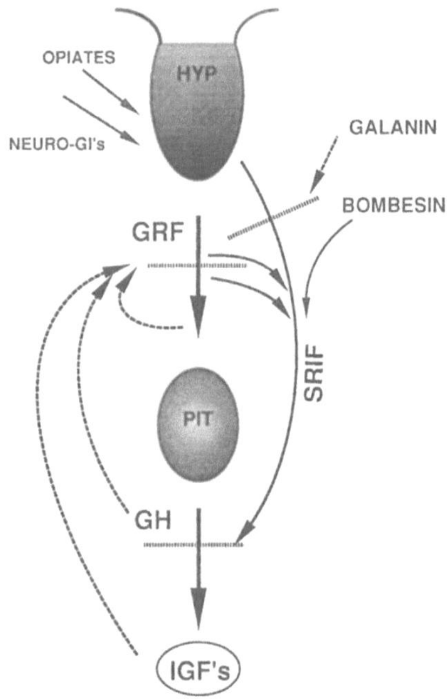

FIGURE 1.1. Stylized representation of some of the peptidic factors influencing the release of pituitary GH secretion.

it was also found to inhibit the secretion of many gastric and pancreatic peptides. The large body of literature that has accumulated on the chemistry and biology of SRIF since its discovery in 1972 (3) attests to its enormous importance in the control of numerous physiological processes. The history of the isolation and the many inhibitory biological activities of somatostatin are well known and have been the subject of several comprehensive reviews (5–8). SRIF pharmacology has also recently received a further stimulus from the cloning and characterization of several additional receptor subtypes (9, 10) coupled to different cellular effector systems.

Short, Potent, Cyclic SRIF Analogs

Its lack of selectivity and short plasma half-life were major obstacles in clinical development of SRIF, but these shortcomings have led to numerous successful structure-activity studies aimed toward their elimination. Most of the more interesting analogs are all shortened, conformationally restricted, cyclic compounds. Studies (11) on the essential nature of each amino acid residue of SRIF quite rapidly revealed that not only could the extracyclic Ala-Gly residues be removed without loss of activity, but also certain residues within the ring system could be deleted with retention of some biological activity. Following this approach to its logical conclusion, Vale et al. (12) demonstrated that the shortest active sequence resided in the octapeptide oxi(Cys-Phe-Phe-D-Trp-Lys-Thr-Phe-Cys). This peptide, while less effective than SRIF itself on inhibition of GH release, had the advantage of being longer acting. These results clearly showed that the 6–11 region, present in the correct conformation, was all that was necessary, at least for pituitary receptor recognition.

A more elegant approach to shorter, active analogs was based on a series of conformational measurements made by Holladay and Puett (13, 14). These measurements suggested the presence of a type II′ β-turn about the 7–11 amino acids. Indeed, it appears that the superior potency of [D-Trp^8]-SRIF (15) results from the stabilization of this turn. Veber et al. (16) realized that this would result in close proximity of the 5–12 and 6–11 Phe amino acid side chains (Fig. 1.2). Bicyclic analogs were than prepared in which disulfide bridges could be formed between Cys residues inserted in these positions with complete retention of activity (17, 18). In fact, the disulfide bridge of one analog could be replaced with 6 methylene groups (present in 7-aminoheptanoic acid or Ahep) and the rest of the molecule dispensed with and still result in an equally active compound (19).

Such small, restricted cyclic peptides offered excellent starting points for computer conformational analysis and modeling, which has resulted in further structural simplification (20) to a simple cyclic hexapetide containing two β-turns, one stabilized by D-Trp in the traditional sense and one by inclusion of a Pro residue, cyclo(Pro-Phe-D-Trp-Lys-Thr-Phe). This hexapeptide was about 5 times more active than SRIF in several test systems. Since the design of this hexapeptide, much research has been done on the optimization of biological activity, most interestingly through use of more superior β-turn stabilizers than Pro. The best appeared to be N-Me-Ala (21), which results in a further 10-fold increase in activity in the hexapeptide cyclo(N-Me-Ala-Tyr-D-Trp-Lys-Val-Phe), which also contains the additional substitutions of Tyr for Phe and Val for Thr (21). This particular analog was also extremely stable metabolically and,

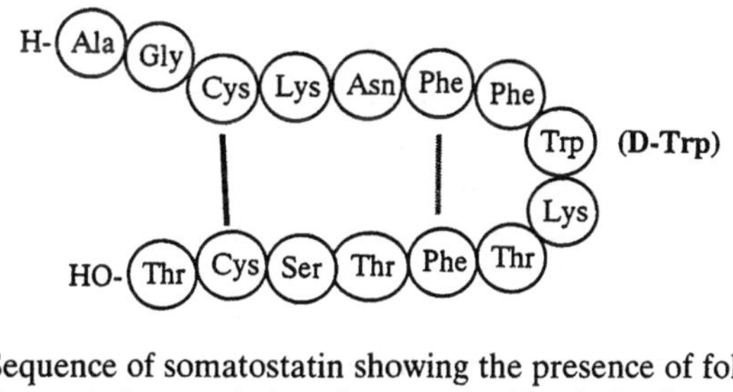

FIGURE 1.2. Sequence of somatostatin showing the presence of folding around the Trp-Lys residue and the proximity of aromatic side chains in positions 6 and 11 later taken advantage of in the design of shorter analogs.

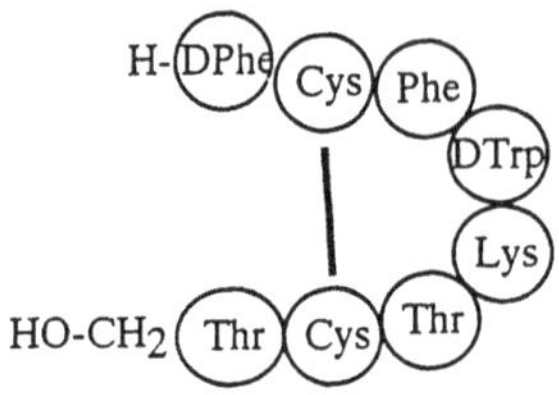

FIGURE 1.3. Sequence of the potent somatostatin analog octreotide that possesses the same type of folding as somatostatin and S-S bridging in the somatostatin position 6 and 11 equivalents.

although slowly well absorbed after oral administration, was subject to rapid biliary excretion. Bioavailability was thus only 1%–3% of administered material.

Along similar lines of development, another attempt at the rational design of short analogs began with the hexapeptide structure oxi(Cys-Phe-D-Trp-Lys-Thr-Cys) (22), in which Cys residues and a covalent S-S bridge essentially replace two proximal Phe residues (Fig. 1.2). Although the hexapeptide itself exhibited only trace SRIF activity, it was found that selective addition of certain amino acid residues to the N- and C-terminus dramatically improved all activity parameters. The best reported structure contained D-Phe in position 1 and threinol in position 8 (Fig. 1.3). This compound was about 20 times more active than SRIF in vivo and, as with other analogs of this type, was long acting. It is usually referred to as SMS 201-995 or octreotide and has undergone extensive clinical investigation. It has also served as a lead structure for the design of additional analogs such as lanreotide, oxi(D-Nal-Cys-Tyr-D-Trp-Lys-Val-Cys-Thr-NH_2), which is an even longer-acting and more receptor-specific analog (23). This type of peptide has been used extensively in oncological studies that have been reviewed (24), and significant binding to and antiproliferative effects against many tumor types have been observed (25–29).

These two families of analogs are still the subjects of in-depth structure-activity studies where the conformations of the various side chains (too flexible to be measured by physicochemical methods in the peptides discussed) are being probed by introducing additional conformational restraint either by substituting a and b positions with bulky Me groups or by tethering amino acid side chains to the peptide backbone. For instance, in our laboratories analogs have been recently synthesized containing D- or L-2,3,4,5,-tetrahydro-β-carboline-4-carboxylic acid (Tcc), which is an analog of Trp in which side chain movement is severely restrained by indole-methylene bridging to the amino group (Fig. 1.4). Neither oxi(D-Nal-Cys-Tyr-D-Tcc-Lys-Val-Cys-Thr) or its L-Tcc counterpart exhibited appreciable biological activity on rat pituitary cells, thus providing useful negative information regarding the desired orientation of the indole side chain for this receptor(s).

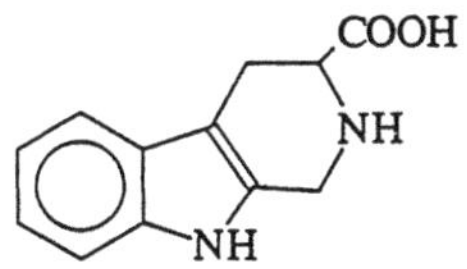

2,3,4,5-tetrahydro-β-carboline-4-carboxylic acid (Tic)

FIGURE 1.4. Structure of the "tethered" amino acid 2,3,4,5-tetrahydro-β-carboline-4-carboxylic acid (Tcc), one of a family of conformationally restrained amino acids proving of great value in analyzing the receptor-bound conformation of somatostatin.

Short, Potent, Linear SRIF Analogs

Recently, further simplifications to the SRIF structure have been developed in our laboratory (30, 31) that make use of the previously discussed proximity of the Phe side chains in positions 6 and 11, together with the conformational stabilization provided by a D-Trp residue in position 8 (Fig. 1.2). Enhancement of π-π electronic interactions between these aromatic rings, rather than the traditional covalent bond between residues in these positions, can result in extremely potent linear octapeptide analogs such as D-Phe-Cpa-Tyr-D-Trp-Lys-Val-Phe-Thr-NH_2 (Cpa = 4-chloro-Phe), which adopts a similar solution conformation to its cyclic counterparts. The Cpa and Phe side chains have been shown by low-temperature NMR studies (31) to stack as expected. This peptide is almost equipotent to SRIF in binding to pancreatic and small-cell lung carcinoma SRIF receptors and inhibiting GH release from rat pituitary cells. Aromatic side chain interactions between N- and C-terminal amino acid residues of these octapeptides could also be utilized to produce potent linear analogs typified by D-Phe-Ala-Tyr-D-Trp-Lys-Val-Ala-D-Nal-NH_2 (D-Nal = D-β-naphthyl-2-Ala), which also has good biological activity and high affinity for SSTR2-type receptors. It is expected that this type of compound, offering ease of synthesis and flexible rather than constrained design possibilities, will be extremely useful for probing the ligand recognition requirements for the numerous SRIF receptor subtypes presently being characterized.

SRIF Analogs that Function as Ligands for Other Peptide Receptors

Folded chain conformations containing hairpin turns, such as the type II′ β-bend in SRIF and its analogs, appear to be a very common motif in neuropeptides and, as such, it is perhaps not surprising that certain SRIF cyclic octapeptide analogs have been found to bind quite well to completely unrelated peptide receptors. In the octreotide series, Maurer et al. (32) found that Tyr-containing sequences displayed affinity for μ opiate receptors and that they were, in fact, opiate antagonists. This work was extended (33) to an investigation of the function of each residue in lanreotide with respect to opiate binding. D-Trp in position 8 was not essential, which differentiates structure-activity requirements totally from those required for SRIF activity. However, the Tyr residue was required, as was a free N-terminal amino group. These are, conversely, not requirements for SRIF activity. It appears, then, that the same backbone conformation is present in all of these peptides, but that entirely different side chains are utilized in binding to SRIF or μ receptors.

Another more recent example of this phenomenon was just discovered in our laboratories during the screening of a library of peptide analogs for displacement of ^{125}I-neuromedin B from specific receptors present on rat olfactory bulb cell membranes. It was found that oxi(D-Nal-Cys-Tyr-D-Trp-Lys-Val-Cys-Nal-NH_2) uniquely exhibited high-affinity binding with an IC_{50} of around 40 nM and, furthermore, that it was devoid of neuromedin B biological activity. This is the first potent competitive receptor antagonist of neuromedin B to be discovered, since the highly potent classical bombesin/gastrin releasing peptide antagonists have little affinity for this receptor (34). The presence of the two Nal residues, presumably occurring in a stacked configuration, was extremely important for binding affinity. This analog had almost no affinity for bombesin/GRP receptors. Thus, once again the value of this folded, constrained backbone as a template for ligand formation for other peptide systems is illustrated.

Receptor Antagonists of SRIF

There are now few small biologically active peptides for which potent competitive receptor antagonists have not been found. Thus, the lack of progress in discovering an SRIF antagonist was discouraging to this and other laboratories, particularly since such a compound might be of therapeutic value in neutralizing the negative effects of SRIF on GH release. In 1982 (35–37), we believed that a major lead had been made with the discovery of a pentapeptide SRIF analog having the curious structure c(Phe-D-Trp-Lys-Thr[Benzyl]-Phe) (the benzyl group [Bzl] is present as a protecting group on the OH group of the Thr side chain), which was closely related to the SRIF hexapeptide agonists already discussed. During routine analog screening in vivo in the rat, it was noticed that this particular compound actually caused release rather than the usual inhibition of release of GH at some doses. A typical dose response is shown in Table 1.1, and it can be seen that at higher doses the effect is lost. A logical explanation for this response was that this compound was acting as a receptor antagonist/partial agonist in blocking the action of endogenous SRIF. Subsequently, this and several other related analogs were indeed

TABLE 1.1. Typical GH dose response obtained with the SRIF antagonist c(Phe-D-Trp-Lys-Thr[Bzl]-Phe) after subcutaneous administration to pentobarbital-anesthetized rats.

Dose (μg/100 g body wt)	Plasma GH (ng/ml)
Saline control	722 ± 101
0.3	801 ± 67
0.6	1082 ± 162
1.2	1070 ± 119
2.4	804 ± 114
4.8	732 ± 220

found to block the effect of injected SRIF on release of GH, insulin, and glucagon (35, 36, 37). However, the analogs had no effect on SRIF inhibition of release of GH from cultured rat pituitary cells. At the time, this result was extremely hard to explain, and no further work on these peptides was performed until quite recently when additional receptor subtypes were characterized and cloned. It now appears that the penta-peptides actually bond to SSTR1 and SSTR3 receptors (38, T. Reisine, personal communication), which appear heavily localized in the brain, but not to pituitary-type SSTR2 receptors. Furthermore, it appears that, as was originally suspected, they are not pure antagonists, being able to activate the receptors on transfected cells at higher dose levels (T. Reisine, personal communication).

In view of the selectivity, it is expected that this class of analog will be of great value in elucidating some of the biological functions of the various SRIF receptors. Further structure-activity studies are now in progress in an effort to generate pure antagonists and to further manipulate selectivity for receptors.

Growth Hormone Releasing Factor

GRF could hardly be more different in terms of its chemistry and physical chemistry than SRIF. GRF has a high degree of sequence homology to the growing family of neuro-GI peptides of the VIP-secretin-glucagon group, present members of which are shown in Table 1.2. In contrast to the other members of this family, however, human growth hormone releasing hormone appears to be highly specific for just pituitary GH release. GRFs have been isolated from human pancreatic tumors (1, 2) and from rat hypothalamic tissue (39). In both rats and humans, the amidated 44-amino acid peptide hpGRF(1-44)-NH_2 (1, 2, 40) and the free acid of the 40-amino acid N-terminal residue (41, 42) have been shown to specifically release GH at physiological doses. GRFs derived from several

TABLE 1.2. Comparison of the amino acid sequences of 8 related peptides of the VIP-secretin-glucagon-GRF family.

Peptide	1 5 10 15 20 25 30 35 40 45
PACAP-38	H-S-D-G-I-F-T-D-S-Y-S-R-Y-R-K-Q-M-A-V-K-K-Y-L-A-A-V-L-G-K-R-Y-K-Q-R-V-K-N-K-NH_2
VIP	H-S-D-A-V-F-T-D-N-Y-T-R-L-R-K-Q-M-A-V-K-K-Y-L-N-S-I-L-N-NH_2
PHI	H-A-D-G-V-F-T-S-D-Y-S-R-L-L-G-Q-L-S-A-K-K-Y-L-E-S-L-I-NH_2
Helodermin	H-S-D-A-I-F-T-E-E-Y-S-K-L-L-A-K-L-A-L-Q-K-Y-L-A-S-I-L-G-S-R-T-S-P-P-P-S
Secretin	H-S-D-G-T-F-T-S-E-L-S-R-L-R-D-S-A-R-L-Q-R-L-L-Q-G-L-V-NH_2
GRF(1-44)	Y-A-D-A-I-F-T-N-S-Y-R-K-V-L-G-Q-L-S-A-R-K-L-L-Q-D-I-M-S-R-Q-Q-G-E-R-N-Q-E-Q-G-A-R-V-R-L-NH_2
Glucagon	H-S-Q-G-T-F-T-S-D-Y-S-K-Y-L-D-S-R-R-A-Q-D-F-V-Q-W-L-M-N-T
GIP	Y-A-E-G-T-F-I-S-D-Y-S-I-A-M-D-K-I-R-Q-Q-D-F-V-N-W-L-L-A-Q-K-G-K-K-S-D-W-K-H-N-I-T-Q

species have now been sequenced with few significant differences apart from His-1 in the rat version, which tends to impart some affinity for such peripheral receptors as the pancreatic VIP receptor. In view of the difficulties and costs associated with the synthesis of peptides of this size, particularly on a commercial scale, much effort has gone into designing shorter analogs with improved potency.

GRF Analogs

We found (43) that full in vivo biological activity is retained by the amidated 1–29 sequence, which has since been used as the basis for many structure-activity studies in numerous laboratories. Also of importance was the discovery (43) that substitution of D-Ala in position 2 of this peptide resulted in a major increase in in vivo potency in the rat, which subsequently was shown to be due to enhanced resistance to proteases rather than increased receptor affinity. This modification has been used in the majority of analogs made since then.

Analog studies generally have tended to focus on the dominating amphiphilic α-helical nature of the 1–29 sequence (Fig. 1.5). Although statistical analysis of its sequence suggests that there might be some

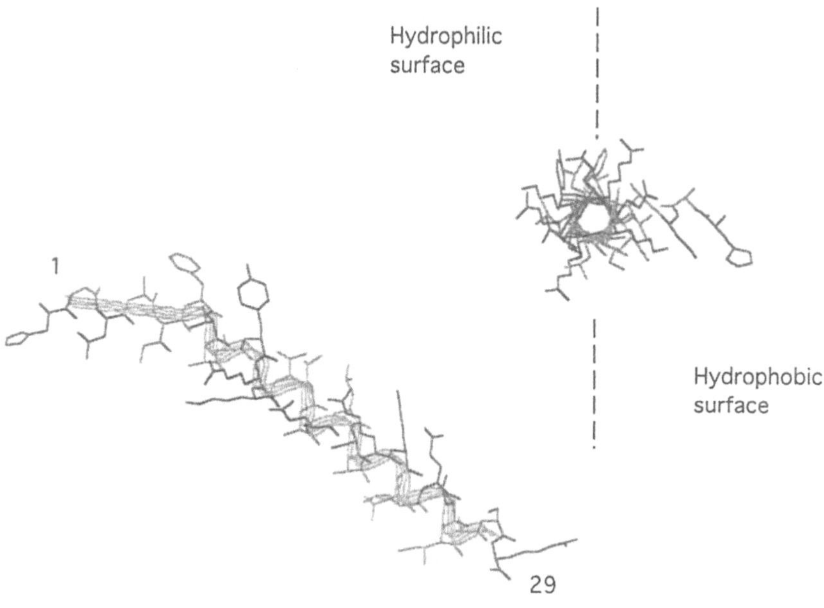

FIGURE 1.5. SYBYL biopolymer modeling of the α-helical, amphiphilic 7–29 region of [His^1,D-Ala^2,$Ala^{8,9,15,27}$,D-Arg^{29}]GRF(1-29)NH_2. Two views are shown with the 1–6 peptide adopting a random structure and hydrophobic residues in a lighter grey tone. The end view from the C-terminus illustrates the well-known occurrence of hydrophilic and hydrophobic side chains on opposite surfaces.

TABLE 1.3. Structures and potencies of some helix-stabilized analogs of $GRF(1\text{-}29)NH_2$ on GH secretion from 4-day cultures of rat pituitary cells.

Analog	Potency ± SE (GRF = 1)
1. $GRF(1\text{-}29)NH_2$	1
2. D-Ala^2	3
3. Ala^{15}	5 ± 1
4. Ala^8	4 ± 1
5. $Ala^{8,15}$	15 ± 2
6. D-Ala^2,$Ala^{8,15}$	27 ± 10
7. D-Ala^2,$Ala^{8,9,15}$	49 ± 14
8. His^1,D-Ala^2,$Ala^{8,9,15}$	2000 ± 542

folded regions (44), conversion to high helix probability sequences using such suitable replacement residues as Ala has resulted in significant improvements to potency. Substitution of Ala for Gly^{15}, for instance, increases in vitro potency about 5-fold (Table 1.3) (45) and for Asn^8 about 4-fold (Table 1.3) (46). Additionally, substitution of Ala for Ser^9 was also very beneficial (analog 6, Table 1.3). Combinations of all of these changes within the same molecule resulted (46) in multiplicative

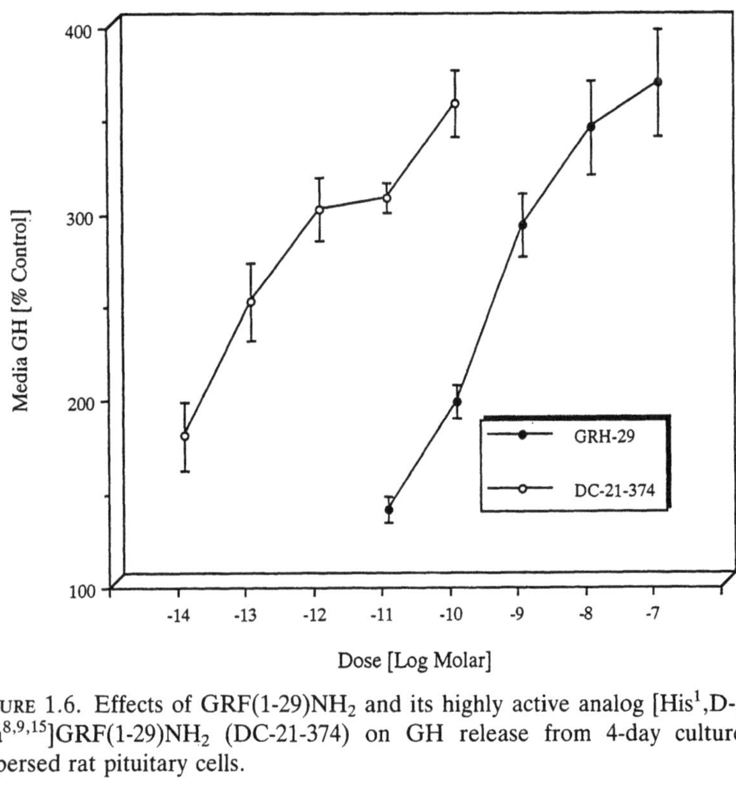

FIGURE 1.6. Effects of $GRF(1\text{-}29)NH_2$ and its highly active analog [His^1,D-Ala^2, $Ala^{8,9,15}$]$GRF(1\text{-}29)NH_2$ (DC-21-374) on GH release from 4-day cultures of dispersed rat pituitary cells.

effects on potency to yield some very active compounds (Table 1.3). Most recently, we found that the presence of His instead of Tyr in position 1 was extremely beneficial, producing a dose response curve (Fig. 1.6) shifted dramatically to the left of GRF itself. This translates into an analog with greater potency than GRF by several orders of magnitude (analog 8, Table 1.3). The reason for the high potency of the His peptide is debatable but could, for instance, be due to the occurrence of His^1 in the rat GRF sequence and the use of rat pituitary cells in this particular assay. Thus, it is not known whether this high level of activity will extend to other animal species. However, His^1-analogs of this type are presently primary candidates for further animal and clinical investigations.

Acknowledgment. Parts of this work were funded by NIH Grant DK-30167.

References

1. Guillemin R, Brazeau P, Bohlem P, Esch F, Ling N, Wehrenberg F. Growth hormone-releasing factor from a human pancreatic tumor that caused acromegaly. Science 1982;218:585–7.
2. Brazeau P, Vale W, Burgus R, Ling N, Butcher M, Guillemin R. Hypothalamic polypeptide that inhibits the secretion of immunoreactive pituitary growth hormone. Science 1973;179:77–9.
3. Rivier JE, Spiess J, Thorner JM, Vale W. Characterization of a growth hormone-releasing factor from a human pancreatic islet tumor. Nature 1982;300:276–8.
4. Magnan E, Cataldi C, Guillaume V, et al. Acute changes in growth hormone-releasing hormone secretion after injection of BIM 23014, a long-acting somatostatin analog, in rams. Life Sci 1992;51:831–8.
5. Meyers CA, Coy DH. Somatostatin, enkephalins, and endorphins. In: Glass GBJ, ed. Gastrointestinal hormones. New York: Raven Press, 1980:363–85.
6. Konturek SJ. Somatostatin and opiate peptides: their action on gastrointestinal secretions. In: Glass GBJ, ed. Gastrointestinal hormones. New York: Raven Press, 1980:693–716.
7. Reichlin S. Somatostatin. N Engl J Med 1983;309:1495–556.
8. Moreau J-P, Defeudis FV. Pharmacological studies of somatostatin and somatostatin analogues: therapeutic advances and perspectives. Life Sci 1987;40:419–37.
9. Yamada Y, Post SR, Wang K, Tager HS, Bell GI, Seino S. Cloning and functional expression of a family of human and mouse somatostatin receptors expressed in brain, gastrointestinal tract and kidney. Proc Natl Acad Sci USA 1992;89:251–5.
10. Yasuda K, Rens-Domiano S, Breder CD, et al. Cloning of a novel somatostatin receptor, SSTR3, coupled to adenylcyclase. J Biol Chem 1992;267.
11. Rivier J, Brown M, Rivier C, Ling N, Vale W. In: Loffet A, ed. Peptides, 1976. Brussels: Universite de Bruxelles, 1976:427.

12. Vale W, Rivier J, Ling N, Brown M. Biological and immunologic activities and applications of somatostatin analogs. Metabolism 1978;27(suppl 1): 1391–401.
13. Holladay LA, Puett D. Conformational studies on somatostatin and analogues. Somatostatin conformation: evidence for a stable intramolecular structure from circular dichroism, diffusion, and sedimentation equilibrium. Proc Natl Acad Sci USA 1976;73:1199–202.
14. Holladay L, Rivier J, Puett D. Conformational studies on somatostatin and analogues. Biochemistry 1977;16:4895–900.
15. Rivier J, Brown M, Vale W. D-Trp^8-somatostatin: an analog of somatostatin more potent than the native molecule. Biochem Biophys Res Commun 1975;65:746–51.
16. Veber DF, Holly FW, Palaveda WJ, et al. Conformationally restricted bicyclic analogs of somatostatin. Proc Natl Acad Sci USA 1978;75:2636–40.
17. Arison BH, Hirschmann R, Veber DF. Inferences about the conformation of somatostatin at a biologic receptor based on NMR studies. Bioorg Chem 1978;7:447–51.
18. Veber DF, Holly FW, Nutt RF, et al. Highly active cyclic and bicyclic somatostatin analogues of reduced ring size. Nature 1979;280:512–4.
19. Veber DF, Friedinger RM, Perlow DS, et al. A potent cyclic hexapeptide analogue of somatostatin. Nature 1981;292:55–8.
20. Freidinger RM, Perlow DS, Randall WC, Saperstein R, Arison BH, Veber DF. Conformational modifications of cyclic hexapeptide somatostatin analogs. Int J Pept Protein Res 1984;23:142–50.
21. Veber DF, Saperstein R, Nutt RF, et al. A superactive cyclic hexapeptide analog of somatostatin. Life Sci 1984;34:1371–8.
22. Bauer W, Briner U, Doepfner W, et al. SMS 201-995: a very potent and selective octapeptide analogue of somatostatin with prolonged action. Life Sci 1982;31:1133–40.
23. Heiman ML, Murphy WA, Coy DH. Differential binding of somatostatin agonists to somatostatin receptors in brain and adenohypophysis. Neuroendocrinology 1987;45:429–36.
24. Lamberts SWJ, Krenning EP, Reubi JC. The role of somatostatin and its analogs in the diagnosis and treatment of tumors. Endocr Rev 1991;12: 450–82.
25. Murphy WA, Taylor JE, Moreau JP, Coy DH. Novel heptapeptide somatostatin analog displays anti-tumor activity independent of effects on growth hormone secretion. Pept Res 1989;2:128–32.
26. Bogden AE, Taylor JE, Moreau JP, Coy DH. Treatment of R-3327 prostate tumors with a somatostatin analogue (somatuline) as adjuvant therapy following surgical castration. Cancer Res 1990;50:2646–50.
27. Taylor JE, Coy DH, Moreau JP. High affinity binding of [^{125}I-Tyr^{11}]-somatostatin-14 to human small cell lung carcinoma (NCI-H69). Life Sci 1988;43:421–8.
28. Taylor JE, Bogden AE, Moreau JP, Coy DH. In vitro and in vivo inhibition of human small cell lung carcinoma (NCI-H69) growth by a somatostatin analogue. Biochem Biophys Res Commun 1988;163:81–6.

29. Murphy WA, Lance VA, Moreau S, Moreau JP, Coy DH. Inhibition of rat prostate tumor growth by an octapeptide analog of somatostatin. Life Sci 1987;40:2515–22.
30. Murphy WA, Taylor JE, Moreau JP, Coy DH. Linear octapeptide analogs of somatostatin retaining biological activity and ability to displace ^{125}I-Tyr^{11}-SRIF binding [Abstract]. 71st Endocr Soc meet, 1990:104.
31. Verheyden PMF, Coy DH, Van Binst G. Conformational study of somatostatin analogues in methanol at low temperature. Magn Reson Chem 1991; 29:607–12.
32. Maurer R, Gaehwiler BH, Buescher HH, Hill RC, Roemer D. Opiate antagonist properties of an octapeptide somatostatin analog. Proc Natl Acad Sci USA 1982;79:4815–7.
33. Walker JM, Bowen WD, Akins ST, Hemstreet MK, Coy DH. μ-Opiate binding and morphine antagonism by octapeptide analogs of somatostatin. Peptides 1987;8:869–75.
34. Jensen RT, Coy DH. Progress in the development of potent bombesin receptor antagonists. Trends Pharmacol Sci 1991;12:13–9.
35. Fries JL, Murphy WA, Sueiraz-Diaz J, Coy DH. Somatostatin antagonist analog increases GH, insulin, and glucagon release in the rat. Peptides 1982;3:811–4.
36. Coy DH, Murphy WA, Fries JL, Sueiras-Diaz J. Potent, short cyclic peptide antagonists of somatostatin. Regul Pept 1983;suppl 2:S101.
37. Coy DH, Murphy WA, Lance VA, Hocart SJ, Sueiras-Diaz J, Mezo I. Somatostatin agonists and antagonists—peptide control of growth hormone secretion. In: Patel YC, Tannenbaum GS, eds. Somatostatin. Advances in experimental medicine and biology. New York: Plenum, 1985;188:325–37.
38. Raynor K, Coy DH, Reisine T. Analogues of somatostatin bind selectively to brain somatostatin receptor subtypes. J Neurochem 1992;59:1241–50.
39. Spiess J, Rivier J, Vale W. Characterization of rat hypothalamic growth hormone-releasing factor. Nature 1983;303:532–5.
40. Wehrenberg WB, Ling N, Brazeau P, et al. Somatocrinin, growth hormone-releasing factor, stimulates secretion of growth hormone in anesthetized rats. Biochem Biophys Res Commun 1982;109:382–6.
41. Rosenthal SM, Schriock EA, Kaplan SL, Guilleman R, Grumbach MM. Effect of growth hormone (GH)-releasing hormone (GRH) on plasma GH in relation to magnitude and duration of GH deficiency in 26 children and adults with isolated GH deficiency or multiple pituitary hormone deficiencies: evidence for hypothalamic GRH deficiency. J Clin Endocrinol Metab 1983;57:677–9.
42. Thorner MO, Rivier J, Spiess J, et al. Human pancreatic growth hormone releasing factor selectively stimulates growth hormone secretion in man. Lancet 1983;i:24–8.
43. Lance VA, Murphy WA, Sueiras-Diaz J, Coy DH. Super-active analogs of growth hormone-releasing factor (1-29)-amide. Biochem Biophys Res Commun 1984;119:265–72.
44. Murphy WA, Coy DH, Lance VA. Superactive amidated COOH-terminal glucagon analogs with no methionine or tryptophan. Peptides 1986;7(suppl 1):69–74.

45. Felix AM, Heimer EP, Mowles TF, et al. Synthesis and biological activity of novel GRF analogs. In: Theodoropoulos D, ed. Proc 19th European Peptide Symposium. Berlin: de Gruyter, 1987:481–4.
46. Coy DH, Hocart SJ, Murphy WA. Human growth hormone-releasing hormone analogues with much improved in vitro growth hormone-releasing potencies in rat pituitary cells. Eur J Pharmacol 1991;204:179–85.

2

Ionic Mechanisms Governing the Control of Growth Hormone Secretion by Somatostatin

JACOB KRAICER AND STEPHEN M. SIMS

Readers are referred to a recent symposium (1) and reviews (2–4) for a comprehensive and broad introduction to *somatostatin* (SS) and its actions. This chapter will be restricted to the mechanisms (primarily ionic) by which SS inhibits *growth hormone* (GH) release. The abundant literature on the actions of SS on other cell types will not be discussed unless directly relevant to ionic mechanisms in somatotrophs. We will begin by summarizing the ionic mechanisms by which *growth hormone releasing factor* (GRF) stimulates the acute release of GH via *cyclic AMP* (cAMP) and cytosolic free Ca^{++} ($[Ca^{++}]_i$) as second messengers (Fig. 2.1). This is discussed in more detail elsewhere in this volume.

GRF, by binding to its receptor on the somatotroph plasma membrane, stimulates the GTP binding protein G_s, which in turn activates adenylate cyclase, resulting in an increase in cAMP formation and an increase in cytosolic cAMP. It is now firmly established that cAMP is a second messenger for GRF and fulfills the following criteria: (i) GRF increases adenylate cyclase activity; (ii) GRF increases cAMP content of the somatotrophs, and the increase in cAMP precedes or is concurrent with the increased release of GH; (iii) exogenous cAMP or analogs stimulate the acute release of GH; and (iv) the GRF-induced release of GH is potentiated by inhibitors of phosphodiesterase activity (5–9).

Using similar criteria, it is also established that $[Ca^{++}]_i$ is a second messenger for GRF, since a rise in $[Ca^{++}]_i$ induced by whatever means will result in the acute release of GH. Conversely, a fall in $[Ca^{++}]_i$ results in a decrease in GH release. Thus, (i) GRF elicits an increase in $[Ca^{++}]_i$ that precedes or is concurrent with the increased release of GH (7); (ii) elevation in $[Ca^{++}]_i$ by any means provokes GH release, even in the absence of GRF (7–9); and (iii) removal of Ca^{++} from the extracellular fluid and/or prevention of the GRF-induced influx of calcium and sub-

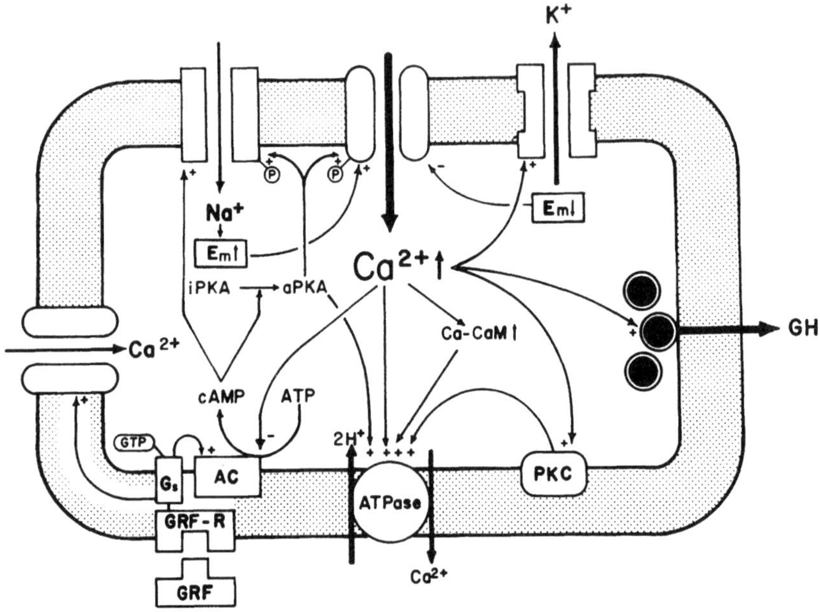

FIGURE 2.1. Model depicting the pathways responsible for the GRF-induced increase in $[Ca^{++}]_i$ and GH release in rat somatotrophs. (AC = adenylate cyclase; E_m = membrane potential [upward arrow, less negative; downward arrow, more negative]; G_s = GTP binding protein G_s; GRF-R = GRF receptor; aPKA = active cAMP-dependent protein kinase; iPKA = inactive cAMP-dependent protein kinase; + = stimulation; − = inhibition; P = channel phosphorylation.) Reprinted with permission from Lussier, French, Moor, and Kraicer (9), © by The Endocrine Society, 1991.

sequent rise in $[Ca^{++}]_i$ results in a loss of responsiveness to GRF (7, 10). The rise in $[Ca^{++}]_i$ induced by GRF results from an influx of Ca^{++} into the somatotrophs from the extracellular fluid and not from an intracellular store or sequestered pool (7, 10).

How are these two second messengers, $[Ca^{++}]_i$ and cAMP, interrelated? The following is a reasonable working model (Fig. 2.1) (9). First, G_s could interact directly with Na^+ channels and/or L-type *voltage-sensitive Ca^{++} channels* (VSCC) in the plasma membrane via a membrane-delimited mechanism. This would cause Na^+ channels to open, resulting in a depolarization with the subsequent opening of VSCC and an influx of Ca^{++} and/or a change in the voltage sensitivity of VSCC, causing them to open at resting membrane potential (E_M) and thus resulting in an influx of Ca^{++}. Second, the GRF-induced increase in cAMP could increase Na^+ conductance directly or via *protein kinase A* (PKA)-induced phosphorylation of Na^+ channels; this increase in Na^+ conductance would lead to depolarization and the opening of VSCC.

Finally, cAMP could increase Ca^{++} conductance directly or via PKA-dependent phosphorylation of Ca^{++} channels, resulting in an influx of Ca^{++}. How the increases in $[Ca^{++}]_i$ and cAMP lead to the acute release of GH by exocytosis is still not understood (see below).

With this introduction, we can now approach the description of the mechanisms by which SS reduces basal and secretagogue-induced GH release. First, we present several key experimental observations. The effect of SS on reducing GH release both in vivo and in vitro is instantaneous (i.e., within seconds). In vitro, SS reduces basal GH release from somatotrophs in a concentration-related manner, and this inhibition persists while SS is maintained (11). Both in vivo and in vitro, SS reduces GRF-induced GH release, again in a reversible and concentration-related manner (5, 6, 12).

Next we present a model for the mechanism of action of SS (Fig. 2.2), followed by the experimental data on which this model is based. A

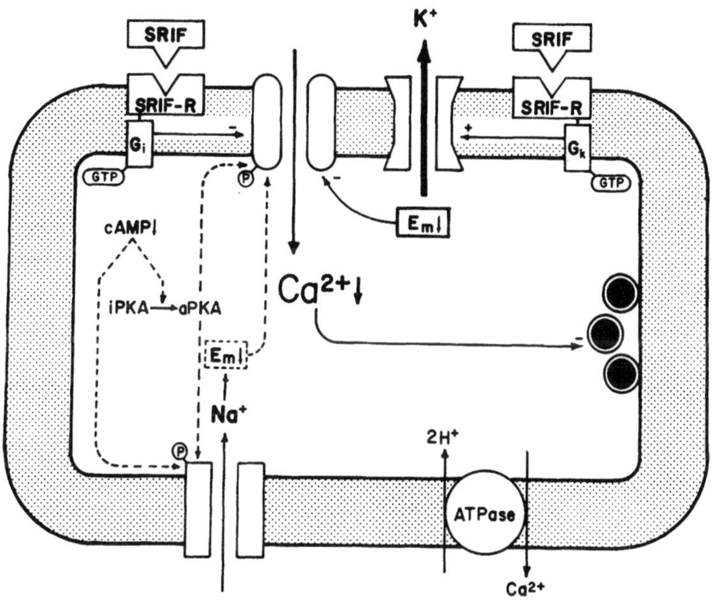

FIGURE 2.2. Model depicting pathways responsible for the SS (SRIF)-induced decrease in $[Ca^{++}]_i$ and inhibition of GH release in rat somatotrophs. (E_m = membrane potential [downward arrow, more negative]; G_i = GTP binding protein G_i; G_k = GTP binding protein G_k; aPKA = active cAMP-dependent protein kinase; iPKA = inactive cAMP-dependent protein kinase; SRIF-R = SRIF receptor; + = stimulation; − = inhibition; P = channel phosphorylation. Dashed arrows denote decreased stimulatory effect due to the moderate reduction in cAMP accumulation seen in GRF-challenged somatotrophs in the presence of SRIF.) Reprinted with permission from Lussier, French, Moor, and Kraicer (9), © by The Endocrine Society, 1991.

brief, albeit simplified, summary of ion channels as they may apply to the actions of SS (2, 4, 13–15) follows. Ion channels may be classified according to the inputs that alter their gating (opening or closing): voltage-gated (in response to a change in membrane potential) and ligand-operated (in response to the binding of a specific ligand to a receptor linked to the channel). Ligand-operated channels can be further subdivided into membrane-delimited and second messenger-mediated. Membrane-delimited channels may be linked to the receptor via G-proteins, or the receptor site may be part of the channel molecule. Second messenger-mediated channels rely on a change in cytosolic second messenger to gate the channel. The following ion channels have been linked to SS action: voltage-gated Ca^{++} channels (L-type), ligand-operated, membrane-delimited K^+ channels, and second messenger-mediated (via cAMP) Ca^{++} channels. A final, but important, concept is that of *modulation*. For example, the second messenger-mediated L-type VSCC, although voltage-gated, can be modulated by cytosolic cAMP in that a rise in cAMP will result in a higher open-state probability so that they will open with a smaller depolarization of the membrane (16).

To return to the model (Fig. 2.2), SS binds to its receptors on the somatotroph plasma membrane. There are multiple receptor subtypes. Some of the receptors are coupled by pertussis toxin-sensitive *guanine nucleotide binding proteins* (G-proteins) to several effector systems via membrane-delimited mechanisms to close VSCC via G_o and/or open K^+ channels via G_k. The closing of VSCC would result in a decrease of Ca^{++} influx into the cells. The opening of K^+ channels would increase K^+ conductance, resulting in hyperpolarization, which would lead to a closing of VSCC. If Ca-ATPase activity is not simultaneously decreased, then the resulting decrease in Ca^{++} influx would lead to a decrease in $[Ca^{++}]_i$ and a subsequent reduction in GH release. While SS does not inhibit basal adenylate cyclase activity, or the increase in adenylate cyclase activity induced by GRF in somatotrophs (6), it does reduce, but only moderately, the cAMP accumulation in GRF-stimulated cells (5, 6). This reduction could lead to a decrease in Ca^{++} and Na^+ channel phosphorylation and/or a decreased direct activation of Ca^{++} and Na^+ channels. Thus Ca^{++} and Na^+ conductances would decrease, causing E_M to become more negative, which would further decrease Ca^{++} influx.

We will now discuss some of the experimental bases for this model. We will attempt to separate observations of single somatotrophs from observations of cell populations. This is essential in order to understand the mechanism of action of SS. Individual somatotrophs exhibit spontaneous action potentials (17, 20) and concurrent $[Ca^{++}]_i$ oscillations (18). The spontaneous activity of the individual cells is hidden when populations of cells are studied, since changes occurring out of phase in individual cells are averaged. Thus, fundamental patterns of activity in individual somatotrophs, which are required in order to understand mechanisms, are lost when cell populations are studied. We will begin with a description of the ionic events in individual somatotrophs during rest, or basal conditions.

Ionic Events in Resting Somatotrophs

Any discussion of ionic events must recognize fundamental electrophysiological properties of somatotrophs. There is a large literature describing the electrophysiology of a number of cloned pituitary cell lines, but these studies will not be discussed in detail because of uncertainty as to their relevance for events occurring in normal somatotrophs and because major alterations in intracellular signaling processes may occur in malignant cells (3, 4). The resting membrane potential of somatotrophs is ~50 mV (17, 19, 20). Somatotrophs exhibit a variety of membrane ionic currents, including Na^+, Ca^{++}, and several types of K^+ currents (20, 21). The activities of the different channels interact to give rise to a regular pattern of action potentials (17, 19, 22) (Fig. 2.3). While the ionic basis for this bursting activity has not been fully characterized, it is likely that voltage-activated Na^+ and/or Ca^{++} currents participate in the rising and plateau phases of the action potentials (20). It is also likely that entry of Ca^{++} during the action potentials gives rise to the oscillations of $[Ca^{++}]_i$ that have been described (3, 23). It remains to be demonstrated if pulses in GH release accompany the action potentials and Ca^{++} oscillations. Voltage-clamp studies have provided a preliminary characterization of the membrane currents involved in the rhythmic bursting patterns. For example, somatotrophs, like some cloned pituitary cell lines, exhibit rapidly inactivating inward current resembling the TTX-sensitive Na^+ current, high-threshold, L-type VSCC, and several K^+ currents, including an outward K^+ current and Ca^{++}-activated K^+ current (19, 20).

The spontaneous bursts of action potentials that we observed in 70% of cells studied with the perforated patch method (20) are quite distinct

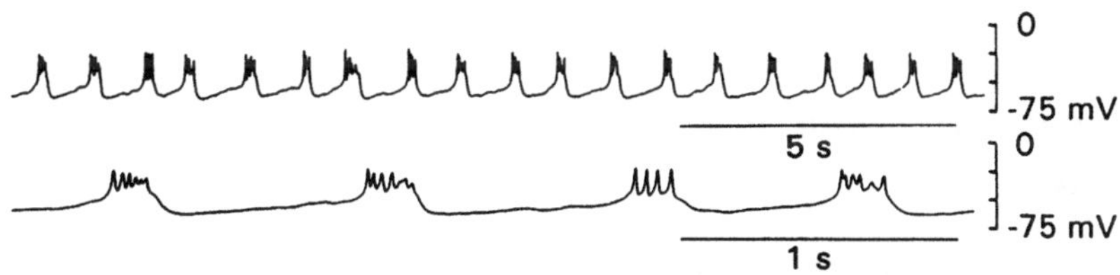

FIGURE 2.3. Spontaneous bursts of action potentials in freshly isolated rat somatotrophs. Membrane potential oscillated between ~−70 mV and −25 mV, with a frequency of $0.9\,s^{-1}$. The first 4 depolarizations at left are shown on an expanded time scale below, revealing small action potentials at the peak of some depolarizations. Recording was obtained using the perforated patch method (20).

from the single, spontaneous action potentials previously described in somatotrophs. Lewis et al. (24) reported that 36% of immunohistochemically identified GH containing rat pituitary cells exhibited irregular spontaneous action potentials. Only 11% of bovine somatotrophs exhibited spontaneous action potentials when recording in whole-cell configuration (25). Chen et al. (26) observed that 44% of 7- to 15-day cultured somatotrophs showed irregular spontaneous action potentials when using whole-cell patch-clamp techniques. In contrast, when recordings were carried out with intracellular microelectrodes, 18% of the cells exhibited regular spontaneous action potentials (27). Brief spontaneous action potentials have also been recorded in clonal pituitary cells (19, 28–30).

Experimental Evidence for the Model Depicting the Mechanism of Action of SS

SS Receptors

SS binds to specific high-affinity saturable receptors in the plasma membrane of somatotrophs (31–33). While it is beyond the scope of this chapter to review the extensive data, multiple subtypes of the SS receptor have been identified, characterized, and purified, and the genes for several SS receptor isoforms have been cloned (32–36).

G-Proteins

The SS receptor is G-protein linked (2). Most of the actions of SS appear to be mediated by pertussis toxin-sensitive G-proteins. A number of G-proteins have been implicated as being coupled to the SS receptor, including G_i (inhibits adenylate cyclase), G_p (activates phosphoinositide-specific phospholipase C), G_o (possibly regulates ion channels including VSCC), G_k (directly activates a K^+ channel), and G_E (mediates exocytosis and is closely linked with the late stages of the exocytotic process) (2, 31, 32). SS receptor-G-protein-linked events are reviewed in detail elsewhere in this volume.

Second Messengers

cAMP

A number of studies have suggested that SS acts, at least in part, via G_i to inhibit basal and the GRF-induced increase in adenylate cyclase activity, subsequently reducing intracellular cAMP accumulation and GH release (31, 32). However, we found that even high concentrations of SS had no effect on basal or the GRF-induced rise in adenylate cyclase activity (6). SS elicited only a small reduction in the GRF-induced increase in cAMP in somatotrophs, even though the GRF-induced increase in GH release was completely blocked (5, 6, 37). Also, since SS completely blocks the increase in GH release induced by cAMP analogs and inhibitors of cAMP phosphodiesterase activity (38–40) and also blocks the increase in GH release induced by high K^+ and calcium ionophores (41), then clearly SS also acts in the release process beyond or independent of cAMP.

cGMP

cGMP is not a second messenger in the action of SS since somatotrophs do not contain measurable guanylate cyclase activity or cGMP, and cGMP analogs do not alter GH release (5, 37, 41).

Inositol Polyphosphate/Diacylglycerol

It has been suggested that SS suppresses GH release by the inhibition of plasma membrane phospholipase C activity, with a subsequent reduction of intracellular inositol polyphosphates (to decrease $[Ca^{++}]_i$) and diacylglycerol (to reduce protein kinase C activity). However, GRF does not act through this second messenger system in somatotrophs (42–44), and SS does not alter phospholipase C activity or protein kinase C activity in somatotrophs (unpublished observations). Therefore, SS does not act through the phosphoinositide second messenger system to inhibit the release of GH.

Leukotrienes

Leukotrienes are products of arachidonic acid metabolism and are synthesized through the action of the enzyme 5-lipoxygenase. It has recently been suggested that SS may act on neurons through the generation of phospholipase A_2 to increase intracellular leukotriene content, which in turn would augment a time- and voltage-dependent K^+ current, the M-current (I_M), and result in hyperpolarization (45). Whether such a mechanism is involved in SS action on somatotrophs awaits study.

$[Ca^{++}]_i$

Basal $[Ca^{++}]_i$ in populations of somatotrophs is between 100 and 300 nM (7–9, 46). As a generalization, it can be stated that a rise in $[Ca^{++}]_i$ is associated with a concurrent concentration-related increase in GH release, and a fall in $[Ca^{++}]_i$ is accompanied by a fall in GH release (see above). A similar range of basal $[Ca^{++}]_i$ is found in individual somatotrophs (18, 46–48). However, regular rhythmic $[Ca^{++}]_i$ oscillations occur in individual somatotrophs with a frequency of up to 15/min (18, 46, 48), and there is a close correlation between the timing of the action potentials and the spikes in $[Ca^{++}]_i$ (18). These Ca^{++} oscillations are due to an influx in Ca^{++}, since they are abolished and basal $[Ca^{++}]_i$ is reduced when extracellular calcium is removed or when calcium channel blockers are added (18, 47, 48). A concurrent pulsatile pattern of GH release would be predicted, but has not yet been reported. It has been reported that the amount of GH released does correlate directly with the amplitude and frequency of calcium oscillation at the level of the single GH cell (18). As discussed in the next section, SS acts primarily, but not exclusively, through altering $[Ca^{++}]_i$ and via membrane-delimited mechanisms.

Membrane-Delimited Actions

There is no compelling evidence that activation/inactivation of a cytosolic second messenger system (other than $[Ca^{++}]_i$) is essential in the mechanism of action of SS. Membrane-delimited mechanisms appear to play key roles in altering $[Ca^{++}]_i$, which in turn alters GH release. The following paragraphs summarize relevant observations.

In cell population studies, SS inhibits basal GH release immediately, reversibly, and in a concentration-related manner. SS also inhibits the increased GH release induced by a number of secretagogues, including GRF, high K^+, calcium ionophores, agents that stimulate adenylate cyclase activity, cAMP analogs, and agents that increase protein kinase C activity (9). Again, the inhibition is immediate, sustained, reversible, and concentration related.

In cell population studies using somatotrophs, SS reduces basal $[Ca^{++}]_i$ and blocks the rise in $[Ca^{++}]_i$ induced by GRF, agents that stimulate adenylate cyclase activity, cAMP analogs, and agents that increase protein kinase C activity (8, 9). This block is immediate, sustained, reversible, and concentration related (8, 9). It should be noted, however, that while SS blocks the increase in GH release induced by high K^+ and calcium ionophores, SS does not reduce the increase in $[Ca^{++}]_i$ induced by these two agents (9).

In cell population studies using somatotrophs, SS reduces basal GH release and the augmented release of GH induced by the agents listed above (except for high K^+ and calcium ionophore) by reducing Ca^{++} influx into the somatotrophs. This conclusion is based on the measurement of Ca^{++} fluxes during steady and nonsteady state, the use of blockers of VSCC, and the use of media free of Ca^{++} (7–9). The pattern of GH release from individual somatotrophs has not yet been reported.

Studies of the Effect of SS on Membrane Potential and Ionic Currents in Single Somatotrophs

We used the perforated patch recording technique (49) to investigate effects of SS on membrane potential and membrane currents of freshly dispersed somatotrophs (20). This technique, in contrast to conventional whole-cell patch-clamp recording, minimizes dialysis of the cytosol with the electrode solution and thus maintains metabolic and second messenger systems intact. Brief application of SS to cells exhibiting spontaneous activity caused hyperpolarization to ~−70 mV and inhibition of the rhythmic bursting pattern. Reappearance of the spontaneous action potentials was apparent after ~1 min (Fig. 2.4A). Other cells that did not show spontaneous action potentials also hyperpolarized in response to SS (Fig. 2.4B). The SS-induced hyperpolarization was always accompanied by a decrease in the amplitude of voltage deflections in response to injected current pulses. This finding indicates that SS caused an increase in the conductance of the membrane for an ion whose equilibrium potential was more negative than the resting membrane potential (likely K^+, see below).

We then studied the actions of SS under voltage clamp to characterize directly the conductance involved. Voltage ramp commands were

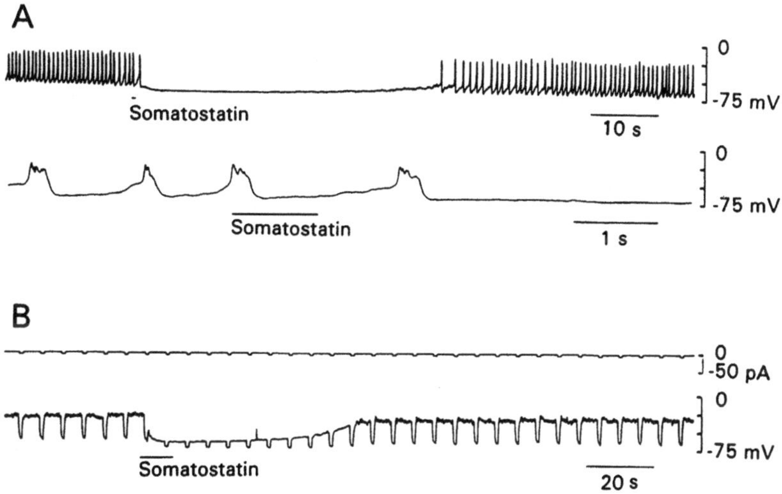

FIGURE 2.4. Somatostatin hyperpolarizes rat somatotrophs. *A:* SS, applied for the time indicated by the bar under the voltage trace by pressure ejection from a pipette (1 s, 100 nM in the application pipette), caused hyperpolarization to −70 mV and inhibited spontaneous action potentials. The portion of the record where SS was applied is shown on an expanded time scale below, revealing a latency of 1–2 s before a response was apparent. Recovery occurred after ~50 s, at right. *B:* Response of a quiescent cell to SS. The upper trace monitors current, with −5 pA pulses being injected into the cell every 6 s. The lower trace displays the membrane potential, with a resting potential of ~−30 mV and an input resistance of ~6 G_Ω. SS (1 μM in the application pipette) caused hyperpolarization to −65 mV, accompanied by a decrease in the amplitude of the voltage deflections in response to injected current pulses, indicating an increase in the conductance of the membrane. Recovery is seen at right. Responses were recorded in perforated patch configuration (20).

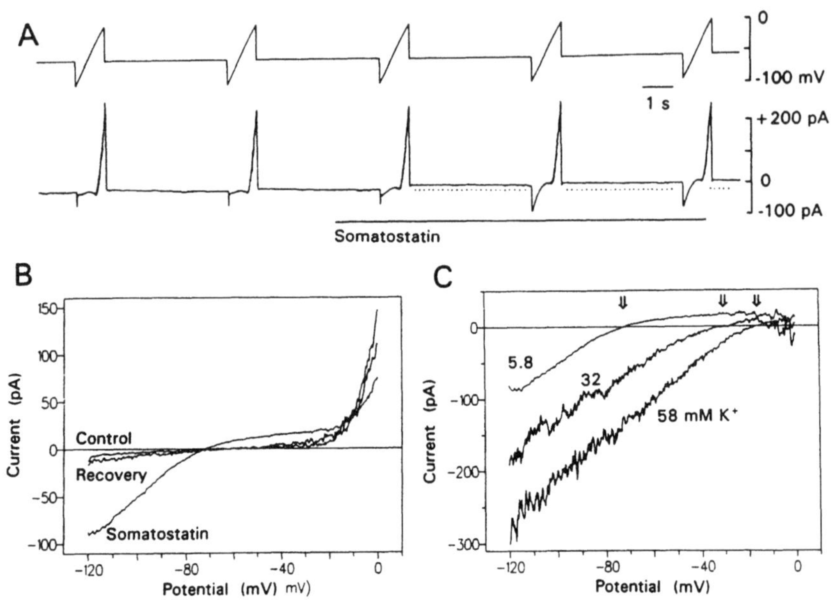

FIGURE 2.5. Current-voltage relationship for SS-activated current. Voltage ramp commands were used to record changes in whole-cell I-V relations. *A:* Cell was held at -60 mV with ramp commands every 5 s from -100 mV to 0 mV over 1 s, as shown in the upper chart recorder trace. SS was applied for the time indicated by the bar underneath the current trace (at bottom) and caused outward current at -60 mV, apparent as the upward deflection of the holding current above the dashed line. Large inward currents were elicited at negative potentials during the ramp commands. *B:* I-V relationships for another cell, determined with ramps from -120 mV to 0 mV. The I-V relation for the control cell, and after recovery, was linear negative to -40 mV. During SS, the I-V showed marked inward rectification. *C:* The SS-induced current was determined by digitally subtracting the two I-V curves. Also shown are SS-induced currents recorded in 32 and 58 mM K^+ for two other cells. The SS-induced current exhibited inward rectification at all levels of $[K^+]_o$. The potential at which the currents reversed direction (arrows) were -72 mV in 5.8 mM K^+, -32 mV in 32 mM K^+, and -19 mV in 58 mM K^+, values close to the equilibrium potentials for K^+. All traces were recorded in perforated patch configuration, and those in *B* and *C* represent averages of two ramps each. SS was 1 μM in the application pipette in *A* and 200 nM for all cells in *B* and *C* (20).

used to record currents over a range of potentials. For cells bathed in 5.8 mM K^+, SS caused outward current at the holding potential and a large inward current at more negative potentials (Fig. 2.5A). The presence of net outward current at the holding potential indicates that, following stimulation by SS, the resting potential was more negative than -60 mV, confirming the current clamp responses shown above. When the currents evoked by the ramp commands were displayed as I-V relations, the I-V was linear at negative potentials before and after recovery, but marked inward rectification was induced by SS (Fig. 2.5B).

The SS-induced current was determined by subtraction of the control and SS I-V curves and is displayed in Figure 2.5C, along with responses of two other cells studied in elevated $[K^+]_o$. The SS-induced current exhibited inward rectification at all concentrations of K^+, apparent by the reduced slope of the I-V curve for outward current (i.e., positive to the reversal potential). Reversal of the SS-induced current occurred close to the theoretical K^+ equilibrium potential and shifted positive along the voltage axis with elevation of $[K^+]_o$. Reversal potentials of the SS-activated current shifted 55 mV positive per 10-fold increase in $[K^+]_o$ (20), close to that predicted for a K^+ current by the Nernst relation.

The inward rectifier induced by SS resembles inward rectifier K^+ currents of other cell types, such as skeletal and cardiac muscle (50, 51). The regulation of an inwardly rectifying K^+ current by SS has also been reported in a variety of other cell types, such as AtT-20 clonal pituitary cells (52), myenteric neurons (53), rat locus coeruleus neurons (54), and cardiac atrial myocytes (55).

The hyperpolarization and inhibition of spontaneous action potentials caused by SS illustrated in Figure 2.4 can be correlated with functional studies on these same cells. By clamping the membrane at ~ -70 mV and abolishing spontaneous action potentials, entry of Ca^{++} through voltage-dependent channels is limited (23). This may explain how SS reduces resting $[Ca^{++}]_i$ in populations of somatotrophs, as demonstrated for these same cells (8, 9). Furthermore, SS and Ca^{++} channel blockers also reduce GH release from somatotrophs (8, 10), leading to the suggestion that suppression of the spontaneous action potentials is accompanied by reduction of basal GH secretion. This proposed mechanism for suppression of GH release from freshly isolated pituitary cells is in accord with models for inhibition of GH release from clonal pituitary cells based on electrophysiological studies (19, 30, 56) and fluorescent indicator dyes (57), as well as for cultured somatotrophs (27) (reviewed in 23). SS has also been shown to reduce calcium influx and basal $[Ca^{++}]_i$ (18, 48) and Co^{++}- and nifedipine-sensitive oscillations in $[Ca^{++}]_i$ in single somatotrophs (18, 46) and clonal pituitary cells (29), suggesting that suppression of rhythmic spontaneous action potentials may be a common response to SS. There is also evidence that SS activates outwardly rectifying K^+ current (58, 59) and directly suppresses voltage-activated Ca^{++} current in pituitary cells (30) and cultured rat somatotrophs (60), actions that would further limit entry of Ca^{++} and reduce secretion of GH.

These data lead to the proposed model for SS action (Fig. 2.2). However, one additional mechanism must be added to take into account the actions of SS that are not simply related to $[Ca^{++}]_i$. The increase in GH release induced by two secretagogues, high $[K^+]_o$ and calcium ionophores, is blocked by SS, but without preventing the increase in $[Ca^{++}]_i$ induced by these two secretagogues. These and other data using permeabilized cells (46) suggest that events distal to $[Ca^{++}]_i$ may also be a site of action of SS. A class of G-proteins, named G_e, has been proposed to interact, possibly via a membrane-limited mechanism, directly with the exocytotic process (46, 61, 62). This pertussis toxin-sensitive G_e would then link the SS receptor with the exocytotic site and provide an additional mechanism for SS action.

Summary

SS binds to specific plasma membrane receptor(s). Then, through the mediation of one or more G-proteins, somatotrophs are hyperpolarized due to an increase in K^+ conductance leading to a decrease in calcium influx. The subsequent reduction in $[Ca^{++}]_i$ results in decreased basal and secretagogue-induced GH release. SS can also reduce GH release without inducing a fall in $[Ca^{++}]_i$. This action is mediated by a G-protein, G_e, that is closely linked with the last stages of the exocytotic process. These mechanisms may all be membrane delimited and may not require the intervention of cytosolic second messengers.

References

1. Patel YC, Tannenbaum GS, eds. Somatostatin: basic and clinical aspects. Metabolism 1990;39(suppl 2):1–191.
2. Inoue M, Yoshii M. Modulation of ion channels by somatostatin and acetylcholine. Prog Neurobiol 1992;38:203–30.
3. Thorner MO, Holl RW, Leong DA. The somatotrope: an endocrine cell with functional calcium transients. J Exp Biol 1988;139:169–79.
4. Mason WT, Rawlings SR, Cobbett P, et al. Control of secretion in anterior pituitary cells—linking ion channels, messengers and exocytosis. J Exp Biol 1988;139:287–316.
5. Sheppard MS, Moor BC, Kraicer J. Release of growth hormone (GH) from purified somatotrophs: interaction of GH-releasing factor and somatostatin and role of adenosine 3′,5′-monophosphate. Endocrinology 1985;117:2364–70.
6. Narayanan N, Lussier B, French M, Moor B, Kraicer J. Growth hormone-releasing factor-sensitive adenylate cyclase system of purified somatotrophs: effects of guanine nucleotides, somatostatin, calcium, and magnesium. Endocrinology 1989;124:484–95.
7. Lussier BT, French MB, Moor BC, Kraicer J. Free intracellular Ca^{2+} concentration ($[Ca^{2+}]_i$) and growth hormone release from purified rat somatotrophs, I. GH-releasing factor-induced Ca^{2+} influx raises $[Ca^{2+}]_i$. Endocrinology 1991;128:570–82.
8. Lussier BT, Wood DA, French MB, Moor BC, Kraicer J. Free intracellular Ca^{2+} concentration ($[Ca^{2+}]_i$) and growth hormone release from purified

rat somatotrophs, II. Somatostatin lowers $[Ca^{2+}]_i$ by inhibiting Ca^{2+} influx. Endocrinology 1991;128:583–91.
9. Lussier BT, French MB, Moor BC, Kraicer J. Free intracellular Ca^{2+} concentration and growth hormone (GH) release from purified rat somatotrophs, III. Mechanism of action of GH-releasing factor and somatostatin. Endocrinology 1991;128:592–603.
10. Lussier BT, Moor BC, French MB, Kraicer J. Release of growth hormone from purified somatotrophs: effects of the calcium channel antagonists diltiazem and nifedipine on release induced by growth hormone-releasing factor. Can J Physiol Pharmacol 1988;66:1373–80.
11. Cowan JS, Moor B, Chow A, Kraicer J. Characteristics of the post-somatostatin rebound in growth hormone secretion from perifused somatotrophs. Endocrinology 1983;113:1056–61.
12. Kraicer J, Cowan JS, Sheppard MS, Lussier B, Moor BC. Effect of somatostatin withdrawal and growth hormone (GH)-releasing factor on GH release in vitro: amount available for release after disinhibition. Endocrinology 1986;119:2047–51.
13. Rudy B. Diversity and ubiquity of K channels. Neuroscience 1988;25:729–49.
14. Castle NA, Haylett DG, Jenkinson DH. Toxins in the characterization of potassium channels. Trends Neurosci 1989;12:59–65.
15. Alberts B, Bray D, Lewis J, Raff M, Roberts K, Watson JD. Molecular biology of the cell. New York: Garland, 1989.
16. Reuter H. Modulation of ion channels by phosphorylation and second messengers. News Physiol Sci 1987;2:168–71.
17. Israel J-M, Denef C, Vincent J-D. Electrophysiological properties of normal somatotrophs in culture. Neuroendocrinology 1983;37:193–9.
18. Holl RW, Thorner MO, Mandell GL, Sullivan JA, Sinha YN, Leong DA. Spontaneous oscillations of intracellular calcium and growth hormone secretion. J Biol Chem 1988;263:9682–5.
19. Yamashita N, Shibuya N, Ogata E. Hyperpolarization of the membrane potential caused by somatostatin in dissociated human pituitary adenoma cells that secrete growth hormone. Proc Natl Acad Sci USA 1986;83:6198–202.
20. Sims SM, Lussier BT, Kraicer J. Somatostatin activates an inwardly rectifying K^+ conductance in freshly dispersed rat somatotrophs. J Physiol (Lond) 1991;441:615–37.
21. DeRiemer SA, Sakmann B. Two calcium currents in normal rat anterior pituitary cells identified by a plaque assay. Exp Brain Res 1986;14:139–54.
22. Ozawa S, Sand O. Electrical activity of rat anterior pituitary cells in vitro. Acta Physiol Scand 1978;102:330–41.
23. Stojilkovic SS, Catt KJ. Calcium oscillations in anterior pituitary cells. Endocr Rev 1992;13:256–80.
24. Lewis DL, Goodman MB, St John P, Barker JL. Calcium currents and fura-2 signals in fluorescence-activated cell sorted lactotrophs and somatotrophs of rat anterior pituitary. Endocrinology 1988;123:611–21.
25. Mason WT, Rawlings SR. Whole-cell recording of ionic currents in bovine somatotrophs and their involvement in growth hormone secretion. J Physiol (Lond) 1988;405:577–93.
26. Chen C, Zhang J, Vincent J-D, Israel J-M. Sodium and calcium currents in action potentials of rat somatotrophs: their possible functions in growth hormone secretion. Life Sci 1990;46:983–9.

27. Chen C, Israel J-M, Vincent J-D. Electrophysiological responses to somatostatin of rat hypophysial cells in somatotroph-enriched primary cultures. J Physiol (Lond) 1989;408:493–510.
28. Dubinsky JM, Oxford GS. Dual modulation of K channels by thyrotropin-releasing hormone in clonal pituitary cells. Proc Natl Acad Sci USA 1985;82: 4282–6.
29. Schlegel W, Winiger BP, Mollard P, et al. Oscillations of cytosolic Ca^{2+} in pituitary cells due to action potentials. Nature 1987;329:719–21.
30. Mollard P, Vacher P, Dufy B, Barker J. Somatostatin blocks Ca^{2+} action potential activity in prolactin-secreting pituitary tumor cells through coordinate actions on K^+ and Ca^{2+} conductances. Endocrinology 1988;123: 721–32.
31. Patel YC, Murthy KK, Escher EE, Banville D, Spiess J, Srikant CB. Mechanism of action of somatostatin: an overview of receptor function and studies of the molecular characterization and purification of somatostatin receptor proteins. Metabolism 1990;39(suppl 2):63–9.
32. Rens-Domiano S, Reisine T. Biochemical and functional properties of somatostatin receptors. J Neurochem 1992;58:1987–96.
33. Bell GI, Reisine T. Molecular biology of somatostatin receptors. Trends Neurosci (in press).
34. Yamada Y, Post SR, Wang K, Tager HS, Bell GI, Seino S. Cloning and functional characterization of a family of human and mouse somatostatin receptors expressed in brain, gastrointestinal tract, and kidney. Proc Natl Acad Sci USA 1992;89:251–5.
35. Kluxen F-W, Bruns C, Lübbert H. Expression cloning of a rat brain somatostatin receptor cDNA. Proc Natl Acad Sci USA 1992;89:4618–22.
36. Yasuda K, Rens-Domiano S, Breder CD, et al. Cloning of a novel somatostatin receptor, SSTR3, coupled to adenylylcyclase. J Biol Chem 1992; 267:20422–8.
37. Sheppard MS, Spence JW, Kraicer J. Release of growth hormone from purified somatotrophs: role of adenosine 3′-5′-monophosphate and guanosine 3′-5′-monophosphate. Endocrinology 1979;105:261–8.
38. Vale W, Brazeau P, Grant G, et al. Premières observations sur le mode d'action de la somatostatine, un facteur hypothalamique qui inhibe la sécrétion de l'hormone de croissance. CR Acad Sci (Paris) 1972;275:2913–6.
39. Sheppard MS, Kraicer J, Milligan JV. Mechanisms governing the release of growth hormone from acutely dispersed purified somatotrophs. In: Jutisz M, McKerns KW, eds. Synthesis and release of adenohypophyseal hormones. Plenum, 1980:495–523.
40. Kraicer J, Chow AEH. Release of growth hormone from purified somatotrophs: use of perifusion system to elucidate interrelations among Ca^{2+}, adenosine 3′,5′-monophosphate, and somatostatin. Endocrinology 1982; 111:1173–80.
41. Kraicer J, Spence JW. Release of growth hormone from purified somatotrophs: use of high K^+ and the ionophore A23187 to elucidate interrelations among Ca^{++}, adenosine 3′,5′-monophosphate, and somatostatin. Endocrinology 1981;108:651–7.
42. French MB, Moor BC, Lussier BT, Kraicer J. Protein kinase C is not essential for growth hormone (GH)-releasing factor-induced GH release from rat somatotrophs. Endocrinology 1989;124:2235–44.

43. French MB, Lussier BT, Moor BC, Kraicer J. Effect of growth hormone-releasing factor on phosphoinositide hydrolysis in somatotrophs. Mol Cell Endocrinol 1990;72:221–6.
44. French MB, Moor BC, Lussier BT, Kraicer J. Growth hormone-releasing factor does not activate protein kinase C in somatotrophs. Mol Cell Endocrinol 1991;79:139–46.
45. Schweitzer P, Madamba S, Siggins GR. Arachidonic acid metabolites as mediators of somatostatin-induced increase of neuronal M-current. Nature 1990;346:464–7.
46. Wollheim CB, Winiger BP, Ullrich S, Wuarin F, Schlegel W. Somatostatin inhibition of hormone release: effects on cytosolic Ca^{++} and interference with distal secretory events. Metabolism 1990;39(suppl 2):101–4.
47. Holl RW, Thorner MO, Leong DA. Intracellular calcium concentration and growth hormone secretion in individual somatotropes: effects of growth hormone-releasing factor and somatostatin. Endocrinology 1988;122:2927–32.
48. Rawlings SR, Hoyland J, Mason WT. Calcium homeostasis in bovine somatotrophs: calcium oscillations and calcium regulation by growth hormone-releasing hormone and somatostatin. Cell Calcium 1991;12:403–14.
49. Horn R, Marty A. Muscarinic activation of ionic currents measured by a new whole-cell recording method. J Gen Physiol 1988;92:145–59.
50. Sakmann B, Trube G. Conductance properties of single inwardly rectifying potassium channels in ventricular cells from guinea-pig heart. J Physiol (Lond) 1984;347:641–57.
51. Hille B. Ionic channels of excitable membranes. 2nd ed. Sunderland, MA: Sinauer, 1992.
52. Pennefather P, Heisler S, MacDonald JF. A potassium conductance contributes to the action of somatostatin-14 to suppress ACTH secretion. Brain Res 1988;444:346–50.
53. Mihara S, North RA, Surprenant A. Somatostatin increases an inwardly rectifying potassium conductance in guinea-pig submucous plexus neurones. J Physiol (Lond) 1987;390:335–55.
54. Inoue I, Nakajima S, Nakajima Y. Somatostatin induces an inward rectification in rat locus coeruleus neurones through a pertussis toxin-sensitive mechanism. J Physiol (Lond) 1988;407:177–98.
55. Lewis DL, Clapham DE. Somatostatin activates an inwardly rectifying K^+ channel in neonatal rat atrial cells. Pflugers Arch 1989;414:492–4.
56. Yamashita N, Shibuya N, Ogata E. Requirement of GTP on somatostatin-induced K^+ current in human pituitary tumor cells. Proc Natl Acad Sci USA 1988;85:4924–8.
57. Koch BD, Blalock JB, Schonbrunn A. Characterization of the cyclic-AMP-independent actions of somatostatin in GH cells, I. An increase in potassium conductance is responsible for both the hyperpolarization and the decrease in intracellular free calcium produced by somatostatin. J Biol Chem 1988;263:216–25.
58. Chen C, Zhang J, Vincent J-D, Israel J-M. Somatostatin increases voltage-dependent potassium currents in rat somatotrophs. Am J Physiol 1990;259:C854–61.
59. White RE, Schonbrunn A, Armstrong DL. Somatostatin stimulates Ca^{2+}-activated K^+ channels through protein dephosphorylation. Nature 1991;351:570–3.

60. Chen C, Zhang J, Vincent J-D, Israel J-M. Two types of voltage-dependent calcium current in rat somatotrophs are reduced by somatostatin. J Physiol (Lond) 1990;425:29–42.
61. Gomperts BD. G_E: a GTP-binding protein mediating exocytosis. Annu Rev Physiol 1990;52:591–606.
62. Luini A, De Matteis MA. Evidence that receptor-linked G protein inhibits exocytosis by a post-second-messenger mechanism in AtT-20 cells. J Neurochem 1990;54:30–8.

3

Biochemical and Functional Properties of Somatostatin Receptors

Terry Reisine, Karen Raynor, Haeyoung Kong, Susan F. Law, Magali Theveniau, Melanie Tallent, Stephanie Rens-Domiano, John Hines, Kazuki Yasuda, and Graeme I. Bell

The tetradecapeptide *somatostatin* (SRIF) exerts diverse biological actions in the endocrine and nervous systems (1, 2). It is the major physiological inhibitor of growth hormone secretion from the anterior pituitary and insulin and glucagon release from the pancreatic islets and also regulates gastric acid secretion from the gut (3–6). SRIF is also expressed in extrahypothalamic locations in the brain, where it has a role in controlling cognitive functions (7, 8). Furthermore, in the striatum, where it has been implicated in modulating locomotor activity (9, 10), SRIF is a major stimulant of dopamine release from nigrostriatal neurons and may interact with dopamine in modulating basal ganglia functions (11).

At the cellular level, SRIF affects multiple signal transduction pathways. It inhibits adenylyl cyclase activity (12, 13) and Ca^{++} conductance (14) and potentiates K^+ currents (15–17). These cellular actions involve pertussis toxin-sensitive G-proteins that functionally couple SRIF receptors to these different effector systems. These biological actions of SRIF have been implicated in the regulation of neuronal firing activity by SRIF and its ability to alter neurotransmitter and hormone release. SRIF has also been reported to modulate Na^+/H^+ exchange to affect intracellular pH (18) and to regulate tyrosine phosphatase activity (19). These effects are pertussis toxin insensitive and have been linked to antiproliferative actions of SRIF on cell growth (20).

SRIF induces its biological actions by interacting with membrane-associated receptors. Functional, pharmacological, and biochemical evidence shows that subtypes of SRIF receptors are expressed in mammalian cells that mediate the diverse biological actions of SRIF (21–27). The recent cloning of five different SRIF receptors (28–36) supports the hypothesis that multiple SRIF receptor subtypes exist and provides the means to study the individual characteristics of each receptor and to develop compounds that are subtype selective, which could be clinically useful. This chapter emphasizes studies that have identified the properties of the cloned receptors and attempts to link that information with our knowledge of the characteristics of SRIF receptor subtypes endogenously expressed in mammalian cells.

The Cloned SRIF Receptors SSTR1, SSTR2, and SSTR3

The first three SRIF receptor genes to be cloned were isolated from mouse and human genomic libraries (28–30), as well as rat cDNA libraries (31–33). They are intronless and encode proteins having the typical 7-transmembrane structure characteristic of other G-protein-linked receptors. The amino acid sequences of the cloned receptors (referred to as SSTR1, SSTR2, and SSTR3) are 45%–50% identical among themselves, but are different from any other neurotransmitter or hormone receptor (see Fig. 3.1). Regions of major sequence divergence among the SRIF receptor subtypes are most apparent in the extracellular N-terminal domain and the intracellular cytoplasmic tail. The receptors are characterized by relatively small third intracellular loops, which have been implicated as the G-protein-coupling domains of other receptors. Each receptor has multiple potential glycosylation sites. In particular, SSTR2 has five potential sites.

The distributions in the body of the first three cloned SRIF receptor mRNAs have been determined by Northern analysis and in situ hybridization. In humans, SSTR1 mRNA is primarily expressed in the stomach and gut, SSTR2 mRNA is expressed in kidney and brain, and SSTR3 mRNA is primarily expressed in brain (28, 30, 36). In rodents, all three receptor mRNAs are expressed in brain, with overlapping but distinct distributions (31–33, 36, 37). For example, all three receptor mRNAs are expressed in the rat hippocampus, but SSTR2 mRNA is mainly expressed in the dentate gyrus, while SSTR3 mRNA is expressed in the dentate gyrus and CA3 region (36, unpublished results). This latter finding is of interest since ^{125}I-CGP 23996, which binds to SSTR3 with high affinity, does not label sites in CA3 region. However, there is a high density of ^{125}I-CGP 23996 binding sites in CA1 region (38), an area of the hippocampus that receives a major input from CA3, suggesting that SSTR3 may be localized presynaptically on CA1 nerve terminals originating from CA3. In contrast, the SSTR2-specific ligand ^{125}I-MK 678 heavily labels the dentate gyrus (38), indicating that this receptor may in part be localized to interneurons or cell bodies in this brain region. Such differential

```
                                                                     <----- M1------
SSTR1 1 MFPNGTASSPSSSPSPSPGSCGEGACSRGPGSGAADGMEEPGRNASQNGTLSEGQGSAILISFIYSVVCLVG  72
SSTR2              1 MEMSSEQLNGSQVWVSSPFDLNGSLGPSNGSNQTEPYYDMTSNAVLTFIYFVVCVVG  57
SSTR3            1 MATVTYPSSEPMTLDPGNTSSTWPLDTTLGNTSAGASLTGLAVSGILISLVYLVVCVVG  59

        ----------->        <-------- M2 -------->                  <---- M3 ----
SSTR1   LCGNSMVIYVILRYAKMKTATNIYILNLAIADELLMLSVPFLVTSTLLRHWPFGALLCRLVLSVDAVNMFTS 144
SSTR2   LCGNTLVIYVILRYAKMKTITNIYILNLAIADELFMLGLPFLAMQVALVHWPFGKAICRVVMTVDGINQFTS 129
SSTR3   LLGNSLVIYVVLRHTSSPSVTSVYILNLALADELFMLGLPFLAAQNALSYWPFGSLMCRLVMAVDGINQFTS 131

        -------->                 <-------- M4 --------->
SSTR1   IYCLTVLSVDRYVAVVHPIKAARYRRPTVAKVVNLGVWVLSLLVILPIVVFSRTAANSDGTVACNMLMPEPA 216
SSTR2   IFCLTVMSIDRYLAVVHPIKSAKWRRPRTAKMINVAVWCVSLLVILPIMIYAGLRSNQWGRSSCTINWPGES 201
SSTR3   IFCLTVMSVDRYLAVVHPTRSARWRTAPVARTVSRAVWVASAVVVLPVVVFSGVP:::RGMSTCHMQWPEPA 200

          <--------- M5 ---------->                                       <----
SSTR1   QRWLVGFVLYTFLMGFLLPVGAICLCYVLIIAKMRMVALKAGW::::::::::QQRKRSERKITLMVMMVV 277
SSTR2   GAWYTGFIIYAFILGFLVPLTIICLCYLFIIIKVKSSGIRVGS::::::::::SKRKKSEKKVTRMVSIVV 262
SSTR3   AAWRTAFIIYMAALGFFGPLLVICLCYLLIVVKVRSTTRRVRAPSCQWVQAPACQRRRRSERRVTRMVVAVV 272

        ------- M6 ------->             <--------- M7 ---------->
SSTR1   MVFVICWMPFYVVQLVNVFA::EQDDATVSQ::LSVILGYANSCANPILYGFLSDNFKRSFQRILCLSWMDN 345
SSTR2   AVFIFCWLPFYIFNVSSVSVAISPTPALKGMFDFVVILTYANSCANPILYAFLSDNFKKSFQNVLCLVKVSG 334
SSTR3   ALFVLCWMPFYLLNIVNVVCPLPEEPAFFGLYFLVVALPYANSCANPILYGFLSYRFKQGFRRILLRPSRRI 344

SSTR1   AAEEPVDYYATALKSRAYSVEDFQPENLESGGVFRNGTCASRISTL - 391
SSTR2   TEDGERSDSKQDKSRLNETTETQRTLLNGDLQTSI - 369
SSTR3   RSQEPGSGPPEKTEEEEDEEEEERREEEERRMQRGQEMNGRLSQIAQAGTSGQQPRPCTGTAKEQQLLPQEA

SSTR3   TAGDKASTLSHL - 428
```

FIGURE 3.1. Predicted amino acid sequences of mouse SSTR1, SSTR2, and SSTR3. Residues in bold are consensus residues. Predicted transmembrane spanning regions are indicated by dashes in the enclosed arrows. Sequences are derived from references 28 and 29.

localization of the SRIF receptor subtypes may have important implications with regard to their different functional roles in brain.

With regard to the localization of SSTR3 mRNA, highest levels of SSTR3 mRNA in rat brain are found in the cerebellum, specifically the granule cell layer (29, 32, 36, unpublished results). This finding is unusual since the adult cerebellum has very low levels of SRIF receptor binding sites compared to most other brain regions (1, 2, 24, 38). However, cerebellum does express relatively moderate levels of SRIF receptors very early in development (39, 40). SRIF receptors are detected just before birth in the cerebellum, peak in the second week after birth, and then fall to just barely detectable levels in the adult. SRIF receptor binding sites are primarily localized to the granule cell layer in the very young cerebellum. The appearance of SSTR3 mRNA in the adult cerebellum may be a vestige of the early developmental appearance of the receptor. However, recent studies by Meyerhof et al. (32) have shown very little detectable SSTR3 mRNA in cerebellum at ages when SRIF receptor binding sites are detected, suggesting little or no relationship between the high levels of SSTR3 mRNA in the adult cerebellum and the early appearance of SRIF receptors in cerebellum. The SSTR3 mRNA in adult cerebellum, therefore, may not be translated, or the expressed protein may not be processed appropriately to bind agonists with high affinity. In contrast, the early appearance of SRIF receptors does correlate with the patterns of expression of SSTR1 and SSTR2 in cerebellum (32, unpublished results), suggesting that these receptor subtypes may be expressed early in development and may be involved in mediating ontogenic effects of SRIF in this brain region.

By Northern analysis of SSTR2 mRNA in rat brain and pituitary, 2 distinct transcripts of 2.3 and 2.8 kb can be detected (36, unpublished results). Interestingly, the relative levels of the two SSTR2 mRNAs differ in brain regions as compared to peripheral tissues such as the pituitary and adrenal cortex. Whether these different mRNAs encode biologically distinct molecules has not been established. These two mRNAs may correspond to splice variants recently described by Vanetti et al. (41), who reported that two forms of SSTR2 mRNA exist with coding regions containing differing C-terminal regions.

Pharmacological Properties of the Cloned SRIF Receptors

In addition to variations in tissue expression, the pharmacological properties of the three cloned receptors are different. Initial characterization of the properties of the receptors was conducted on *Chinese hamster ovary* (CHO) cells stably expressing each cloned receptor (29, 36, 42). All three receptors bound SRIF and its N-terminally extended peptide SRIF-

28 with high and similar affinity. Two rigid analogs of SRIF, the cyclohexapeptide MK 678 and the octapeptide SMS-201-995, have high potency for SSTR2, but low affinity for SSTR3, and do not interact with SSTR1 at 1 μM concentrations. Furthermore, ^{125}I-MK 678 selectively labels SSTR2, indicating that this peptide is a specific SSTR2 ligand. In contrast, the peptide c(Ahep-Phe-D-Trp-Lys-Thr[Bzl]) and several structural analogs inhibit ^{125}I-Tyr^{11} SRIF binding to SSTR1 and SSTR3, but not SSTR2, further indicating differences in peptide specificities of the cloned receptors. Interestingly, the peptide CGP 23996 potently binds to SSTR2 and SSTR3 with low nanomolar IC_{50} values, but is impotent at binding to SSTR1 (36). However, iodinated CGP 23996 potently binds to SSTR1, as well as the other cloned SRIF receptors. This finding suggests that iodination of CGP 23996 greatly increases its affinity for SSTR1, possibly by modifying the conformation of the peptide.

In addition to differences in peptide specificities, the cloned receptors vary in their coupling to G-proteins. In the presence of GTP or its stable analogs, G-protein coupling to receptors is diminished and has been shown in many cases to decrease affinity of the receptors for agonist. GTPgS, a highly stable analog of GTP, has been reported to diminish SRIF receptor/G-protein coupling and cause a loss of high-affinity agonist binding to the receptor (1, 2, 13, 24, 25, 27). GTPgS diminishes agonist binding to SSTR2 and SSTR3, but not to SSTR1 when expressed in CHO cells (29, 42). Pertussis toxin, an agent that catalyzes the ADP-ribosylation of G_i and G_o, thereby inactivating these G-proteins and uncoupling them from receptors and reducing the affinity of the receptors for agonists (43), has been used to probe G-protein coupling to the cloned SRIF receptors. Pretreatment of CHO cells expressing SSTR2 or SSTR3 with pertussis toxin abolishes subsequent high-affinity agonist binding to these cloned receptors (29, 42). Similar treatment of CHO cells expressing SSTR1 with pertussis toxin did not alter the binding of agonists to this receptor (42). This finding suggests that SSTR1 expressed in CHO (DG44) cells does not efficiently couple to G-proteins, whereas both SSTR2 and SSTR3 associate with pertussis toxin-sensitive G-proteins.

To directly identify the G-proteins coupled to SSTR2, we have solubilized SSTR2 from CHO cells and immunoprecipitated or uncoupled SSTR2/G-protein complexes using peptide-directed antisera against the different pertussis toxin-sensitive G-proteins (44). Using this approach previously to study SRIF receptor/G-protein interactions in rat brain and the pituitary-derived tumor cell line AtT-20, we reported that SRIF receptors interact with G_{ia1}, G_{ia3} and G_{oa}, but not G_{ia2} or G_{sa} (45, 46). The results of studies with the cloned SSTR2 in CHO cells suggest that SSTR2 primarily associates with G_{ia3} and G_{oa} (44). Furthermore, in the human kidney-derived cell line HEK 293, which endogenously expresses SSTR2 and similar levels of G_{ia1} and G_{ia3} but not G_{oa}, SSTR2 was observed to associate primarily with G_{ia3}. These findings are of interest because SSTR2 did not mediate SRIF inhibition of cAMP formation when expressed in COS cells that express all forms of G_{ia} (36, 42) or in 293 cells (44). In previous studies, we have suggested that G_{ia1} is critical in coupling SRIF receptors to adenylyl cyclase since antiserum directed against G_{ia1} blocked the ability of SRIF to inhibit forskolin-stimulated adenylyl cyclase activity in membranes from AtT-20 cells, whereas antisera directed against G_{ia2}, G_{ia3} or G_{oa} did not (47). The inability of SSTR2 to associate effectively with G_{ia1} may explain its lack of functional coupling to adenylyl cyclase.

While SSTR2 may not efficiently couple with adenylyl cyclase, its association with G_{ia3} and G_{oa} raises the possibility that it may mediate effects of SRIF on ionic conductances. SRIF inhibits voltage-dependent Ca^{++} currents in brain neurons and pituitary cells (14, 48, 49). Furthermore, it has been shown to potentiate a delayed rectifying K^+ current in brain neurons and an inward rectifying current in pituitary cells (15, 16, 48, 49). Both actions of SRIF are pertussis toxin sensitive and G-protein mediated. Recent "mRNA knockout" studies of Kleuss et al. (50) indicate that G_{oa} selectively couples SRIF receptors to Ca^{++} channels in rat pituitary GH_3 cells. Similarly, reconstitution studies of Yatani et al. (51) show that G_{ia3} couples SRIF receptors to K^+ channels. Furthermore, MK 678, a specific SSTR2 agonist, inhibits Ca^{++} conductance and potentiates K^+ currents in both brain and pituitary-derived cells (48, 49). Therefore, SSTR2 may be able to associate with G_{ia3} and G_{oa} to mediate effects of SRIF on K^+ and Ca^{++} currents, respectively.

In contrast to SSTR2, SSTR3 mediates the inhibitory effects of SRIF on adenylyl cyclase (29, 36) and appears to be able to associate with G_{ia1}. This is suggested by recent studies (Law, Reisine, unpublished results) that showed that SSTR3 in CHO cells lacking significant levels of G_{ia1} did not functionally couple to adenylyl cyclase, but in cells coexpressing both G_{ia1} and SSTR3, SRIF was able to inhibit adenylyl cyclase activity. This finding supports our hypothesis that G_{ia1} is involved in coupling SRIF receptors to adenylyl cyclase.

The third intracellular loops of many neurotransmitter receptors have been proposed to contain G-protein-coupling domains of the receptors and are critical regions in the association of receptors with effector systems. The third intracellular loops of the cloned SRIF receptors are fairly divergent, and the size of the loop of SSTR3 is longer than the other two subtypes. Determination of whether this part of the receptor confers specific G-protein coupling would be of great interest.

Physical Analysis of SSTR2

To investigate the biochemical properties of SSTR2, we have recently developed peptide-directed antibodies against this receptor (52). Two antibodies were generated, one against the third extracellular loop of the receptor (2e3) and the other directed against the cytoplasmic tail of the receptor (2i4). The 2e3 was able to specifically detect SSTR2 by immunoblotting. The antibody reacted with two proteins of 93 and 148 kd from CHO cells expressing SSTR2. The predominant immunoreactive material was 93 kd. Similarly, 2e3 detected 87 and 148 kd proteins from AtT-20 cells, a mouse pituitary-derived cell line expressing high levels of SSTR2 mRNA. Consistent with these findings, both 2e3 and 2i4 immunoprecipitated 93 and 148 kd ^{35}S-methionine-labeled proteins from CHO cells expressing SSTR2, and 2i4 specifically immunoprecipitated high-affinity ^{125}I-MK 678 binding sites from both CHO cells expressing SSTR2 and AtT-20 cells.

Photocrosslinking studies with iodinated SRIF radioligands revealed a prominent labeled SRIF receptor of approximately 90 kd in both CHO cells expressing SSTR2 and AtT-20 cells. Furthermore, in AtT-20 cells a higher molecular weight protein of approximately 150 kd was detected. These sizes correspond to the masses of the proteins specifically detected by immunoblotting and immunoprecipitated by the SSTR2 antibodies, indicating that these two proteins are both SRIF receptors. Interestingly, 2e3 only detected a 148-kd protein from rat brain and pituitary, as well as the cell lines AR42J and HEK 293, all of which express SSTR2 mRNA. This protein is a SRIF receptor, since both 2e3 and 2i4 immunoprecipitated high-affinity solubilized ^{125}I-MK 678 binding sites from rat brain. The expression of only a single isoform of SSTR2 in these tissues suggests that the expression of the 90 kd SSTR2 is selectively regulated.

The predicted size of SSTR2 from its amino acid sequence is 42 kd (28). The large size of the SSTR2 isoforms detected with the antibodies suggests this receptor undergoes significant posttranslation processing. This is further indicated by the finding that 2e3 immunoprecipitated 42-kd in vitro translation products from mRNA extracted from CHO cells expressing SSTR2.

Previous studies have indicated that SRIF receptors from rat brain and AtT-20 cells, as well as pancreas and pituitary, are heavily glycosylated, and oligosaccharides associated with SSTR2 may significantly contribute to its mass (27, 53). In fact, SSTR2 has 5 potential glycosylation sites in its sequence that may serve as acceptors of various sugars. In addition to contributing to its mass, oligosaccharides in SSTR2 may be important for the functioning of this receptor, since cleavage of selective oligosaccharide from brain and AtT-20 cell SRIF receptors, as well as SSTR2, greatly reduces high-affinity binding of ^{125}I-MK 678 to the receptor (53, 54).

To investigate further the role of oligosaccharides in the functioning of SSTR2, we have used site-directed mutagenesis to convert the asparagines of each potential glycosylation site in SSTR2 to glutamines so as to prevent the incorporation of carbohydrates into the protein (54, Yasuda, Law, Bell, Reisine, unpublished results). Each mutant receptor was then transiently expressed in COS cells and labeled with either the SSTR2-selective ligand ^{125}I-MK 678 or ^{125}I-CGP 23996, which labels all SRIF receptors. Mutations at each individual glycosylation site did not alter the binding of either ligand to the receptor. However, consecutive mutations of the N-terminal, 4 glycosylation sites, or all the glycosylation sites greatly reduced the affinity of the receptor for ^{125}I-MK 678, but did not appreciably affect the binding of the larger peptide, ^{125}I-CGP 23996. This finding indicates the glycosylation of SSTR2 was necessary for the receptor to bind the small peptide, ^{125}I-MK 678, with high affinity, but was not required for larger peptides to interact with the receptor. Since SSTR2 is the only cloned SRIF receptor that has very high affinity for MK 678, glycosylation of this receptor subtype may create the pharmacological selectivity by imposing specific confirmational constraints on the receptor. Further mutational analysis of the sites involved in the selective binding of MK 678 to SSTR2 suggests that the first glycosylation site at residue 9 of the receptor is likely to be glycosylated and involved in ligand binding, since mutations of the second, third, and fourth potential glycosylation sites did not reduce the affinity of the receptor for ^{125}I-MK 678. Since an individual mutation of the first glycosylation site did not reduce agonist binding, sugars incorporated into this site must act in concert with oligosaccharide at other sites within the receptor to affect agonist binding. Further mutational analysis should reveal the location of those other sites.

In addition to glycosylation, another posttranslational modification that may affect the functioning of SSTR2 is phosphorylation. SRIF receptors desensitize following continuous exposure to agonist (2, 16, 55, 56). Prolonged treatment of SSTR2 expressed in CHO cells results in a decrease in the affinity of the receptor for agonists, which would be consistent with desensitization (42). SRIF receptors in brain neurons in culture with similar pharmacological characteristics as SSTR2 desensitize when continuously treated with SRIF and have a diminished capability to mediate the potentiation of K^+ currents by SRIF (16). Receptors with similar characteristics to SSTR2 in AtT-20 and GH_3 cells, when exposed to SRIF or the SSTR2 selective agonist MK 678, have a reduced affinity for agonists (55, 56). The cellular mechanisms involved in SRIF receptor desensitization have been proposed to involve the phosphorylation of the receptor catalyzed by the enzyme *beta-adrenergic receptor kinase* (BARK). Mayor et al. (57) reported that continuous exposure of SRIF receptors in S49 lymphoma cells to SRIF diminished the ability of SRIF to inhibit adenylyl cyclase activity and induced the translocation of BARK activity in the cells. Furthermore, exposure of AtT-20 cell membranes to SRIF, BARK, and ATP caused a decrease in the specific binding of the SSTR2-selective agonist ^{125}I-MK 678 (58). This effect was blocked by the BARK inhibitor dextran and did not occur in the absence of SRIF, consistent with previous experimental evidence that BARK only modulated the activity of the agonist-occupied receptor.

To test directly whether SSTR2 is phosphorylated and regulated by agonist treatment, we (Hines, Theveniau, Benovic, Reisine, unpublished results) showed that pretreatment of CHO cells expressing SSTR2 with 10 nM SRIF caused a decrease in high-affinity agonist binding that was reversible. Using the SSTR2-directed antibody 2e3, it was shown that SRIF treatment of the CHO cells expressing SSTR2 increased the incorporation of ^{32}P-phosphate into the receptor. In contrast, prolonged treatment of the cells with SRIF did not modify the levels of immunoreactive SSTR2, suggesting that the receptor was not down-regulated, but that instead the decrease in high-affinity agonist binding is due to its phosphorylation. These CHO cells express BARK immunoreactivity, and in preliminary studies, exposure of CHO cell membranes to SRIF and purified BARK caused a specific decrease in high-affinity ^{125}I-MK 678 binding, indicating that BARK may have catalyzed the in vitro phosphorylation of the agonist-occupied SSTR2 to decrease the affinity of the receptor for agonist. These preliminary findings support the hypothesis that phosphorylation may be involved in the desensitization of SSTR2.

Cloned SRIF Receptor SSTR4 Is Pituitary Specific and May Be Directly Linked to Inhibition of Growth Hormone Release

Following the cloning of the first three SRIF receptors, O'Carroll et al. (34), identified a fourth SRIF receptor whose amino acid sequence and characteristics were distinct from the other receptor subtypes. SSTR4 is a 7-transmembrane-spanning, G-protein-linked receptor consisting of 383 amino acids. It has considerable sequence similarity to the first three cloned SRIF receptors. Northern analysis reveals high levels of expression of SSTR4 mRNA in the pituitary, but no mRNA is detected in the rat brain, pancreas, or kidney. The expression of SSTR4 mRNA only in the pituitary is unique among the cloned SRIF receptors and suggests that this subtype may preferentially mediate effects of SRIF on pituitary hormone secretion. Pharmacological analysis of the cloned SSTR4 expressed in CHO cells reveals a higher affinity of the receptor for SRIF-28 than SRIF, further distinguishing this receptor from the first three cloned SRIF receptor subclasses. Previously, a SRIF-28-preferring receptor had been identified in pancreas and was shown to be coupled to the inhibition of insulin release (21), and SRIF-28 was shown to inhibit a delayed rectifier K^+ current in rat brain neurons in culture (15, 16), whereas SRIF potentiated this same current. Furthermore, receptor binding studies have shown that the rat pituitary preferentially expresses a receptor with higher affinity for SRIF-28 than SRIF (59). SSTR4 may be the receptor subtype described in these previous studies. Interestingly, SSTR4 has very high affinity for SMS-201-995, but over 50-fold lower affinity for MK

678. This is the first receptor that has such differing affinities for these two rigid analogs of SRIF. SSTR4 can mediate the inhibition of cAMP by SRIF when expressed in CHO cells, indicating that it couples to G-proteins. The unique tissue localization of SSTR4 mRNA and pharmacological characteristics of the receptor suggest that this receptor is likely to be the classical SRIF receptor involved in mediating the inhibition of growth hormone secretion by SRIF.

Other SRIF Receptor Subtypes

Pharmacological and biochemical evidence suggests that other SRIF receptors that have not yet been cloned are expressed in mammalian cells. In particular, the striatum expresses moderate densities of SRIF receptors (24, 38) that mediate effects of SRIF on locomotor activity and stereotypy (9, 10) and the stimulation of dopamine release (11). Northern analysis reveals very little, if any, expression of the recently cloned SRIF receptor mRNAs in rat striatum (unpublished results), suggesting that either the striatal SRIF receptors are presynaptic in this brain region and the receptor mRNA is localized to other brain regions, that there is no relationship between mRNA levels and receptor protein expression, or that another SRIF receptor subtype is expressed in this tissue. Consistent with the latter possibility, Theveniau et al. (60) developed an antibody against a brain SRIF receptor that detects SRIF receptors of 60 kd in rat brain and the pituitary cell line AtT-20, but does not interact with rat pituitary SRIF receptors. The striatum expresses the highest levels of immunoreactivity detected by this antibody. This antibody does not crossreact with either SSTR1, SSTR2, or SSTR3 by immunoblotting (61), indicating that it recognizes a protein distinct from the cloned SRIF receptors. This striatal SRIF receptor does not appear to correspond to a recently cloned fifth SRIF receptor described by Bruno et al. (35) since that receptor has no affinity for SMS-201-995, whereas rat striatal SRIF receptors have high affinity for this peptide. Presently, studies are under way to screen a rat striatal cDNA library with the antibody recognizing this unique SRIF receptor subtype to clone this receptor.

Conclusion

The recent cloning of multiple SRIF receptors has now made it possible to study this receptor family in greater detail than previously possible. The cloning of these receptors is greatly facilitating our understanding of their unique functional and biochemical properties. In preliminary studies, we have identified subtype-specific SRIF analogs for each SRIF receptor subtype. These compounds will be employed in examining the biological properties of each receptor subtype and could be clinically useful. This latter possibility may be further potentiated by the apparent tissue-specific expression of some of these receptor subclasses in humans. Such compounds could be effective in treating various endocrine disorders and as antiproliferative agents. Furthermore, development of SRIF receptor-selective analogs that can cross the blood brain barrier may also find use in treatment of central disorders with altered SRIF transmission, such as Alzheimer's disease and epilepsy.

Acknowledgments. This work was supported by the Howard Hughes Medical Institute and NIH Grants MH-45533, MH-48518, DK-20595, DK-42086, and HD-07029.

References

1. Epelbaum J. Somatostatin in the central nervous system: physiology and pathological modifications. Prog Neurobiol 1986;27:63–100.
2. Raynor K, Reisine T. Somatostatin receptors. Crit Rev Neurobiol 1992; 16:273–89.
3. Brazeau P, Vale W, Burgus R, Ling N, Rivier J, Guillemin R. Hypothalamic polypeptide that inhibits the secretion of immunoreactive pituitary growth hormone. Science 1972;129:77–9.
4. Patel Y, Sirkant C. Somatostatin mediation of adenohypophysial secretion. Annu Rev Physiol 1986;48:551–67.
5. Koerker D, Ruch W, Chickadel E, et al. Somatostatin: hypothalamic inhibitor of the endocrine pancreas. Science 1974;184:482–4.
6. Chickadel E, Palmer J, Koerker D, Ensinck J, Davidson M, Goodner C. Somatostatin blockade of acute and chronic stimuli of the endocrine pancreas and consequences of this blockade on glucose homeostasis. J Clin Invest 1975;55:754–7.
7. Haroutunian V, Mantin G, Campell G, Tsuboyama G, Davis K. Cysteamine-induced depletion of central SRIF-like immunoreactivity: effects on behavior, learning, memory, and brain neurochemistry. Brain Res 1987;403:234–42.
8. DeNoble V, Hepler D, Barto R. Cysteamine-induced depletion of somatostatin produces differential cognitive deficits in rats. Brain Res 1989;482:42–8.
9. Rezek M, Havlicek V, Hughes K, Friesen H. Neostriatal administration of somatostatin: differential effect of small and large doses on behavior and motor control. Can J Physiol Pharmacol 1977;55:234–42.
10. Raynor K, Lucki I, Reisine T. $Somatostatin_1$ receptors in the nucleus accumbens selectively mediate the stimulatory effect of somatostatin on locomotor activity in rats. J Pharmacol Exp Ther (in press).
11. Chesselet MF, Reisine T. Somatostatin regulates dopamine release in rat striatal slices and cat caudate nuclei. J Neurosci 1983;3:232–6.
12. Jakobs K, Akotories K, Schultz G. A nucleotide regulatory site for somatostatin inhibition of adenylate cyclase in S49 lymphoma cells. Nature 1983; 303:177–8.

13. Mahy N, Woolkalis M, Thermos K, Carlson K, Manning D, Reisine T. Pertussis toxin modifies the characteristics of both the inhibitory GTP binding proteins and the somatostatin receptor in anterior pituitary tumor cells. J Pharmacol Exp Ther 1988;246:779–85.
14. Wang H, Reisine T, Dichter M. Somatostatin-14 and somatostatin-28 inhibit calcium currents in rat neocortical neurons. Neuroscience 1990;342: 335–42.
15. Wang H, Bogen C, Reisine T, Dichter M. Somatostatin-14 and somatostatin-28 induce opposite effects on potassium currents in rat neocortical neurons. Proc Natl Acad Sci USA 1989;86:9616–20.
16. Wang H, Dichter M, Reisine T. Lack of cross-desensitization of somatostatin-14 and somatostatin-28 receptors coupled to potassium channels in rat neocortical neurons. Mol Pharmacol 1990;38:357–61.
17. Jacquin T, Champagnat J, Madamba S, Denavit-Saubie M, Siggins G. Somatostatin depresses excitability in neurons of the solitary tract complex through hyperpolarization and augmentation of I_M, a non-inactivating voltage-dependent outward current blocked by muscarinic agonist. Proc Natl Acad Sci USA 1988;85:948–52.
18. Barber D, McGuire M, Ganz M. Beta-adrenergic and somatostatin receptors regulate Na^+/H^+ exchange independent of cAMP. J Biol Chem 1989;264: 21038–42.
19. Tahiri-Jouti N, Cambillau C, Viguerie N, et al. Characterization of a membrane tyrosine phosphatase in AR42J cells: regulation by somatostatin. Am J Physiol 1992;262:G1007–14.
20. Viguerie N, Tahiri-Jouti N, Ayral A, et al. Direct inhibitory effects of somatostatin analog SMS-201-995 on AR4-J cell proliferation via pertussis toxin-sensitive GTP binding protein-independent mechanism. Endocrinology 1989;124:1017–25.
21. Mandarino L, Stenner D, Blanchard W, et al. Selective effects of somatostatin-14, -25, and -28 on in vitro insulin and glucagon secretion. Nature 1981; 291:76–8.
22. Reubi JC. Evidence for two somatostatin-14 receptor subtypes in rat brain cortex. Neurosci Lett 1984;49:259–63.
23. Tran V, Beal F, Martin J. Two types of somatostatin receptors differentiated by cyclic somatostatin analogs. Science 1985;228:492–5.
24. Raynor K, Reisine T. Analogs of somatostatin selectively label distinct subtypes of somatostatin receptors in rat brain. J Pharmacol Exp Ther 1989; 251:510–7.
25. Raynor K, Coy D, Reisine T. Analogues of somatostatin bind selectively to brain somatostatin receptor subtypes. J Neurochem 1992;59:1241–50.
26. Sakamoto C, Nagao M, Matozaki T, Nishizaki H, Konda Y, Baba S. Somatostatin receptors on rat cerebrocortical membranes. J Biol Chem 1988;263: 14441–5.
27. Rens-Domiano S, Reisine T. Biochemical and functional properties of somatostatin receptors. J Neurochem 1992;58:1987–96.
28. Yamada Y, Post S, Wang K, Tager H, Bell GI, Seino S. Cloning and functional characterization of a family of human and mouse somatostatin receptors expressed in brain, gastrointestinal tract and kidney. Proc Natl Acad Sci USA 1992;89:251–5.

29. Yasuda K, Rens-Domiano S, Breder C, et al. Cloning of a novel somatostatin receptor, SSTR3, coupled to adenylyl cyclase. J Biol Chem 1992;267:20422–8.
30. Yamada Y, Reisine T, Law SF, et al. Somatostatin receptors, an expanding gene family: cloning and functional characterization of human SSTR3, a protein coupled to adenylyl cyclase. Mol Endocrinol (in press).
31. Li XJ, Forte M, North R, Ross C, Snyder S. Cloning and expression of a rat somatostatin receptor enriched in brain. J Biol Chem 1992;267:21307–12.
32. Meyerhof W, Wulfsen I, Schonrock C, Fehr S, Richter D. Molecular cloning of a somatostatin-28 receptor and comparison of its expression pattern with that of a somatostatin-14 receptor in rat brain. Proc Natl Acad Sci USA 1992;89:10267–71.
33. Kluxen F, Bruns C, Lubbert H. Expression cloning a rat brain somatostatin receptor cDNA. Proc Natl Acad Sci USA 1992;89:4618–22.
34. O'Carroll AM, Lolait S, Konig M, Mahan L. Molecular cloning and expression of a pituitary somatostatin receptor with preferential affinity for somatostatin-28. Mol Pharmacol (in press).
35. Bruno J, Xu Y, Song J, Berelowitz M. Molecular cloning and functional expression of a brain-specific somatostatin receptor. Proc Natl Acad Sci USA (in press).
36. Bell GI, Reisine T. Molecular biology of somatostatin receptors. Trends Neurosci (in press).
37. Breder C, Yamada Y, Yasuda K, Seino S, Saper C, Bell GI. Differential expression of somatostatin receptor subtypes in brain. J Neurosci 1992;12:3920–34.
38. Martin JL, Chesselet MF, Raynor K, Gonzales C, Reisine T. Differential distribution of somatostatin receptor subtypes in rat brain revealed by newly developed somatostatin analogs. Neuroscience 1991;41:581–93.
39. Gonzalez B, Leroux P, Bodenant A, Laquerrier A, Coy D, Vaudry H. Ontogeny of somatostatin receptors in rat brain: biochemical and autoradiographic study. Neuroscience 1989;29:629–44.
40. Gonzalez B, Leroux P, Lamacz M, Bodenant C, Balazs R, Vaudry H. Somatostatin receptors are expressed by immature cerebellar granule cells: evidence for a direct inhibitory effect of somatostatin on neuroblast activity. Proc Natl Acad Sci USA 1992;89:9627–31.
41. Vanetti M, Kouba M, Wang X, Vogt G, Hollt V. Cloning and expression of a novel mouse somatostatin receptor (SSTR2B). FEBS Lett 1992;311:290–4.
42. Rens-Domiano S, Law SF, Yamada Y, Seino S, Bell G, Reisine T. Pharmacological properties of two cloned somatostatin receptors. Mol Pharmacol 1992;42:28–34.
43. Sekura R, Moss J, Vaughan M. Pertussis toxin. New York: Academic Press, 1985.
44. Law SF, Manning D, Reisine T. Identification of the subunits of GTP binding proteins coupled to somatostatin receptors. J Biol Chem 1991;266:17885–97.
45. Law SF, Reisine T. Agonist binding to rat brain somatostatin receptors alters the interaction of the receptors with guanine nucleotide-binding regulatory proteins. Mol Pharmacol 1992;42:398–402.
46. Law SF, Reisine T. The somatostatin receptor expressed in 293 cells and the cloned SRIF receptor SSTR2 selectively couple to G_{ia3}. Mol Biol Cell 1992;3:280a.

47. Tallent M, Reisine T. G_{ia1} selectively couples somatostatin receptors to adenylyl cyclase in AtT-20 cell membranes. Mol Pharmacol 1992;41:452–5.
48. Tallent M, Raynor K, Dichter M, Reisine T. Properties of $somatostatin_1$ receptor subtypes in AtT-20 cells. Soc Neurosci Abst 1992;18:449a.
49. Raynor K, Wang HL, Dichter M, Reisine T. Subtypes of somatostatin receptors couple to multiple cellular effector systems. Mol Pharmacol 1991; 40:248–53.
50. Kleuss C, Hescheler J, Ewel C, Rosenthal W, Schultz G, Wittig B. Assignment of G protein subtypes to specific receptors inducing inhibition of calcium currents. Nature 1991;353:43–8.
51. Yatani A, Codina J, Sekura R, Birnbaumer L, Brown A. Reconstitution of somatostatin and muscarinic receptor mediated stimulation of K^+ channels by G_k protein in clonal rat anterior pituitary cell membranes. Mol Endocrinol 1987;1:283–93.
52. Theveniau M, Raynor K, Yasuda K, Bell G, Reisine T. Identification of two isoforms of the cloned somatostatin receptor, SSTR2 (submitted).
53. Rens-Domiano S, Reisine T. Structural analysis and functional role of the carbohydrate component of somatostatin receptors. J Biol Chem 1991;266: 20094–102.
54. Reisine T, Rens-Domiano S, Law SF, Takeda J, Yasuda K. Location of carbohydrates in the somatostatin receptor SSTR2 involved in high affinity agonist binding. Mol Biol Cell 1992;3:299a.
55. Raynor K, Reisine T. Subtypes of somatostatin receptors are expressed in the anterior pituitary cell line GH_3. J Pharmacol Exp Ther (in press).
56. Mahy N, Woolkalis M, Manning D, Reisine T. Characteristics of somatostatin desensitization in the pituitary tumor cell line AtT-20. J Pharmacol Exp Ther 1988;247:390–6.
57. Mayor F, Benovic J, Caron M, Lefkowitz R. Somatostatin induces translocation of the beta-adrenergic receptor kinase and desensitizes somatostatin receptors in S49 lymphoma cells. J Biol Chem 1987;262:6468–71.
58. Delfs J, Rens-Domiano S, Benovic J, Reisine T. Regulation of somatostatin receptors by beta-adrenergic receptor kinase. Soc Neurosci Abst 1991;17: 319–31.
59. Srikant C, Patel Y. Receptor binding of somatostatin-28 is tissue specific. Nature 1981;294:259–60.
60. Theveniau M, Rens-Domiano S, Law SF, Rougon G, Reisine T. Development of antibodies against the rat brain somatostatin receptor. Proc Natl Acad Sci USA 1992;89:4314–8.
61. Kong H, Theveniau M, Spencer C, Murray C, Eberwine J, Reisine T. Progress towards cloning of a novel somatostatin receptor. Soc Neurosci Abst 1992;18:448a.

4

Regulation of Growth Hormone Output: The GRF Signal

I.C.A.F. Robinson

Like most anterior pituitary hormones, *growth hormone* (GH) is secreted into the bloodstream in a highly episodic manner. In recent years, it has become clear that both the amount of GH and the pattern in which it is secreted are important in determining its biological effects. The pattern of GH release is in turn governed primarily by output of hypothalamic hypophysiotropic peptides, *somatostatin* (SS) and *GH releasing factor* (GRF), integrated with input from peripheral endocrine (adrenal, thyroid, gonadal, and pancreatic hormones) and nutritional factors at both hypothalamic and pituitary levels. Furthermore, GH itself and the IGF-I that it generates both can feed back to inhibit production and/or release of GH into peripheral blood.

Although ten years have elapsed since the discovery and characterization of GRF (1, 2), many questions remain concerning the timing and amount of GRF release. Furthermore, despite its potent and specific actions in man, the clinical use of this peptide or its analogs has not yet been unequivocally established. In this chapter, we review some studies of our own and other laboratories in an attempt to highlight a few of the newer developments and outstanding questions in the physiology of GRF and SS in controlling the activity of the GH system.

GRF and SS: The GH Secretory Signal

Much of what has been learned about the detailed control of the GH secretory pattern is interpreted in light of the fundamental observations of Tannenbaum and Martin (3), who were the first to describe the ultradian rhythm of GH secretion in the conscious rat. Subsequent studies from Tannenbaum and her colleagues have clearly shown that both GRF and SS are necessary for a coordinated, rhythmic GH secretion and that pulsatile GH requires an intact GRF system (4). It is likely that similar interactions occur in most species, but the rat is particularly useful since the episodic secretory pattern in the male rat is regular and pronounced. Blockade of GRF by antibodies blocks the incidence of spontaneous GH pulses, whereas SS blockade increases basal GH levels, but does not prevent spontaneous peaks of GH secretion (5, 6). Since it was established soon after GRF discovery that simultaneous SS exposure could inhibit the GH releasing effects of GRF (7), it is not surprising that the in vivo effects of single doses of GRF are highly variable in conscious animals and man and that the magnitude of the GH release in response to GRF is heavily dependent on the prevailing SS tone.

This fact makes it difficult to infer the patterns of GRF and SS release from any given pattern of GH release, unless one or another of the GRF or SS components is given exogenously in a systematic manner so as to

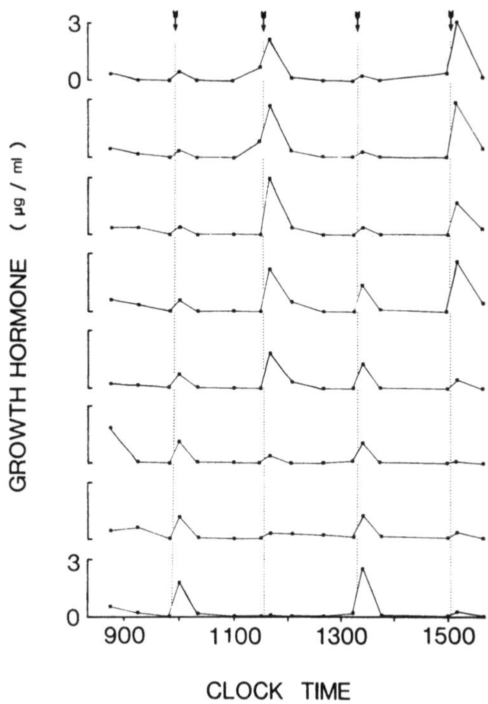

FIGURE 4.1. Intermittent responsiveness to GRF. Male rats secrete GH in episodic pulses at intervals of approximately 3–3.5 h. If they are given a series of GRF pulses (1 µg IV, arrows) at twice that frequency, they show an intermittent pattern of responsiveness (10). The figure shows 8 individual male rats given 4 GRF injections every 100 min, and the animals are ranked according to the size of the first response. Note that the animals are not synchronized with respect to clock time, and some are completely out of phase with others. Data from Robinson and Clark (63).

swamp out endogenous variations. Most indirect evidence gained from studies of this type is consistent with a model in which cyclic variation in SS tone dominates the secretory output of GH in response to GRF release (8). Using methods that permitted chronic venous access and automatic blood sampling and infusions in conscious rats (9), we have performed several studies of this type. Multiple injections of GRF in conscious male rats showed that there is an underlying rhythm of responsiveness and refractoriness to exogenous GRF (Fig. 4.1) similar to the rhythm of endogenous GH secretion (10). This is seen in males, but not females, and can be eliminated by a number of treatments that render GRF responsiveness uniform (5, 8), showing that it is not simply due to a waxing and waning of pituitary GH stores available for release, although the magnitude of any single burst of secretion will clearly be influenced by both the immediate history of secretory activity as well as the prevailing level of synthetic activity.

Similarly, experiments with patterned infusions of SS show that cyclic responsiveness can be mimicked experimentally, both with sinusoidal infusions of SS and even more dramatically with intermittent square wave infusions, the latter inducing a sufficiently pulsatile GH pattern in normal female rats to produce a paradoxical increase in growth in these animals (11). Intermittent infusions of SS produce large rebound releases of GH (Fig. 4.2). A small component of this response must lie at the somatotroph level, since rebound GH release can be demonstrated in vitro (12), but the magnitude of the response is much greater in vivo, particularly in the conscious animal. Experiments by ourselves and others using passive immunization against endogenous GRF (Fig. 4.2) suggest that the main reason for the dramatic and abrupt release of GH is that GRF release is associated with SS withdrawal (13, 14).

What is not yet clear is whether SS infusions stimulate high GRF secretion that is left unopposed when SS is rapidly withdrawn or whether the fall in SS tone precipitates an increase in GRF. There is anatomical evidence for both SS projections and SS receptors on or close to many arcuate GRF neurons (15). These arcuate SS receptors could well mediate the rebound GRF release induced by peripheral withdrawal of infused SS and be the primary targets for an endogenous system whereby the activity of SS and GRF systems are coordinated. However, it remains to be determined whether the source of endogenous SS impinging on GRF neurons derives from the periventricular SS system or from intra-arcuate SS-containing interneurons (16).

These observations suggest that exogenous intermittent infusions of SS might be one way of examining the endogenous GRF release and by swamping endogenous SS, evening out the variability in the pituitary response to exogenous GRF injections. We have tried to take this model from animals into man. In a balanced experimental design, the effects of SS-induced GH secretion with or without concomitant GRF exposure

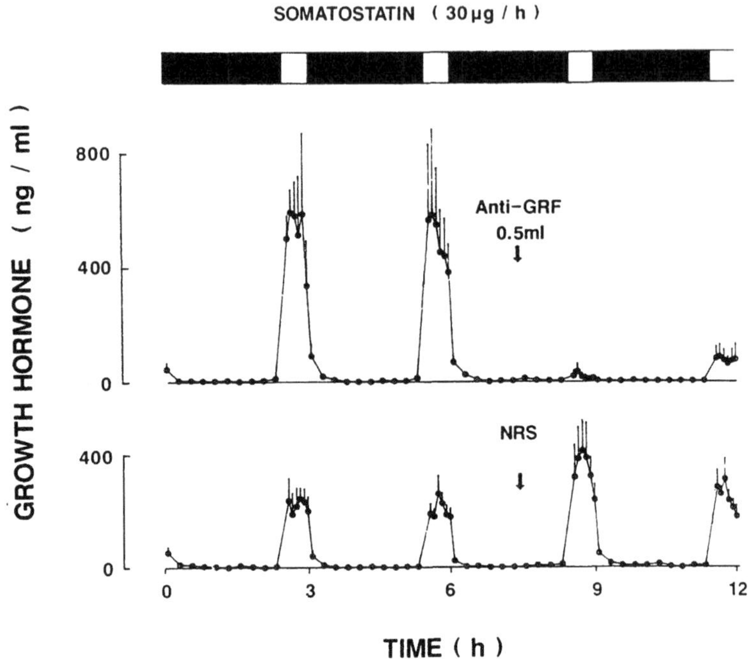

FIGURE 4.2. Somatostatin-induced rebound GH secretion depends on GRF. Regular rebound GH secretion is induced by intermittent infusions of SS in conscious rats. Two groups of 4 rats were given SS IV in a pattern 2.5 h on and 0.5 h off. After 2 rebounds, half the animals (upper panel) were passively immunized with 0.5 ml anti-GRF serum, which blunted the subsequent GH rebounds. Animals given nonimmune rabbit serum continued to show GH rebounds on SS withdrawal (lower panel). Data are mean ± SEM. Reprinted with permission from Clark, Carlsson, Rafferty, and Robinson (13).

were compared (17). The GH responses to repeated SS withdrawal were not as regular as in the animal studies, but GRF injections combined with SS withdrawal produced more regular GH responses than when GRF injections alone were given. One problem with the interpretation of such studies in man is that SS infusions also cause suppression and rebound release of insulin, which itself stimulates GH output (unlike in the rat), which exemplifies some of the difficulties in extrapolating experimental maneuvers between these species.

Since GH pulses are blocked when GRF action is inhibited by passive immunization (5, 6), or GRF cells destroyed by neonatal monosodium glutamate treatment (18), it would seem reasonable to assume that, in common with other neuroendocrine neuronal systems, the release of GRF is likely to be episodic and that episodes of GRF release would induce GH release in periods of trough secretion of SS. In order to determine whether this is so, it is necessary to have a more direct measurement of GRF secretory activity, but this has proved difficult technically.

The classic endocrine approach to study this directly is to gain access to the hypophysial portal circulation and measure the levels of SS and GRF in portal blood. The hypophysial portal system is relatively inaccessible, and ideally the sampling should be carried out in conscious unstressed animals. It requires highly sensitive GRF assays, given the low levels of GRF in portal blood and the small volume of blood available. In the rat, portal blood sampling is only possible with major surgery under anesthesia, and while there is evidence for an inverse correlation of GRF and SS secretion (6), it is extremely difficult to relate this to normal physiological pulsatility.

Portal sampling has also been carried out in the conscious sheep (19, 20). The GH pulses are of smaller magnitude in this species than in the rat, and while GRF pulses are sometimes associated with GH release, there are GRF pulses that do not cause GH release and there is not always a relationship between falls in SS levels and the onset of GH release. However, this approach is much more direct, and it is likely that

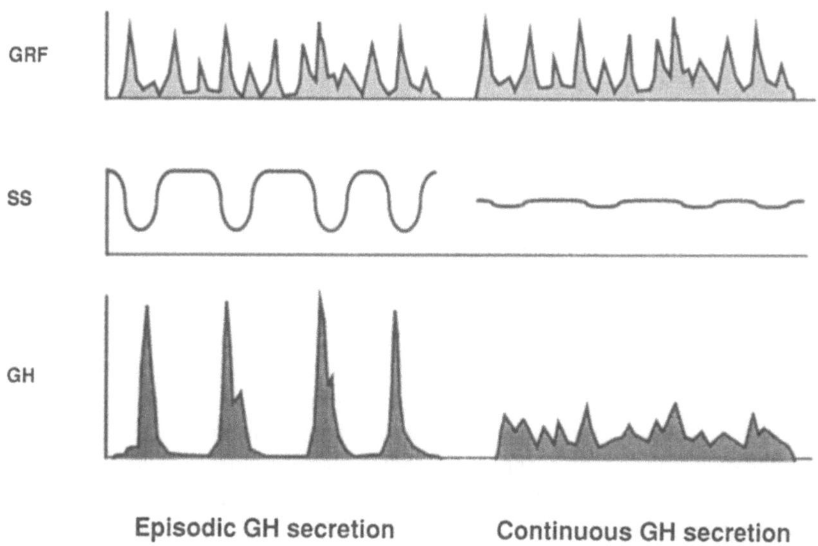

FIGURE 4.3. GRF and SS interactions produce both episodic or continuous GH output. Recent data from direct measurements in hypophysial portal blood, together with effects of patterned GRF or SS infusions, suggest that GRF and GH output are not tightly correlated and that it is largely the variations in SS tone that determine both endogenous GH secretion and the response to exogenous GRF. In this model, an identical GRF secretory pattern can give rise to both pulsatile or continuous GH profiles (typical of male or female rats, respectively), depending on the pattern of SS release.

further studies in conditions of more marked pulsatility or using other stimuli to induce GH secretion will enable the relationship between portal levels of GRF and SS and the pattern of GH release to be elucidated, at least in this species.

The general notion that GRF pulses may be dissociated from GH release and that differences in SS secretion may have an overriding influence on acute GH output leads to a modification of the original model (8). Our current working model for the rat (Fig. 4.3) suggests that GRF is secreted in many episodes in both males and females and is required for GH release. A difference in SS secretion between the sexes results in the different patterns of GH output, but the repeated exposure to GRF builds up GH stores between pulses and is required for the maintenance of the somatotroph population. It is possible that GRF secretory patterns may also differ between the sexes, but it is not a requirement of this model. GRF release is shown as episodic, rather than continuous secretion, since peptidergic neurons tend to fire in an episodic fashion (see below), though whether the net output of activating GRF cells will result in a pulse or a more gradual increase in concentration will depend on the degree to which the GRF population response is synchronized. Indeed, the main function of rapid SS withdrawal from arcuate GRF cells might be to ensure that their rebound activation is synchronous.

GRF: The Neuronal Signal

We have recently tried to obtain more direct information about the neuronal control of GRF output. Like other hypothalamic releasing factors, GRF is the product of neurosecretory neurons, and the output of these neurons from the terminals in the median eminence will largely reflect the electrical activity generated by afferent input at the cell body. Much has been learned about the patterns of episodic peptide secretion from the hypothalamo-neurohypophysial *oxytocin* (OT) and *vasopressin* (AVP) systems, principally because the terminals are accessible in the posterior pituitary and the cell bodies can be located and identified antidromically. The main properties of these neuroendocrine neurons can be summarized as follows.

Although both cell types show episodic activation, the patterns are very different and lead to quite different patterns of hormone release under different circumstances (21). In response to suckling, OT is released in episodic large pulses, causing milk ejection—continuous OT release would be much less effective to stimulate myoepithelial cell contractions necessary for milk letdown. Electrical recordings during milk ejection show a large acceleration in firing (22), which is synchronized between different OT cells (23), and it is the synchronous episodic activation that leads to a burst of peptide release. The very similar AVP cells show a different pattern: They show increases and decreases in firing activity in a phasic pattern that increase during osmotic stimuli, but these are not synchronized between cells so that AVP secretion is more continuous. Furthermore, the peptides can be released from OT or AVP terminals by direct electrical stimulation, and their output is highly frequency dependent (24). These techniques should be applicable to the GH controlling systems, and we have previously argued (25) that GRF neurons should show similar generalized properties of these peptidergic neurosecretory neurons.

With Dickson and Leng, recordings have been made from arcuate neurons in the hope of being able to characterize the electrophysiological activity of identified GRF neurons. Candidates for GRF neurons have been identified on the basis of several criteria, albeit indirect. They should project to the median eminence, and this can be tested by antidromic activation following median eminence stimulation. While such direct stimulation of the median eminence has little effect on net GH secretion (probably because of concomitant SS release that overrides the effects of GRF on GH output), appropriate electrical stimulation of the arcuate nucleus elicits a pronounced GH release that is relatively independent of the frequency of stimulation, but very sensitive to the duration of the stimulus train (26).

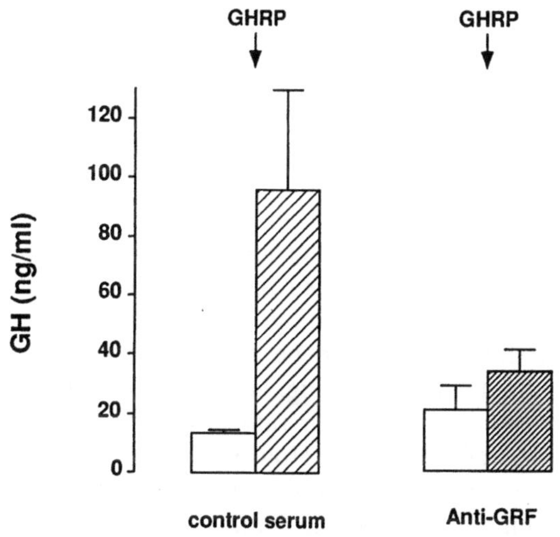

FIGURE 4.4. Endogenous GRF is involved in GH release induced by GH releasing peptide in conscious rats. Mean (±SEM) plasma GH levels (hatched bars) achieved by an intravenous infusion of GHRP (20 µg/h) are shown. Four animals (right panel) were pretreated with anti-GRF serum, while another four received control serum (left panel). Blockade of endogenous GRF significantly ($P < 0.05$) reduced the amount of GH released by GHRP in this model (30).

Preliminary results suggest that this electrophysiological approach can be used to identify neuronal and pharmacological inputs into putative GRF cells (27). For example, stimulation of the periventricular nucleus inhibits and stimulation of the basolateral amygdala stimulates neuronal activity in candidate GRF cells, as well as GH release. Furthermore, we have suggested that the potent effects of the synthetic GH releasing hexapeptide GHRP (28, 29) in conscious animals reflects primarily its hypothalamic action, either to inhibit SS or stimulate GRF, with which it synergizes to release GH. GHRP infusions enhance GRF-induced GH responses (30), and blockade of GRF with antibodies reduces the effects of GHRP in the rat (Fig. 4.4), but these experiments do not identify the site(s) of their interaction. We have recently obtained direct evidence that GHRP injections activate both neuronal firing and gene expression in a population of arcuate neurons that are close to, or identical with, GRF cells (Dickson, Leng, Robinson, unpublished results).

GRF: The Trophic Signal

If recent evidence indicates that GRF pulses may occur more frequently than GH pulses, this does not mean that "unsuccessful" GRF pulses are without effect. GRF stimulates GH gene transcription, and this effect of GRF is not blocked by SS (31). Furthermore, GRF stimulates the proliferation of somatotrophs, and transgenic mice overexpressing the GRF gene in an unregulated fashion have a dramatically increased pituitary cell size due to a specific increase in somatotroph number (32, 33). Although a transient desensitization can be demonstrated with GRF in vivo (34) and in vitro (35), it is difficult to maintain a significant functional down-regulation for longer periods of time. On the contrary, the marked pituitary enlargement and somatotroph hyperplasia, causing gigantism or acromegaly in man, clearly is not overridden by endogenous SS secretion, even if this is increased as a feedback response to the excess GH secretion.

In our early studies with GRF and growth promotion, we showed that an acute growth stimulation could be obtained if GRF was given in a pulsatile pattern in female rats (Fig. 4.5) to generate a pulsatile GH pattern that increased their growth rate (36). The same dose of GRF given continuously was ineffective. However, in both male rats and man, provided there is an endogenous cyclic variation in SS tone, pulsatile GH output can be maintained despite constant GRF exposure. Experimental studies with constant intravenous infusions of GRF show that the endogenous rhythm is greatly amplified in the short term (37, 38). Even if GRF exposure is greatly elevated and prolonged, as in patients with ectopic GRF production, GH release remains episodic despite continuous high GRF tone. Under these circumstances at least, GRF patterns do not have to be pulsatile in order to achieve growth.

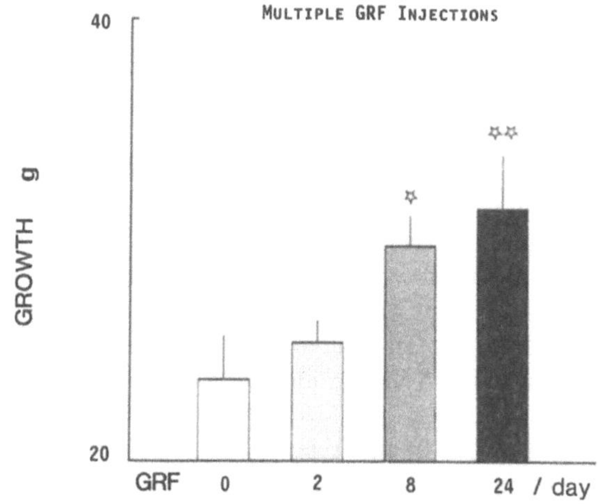

FIGURE 4.5. Multiple injections of GRF stimulate growth in female rats. Groups of normal female animals were given a total of 24 μg GRF/day IV, divided as 2, 6, or 24 pulses per day, for 7 days. Compared to saline-infused controls, significant growth was obtained at the 2 higher frequencies. Data shown are mean ± SEM and are redrawn from Robinson and Clark (63).

The potential therapeutic usefulness of this observation has been tested in man by giving continuous subcutaneous infusions of low doses of GRF for up to one year in a number of GH-deficient (but GRF-responsive) children, who showed both an increase in the amplitude of their spontaneous GH pulses and stimulated growth (39). This opens the way for development of long-acting GRF preparations that should accomplish an increased GH pulsatile profile by both stimulating the number of somatotrophs and their GH synthesis and amplifying the residual GH pulses that may be detected in these patients. However, it should be recognized that this treatment is essentially employing the trophic effects of GRF on GH synthesis and somatotroph proliferation, rather than the acute secretagogue action of GRF to release a pulse of GH. It may be that much less GRF is required over such long periods, and it is uncertain what proportion of patients will benefit from a regime of continuous GRF.

Faults in Receiving the GRF Signal

There is now considerable evidence implicating the adenyl cyclase system in the trophic actions of GRF. For example, experiments with transgenic mice in which cAMP levels are elevated by targeting cholera toxin to the somatotroph with cell-specific elements of the GH promoter show that cyclase activation in the somatotroph leads to a dramatic but selective increase in this cell population (40), extending the important clinical observations that some GH-secreting pituitary adenomas result from somatic mutations in the Gs_{α} regulatory subunit of adenyl cyclase, leading to excess activity (41). Conversely, impairment in the function of the somatotroph adenyl cyclase system leads to a reduced pituitary size and a specific reduction in somatotroph cell number, with a consequent diminution in GH output and dwarfism. One elegant example of this is the transgenic mouse model in which the transcriptional stimulation of cyclic AMP is blocked by targeted expression of a cAMP response element binding protein mutated so that it cannot be activated by cAMP-dependent protein kinase: There is a 5-fold reduction in GH-producing cells, and the mice are dwarfed (42).

Another animal model of this type is the dwarf (dw/dw) rat (43); this animal shows a significantly reduced adenyl cyclase response to GRF (44) and a specific reduction in somatotroph number and GH output (45). Although the nature of the defect in this mutant rat remains to be elucidated, it is not a defect in the GH gene itself, unlike in the Japanese dwarf rat, which cannot synthesize a normal GH molecule (46). Also, GRF is effective to release GH both in vitro and in vivo in the dwarf rat, unlike the phenotypically similar mutation in the Little mouse (47, 48). Our dwarf mutant produces normal GH, but at a greatly reduced level so that the peak GH levels achieved in response to GRF are roughly proportional to the pituitary GH contents (49), presumably relating to the smaller pool of readily releasable GH in this mutant.

When compared with normal rats from the same strain, Dw rats were slightly less sensitive to GRF injections IV, and this reduced sensitivity occurs over the period of 20–40 days postnatally (49), which coincides with the onset of GH-dependent growth in this species and an activation of the GH system generally. Whereas the GH response to a low dose of GRF increases 7-fold between 20 and 40 days in normal animals, the response actually diminishes over this period in dwarf rats (Fig. 4.6). This is not due to a failure to increase GH synthesis during this period; pituitary GH contents increase in both wild-type and mutant strains over this period (49) so that the relative deficit remains roughly constant.

This dwarf rat is a unique model in which one can study the effects of early specific GH insufficiency on the development of subsequent GH responsiveness to GRF. So far the molecular nature of the defect has not been established, though it seems likely to lie somewhere in the signal transduction pathway for the trophic effects of GRF, which can therefore be separated from the pathway mediating acute secretagogue effects of GRF on GH secretion. Our attempts to quantify GRF receptors by binding assays on dwarf pituitaries have been unsuccessful. However, the recent cloning of the GRF receptor (50) in both rat and man is an important development that will make it possible to study possible receptor

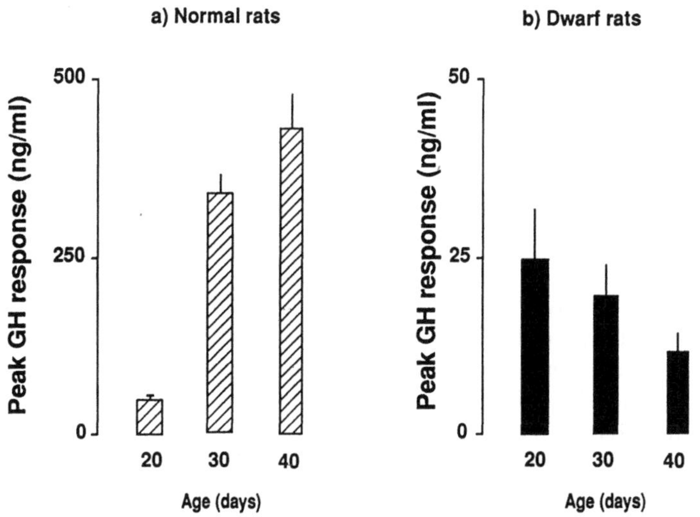

FIGURE 4.6. GRF responsiveness differs with maturation in normal and dwarf rats. Plasma GH responses 5 min after intravenous injection of a moderate dose of GRF (31.25 ng) in groups of anesthetized normal (*a*) or dwarf (*b*) rats at 20, 30, or 40 days of age. Normal rats show a 7-fold increase in their plasma GH response to GRF over this period, whereas the GH response in dwarf rats declines (49). Data shown are mean ± SEM for 5–8 animals in each group.

deficits in this and other models, as well as to investigate the regulation of the GRF receptor gene under different circumstances involving chronic hypo- or hypersecretion of GRF.

What Signals Control Basal GH Levels?

Thus far we have considered the interaction between GRF and SS in terms of optimizing an episodic GH secretory pattern. However, an equally fundamental aspect of the difference between a pulsatile and continuous GH pattern (as in normal male and female rats) is the marked difference in basal GH levels. While falls in SS may initiate GRF and GH output, it is the extent of secretion of SS between pulses of GH that determines the basal secretion of GH. This is significant in two respects. By conserving the releasable pool of GH, large GH pulses can be produced and hence achieve higher peak plasma GH levels during secretory bursts. A second effect of SS, however, is to lower the circulating GH levels between secretory episodes, and the expression of a number of GH-sensitive proteins is known to be sensitive to the presence or absence of low levels of GH in the trough periods between secretory bursts (51–53).

In this context, it is important to recall that there are other mechanisms that can regulate the basal plasma GH levels. While the entry rate of GH to the circulation can be limited by SS, the basal levels following secretory episodes are largely determined by clearance mechanisms. Many species have considerable quantities of circulating *GH binding proteins* (GHBPs) whose principle component corresponds to the extracellular domain of the GH receptor, demonstrating a similar avidity and specificity (54, 55). In man, it appears that the extracellular domain of the GH receptor is cleaved, and this forms the circulating GH binding protein (56). In the rat, an alternatively spliced messenger RNA has been identified that gives rise to a truncated form of the receptor, with a hydrophilic domain replacing the transmembrane hydrophobic sequence in the full-length receptor, and there is good evidence that this is the major source of circulating GHBP in this species (57, 58). The prospect that basal plasma GH levels could be effectively regulated by a component of the GH receptor system prompted us to investigate the role of GHBP in modifying the pulsatile GH pattern in experimental animals.

A direct RIA for GHBP, capable of measuring rat GHBP in 10-μl plasma samples and unaffected by the presence or absence of GH, has enabled us (59) to compare the plasma profiles of GHBP and GH in the rat (Fig. 4.7). While GHBP levels fluctuate to a minor extent, we found

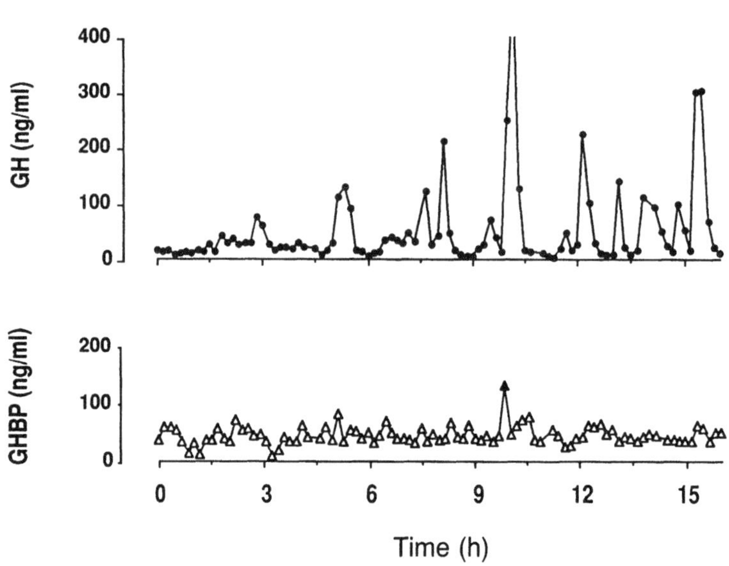

FIGURE 4.7. Serial microsamples of blood were withdrawn automatically from conscious rats (9), subdivided, and assayed for both rat GH and rat GHBP by specific RIAs (59). One example is illustrated. Despite large episodes of GH release, blood levels of GHBP in the same samples were relatively stable, and the minor fluctuations did not correlate with GH pulsatility.

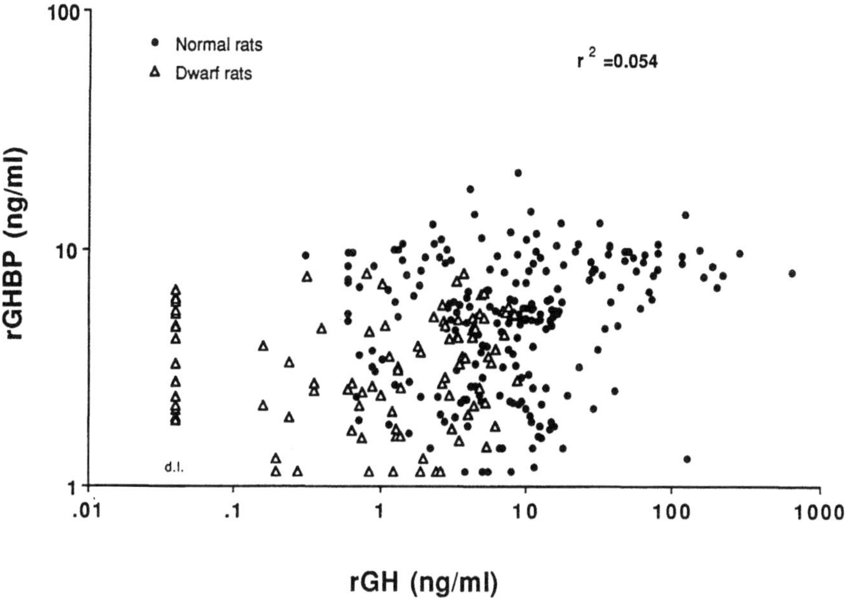

FIGURE 4.8. Plasma GHBP and GH levels vary independently. In a series of sampling studies, plasma GH and GHBP levels were assayed in conscious male dwarf (open triangles) or normal (solid circles) rats (59). Despite a 1000-fold variation in plasma GH values, there was no correlation between the GHBP or GH values measured in each sample (n = 317). Furthermore, GHBP levels in dwarfs were similar to those in normal males, despite the much lower GH levels in the dwarf rats.

no evidence for any correlation between GHBP and GH in serial samples from normal or dwarf male rats showing regular GH pulses (Fig. 4.8). While this appears at odds with an earlier study (60) that reported GH binding activity inversely correlated with GH in single samples from groups of animals, the amount of "free" GHBP (i.e., unoccupied by GH) will indeed vary inversely with GH secretion if GH pulses encounter a constant amount of GHBP in the blood.

In the context of a pulsatile secretion of GH, the role of GHBP is difficult to establish. The amount of circulating GHBP is sufficient to complex a large proportion of "basal" circulating GH levels, but peak secretory rates generate concentrations of GH that far exceed GH binding saturation and, thus, free GH is largely present. However, as GH is cleared, it may well be that a significant fraction of lower concentrations of GH remains associated with the binding protein. Formation of GH/GHBP complexes could occur within the plasma compartment or within the *extracellular fluid* (ECF). Circulating GH/GHBP complexes are cleared at a significantly slower rate than free GH or free GHBP (61) and could act

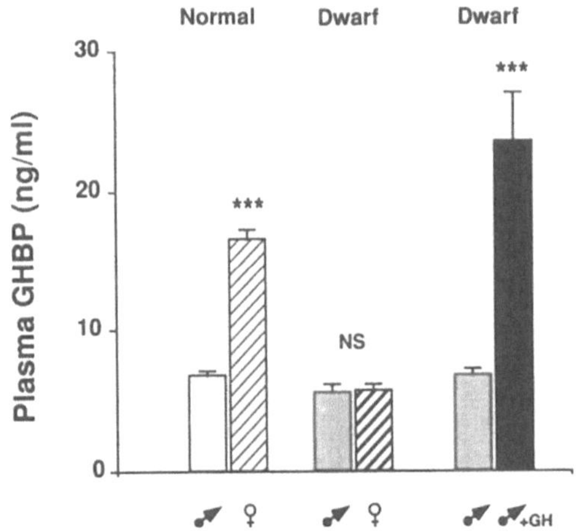

FIGURE 4.9. Plasma GHBP levels in normal and dwarf rats. In normal rats, plasma GHBP levels are higher in females than males. This difference is not seen in dwarf rats. Treatment of a group of dwarf males with a continuous subcutaneous infusion of hGH (200 μg/day) for 7 days raised plasma GHBP levels significantly. Data shown are mean ± SEM. (*** = $P < 0.001$, NS = not significant.) Reprinted with permission from Carmignac, Wells, Carlsson, Clark, and Robinson (59).

to trap low concentrations of GH so that, effectively, once a concentration has fallen below a given level, it is no longer available to stimulate tissue receptors. Alternatively, any GH/GHBP complex formed in the ECF will not be cleared rapidly from the vicinity of the cell and, depending on the amounts of GHBP relative to GH receptor, could either inhibit GH action or, conversely, preserve a reservoir of GH long after the peak of GH has been cleared from the circulation.

Although acute changes in GH do not affect GHBP levels, these are regulated in the longer term by several factors, including gonadal steroids and GH itself. Normal female rats have higher GHBP levels than males (59, 62), but this sexual dimorphism in plasma GHBP levels is not seen in GH-deficient dwarf animals (Fig. 4.9): Female dwarfs have the same low levels as normal or dwarf males (59). This suggested that it might be the high basal GH levels in normal female rats that elevated their plasma GHBP. Accordingly, dwarf males were given hGH (200 μg/day) by continuous subcutaneous infusion for 7 days. This treatment also raised plasma GHBP levels to normal female levels (Fig. 4.9). These and other studies suggest that GHBP levels, like GH receptors, are sensitive to the pattern of GH exposure, and we are currently investigating the role of basal GH levels, regulated by SS, in influencing both metabolic and growth-promoting actions of GH.

Conclusions

These studies suggest that the hypothalamus has far wider effects on the GH system than the acute regulation of anterior pituitary hormone output. By regulation of hormone synthesis and GH cell number, even in adult life, GRF can quantitatively determine the activity of the GH system in the longer term. By control over basal GH output between bursts, SS can influence GHBP to affect GH clearance and availability to GH receptors and ultimately change the effective GH responsiveness of peripheral target tissues to any given pattern of GH exposure. The development of methods to study GRF neuronal activity directly should help us to identify the many diverse inputs to this system that are integrated to produce an appropriate signal for growth.

Acknowledgments. I would like to thank my past and present colleagues for discussion, for their help with the provision of reagents, and for permission to quote some of their unpublished data.

References

1. Guillemin R, Brazeau P, Böhlen P, Esch F, Ling N, Wehrenberg WB. Growth hormone-releasing factor from a human pancreatic tumour that caused acromegaly. Science 1982;218:585–7.
2. Rivier J, Spiess J, Thorner M, Vale W. Characterization of a growth hormone-releasing factor from a human pancreatic islet tumour. Nature 1982;300:276–8.
3. Tannenbaum GS, Martin JB. Evidence for an endogenous ultradian rhythm governing growth hormone secretion in the rat. Endocrinology 1976;98: 562–70.
4. Tannenbaum GS. Interrelationship of somatostatin and growth hormone-releasing hormone in the genesis of the rhythmic secretion of growth hormone. Acta Paediatr Scand Suppl 1990;367:76–80.
5. Wehrenberg WB, Ling N, Böhlen P, Esch F, Brazeau P, Guillemin R. Physiological roles of somatocrinin and somatostatin in the regulation of growth hormone secretion. Biochem Biophys Res Commun 1982;109:562–5.
6. Plotsky PM, Vale W. Patterns of growth hormone releasing factor and somatostatin secretion into the hypophysial portal circulation of the rat. Science 230:461–3.
7. Guillemin R. Hypothalamic control of pituitary functions: the growth hormone releasing factor. The XVIII Sherrington Lectures. Liverpool: Liverpool University Press, 1986.
8. Tannenbaum GS, Ling N. The interrelationship of growth hormone (GH)-releasing factor and somatostatin in the generation of the ultradian rhythm of GH secretion. Endocrinology 1984;115:1952–7.
9. Clark RG, Chambers G, Lewin J, Robinson ICAF. Automated repetitive microsampling of blood: growth hormone secretion in conscious male rats. J Endocrinol 1986;111:27–35.

10. Clark RG, Robinson ICAF. Growth hormone (GH) responses to multiple injections of a fragment of human GH-releasing factor in conscious male and female rats. J Endocrinol 1985;106:281–9.
11. Clark RG, Robinson ICAF. Paradoxical growth promoting effects induced by patterned infusions of somatostatin in female rats. Endocrinology 1988;122: 2675–82.
12. Kraicer J, Cowan JS, Sheppard B, Lussier B, Moor BC. Effect of somatostatin withdrawal and growth hormone (GH)-releasing factor on GH release in vitro—amount available for release after disinhibition. Endocrinology 1986; 119:2047–51.
13. Clark RG, Carlsson LMS, Rafferty B, Robinson ICAF. The rebound release of growth hormone (GH) following somatostatin infusion in rats involves hypothalamic GH-releasing factor release. J Endocrinol 1988;119:397–404.
14. Miki N, Ono M, Shizume K. Withdrawal of endogenous somatostatin induces secretion of growth hormone-releasing factor in rats. J Endocrinol 1988;117: 245–52.
15. Epelbaum J, Moyse E, Tannenbaum GS, Kordon C, Beaudet A. Combined autoradiographic and immunohistochemical evidence for an association of somatostatin binding sites with growth-hormone-releasing factor-containing nerve cell bodies in the rat arcuate nucleus. J Neuroendocrinol 1989;1:109–15.
16. Willoughby J, Brogan M, Kapoor R. Hypothalamic interconnections of somatostatin and growth hormone releasing factor on growth hormone secretion. Neuroendocrinology 1989;50:584–91.
17. Hindmarsh PC, Brain CE, Robinson ICAF, Matthews DR, Brook CGD. The interaction of growth-hormone releasing hormone and somatostatin in the generation of a GH pulse in man. Clin Endocrinol (Oxf) 1991;35:353–60.
18. Bloch B, Ling N, Benoit R, Wehrenberg WB, Guillemin R. Specific depletion of immunoreactive growth hormone-releasing factor by monosodium glutamate in rat median-eminence. Nature 1984;307:272–3.
19. Thomas GB, Cummins JT, Francis H, Sudbury AW, McCloud PI, Clarke IJ. Effect of restricted feeding on the relationship between hypophysial portal concentrations of growth hormone (GH)-releasing factor and somatostatin, and jugular concentrations of GH in ovariectomized ewes. Endocrinology 1991;128:1151–8.
20. Frohman L, Downs T, Clark I, Thomas GB. Measurement of growth hormone-releasing hormone and somatostatin in hypophysial portal plasma of unanaesthetized sheep: spontaneous secretion and response to insulin-induced hypoglycaemia. J Clin Invest 1990;86:17–24.
21. Leng G, Bicknell RJ. The neurohypophysis. In: Lightman SL, Everitt BJ, eds. Neuroendocrinology. Oxford: Blackwell, 1986:177–96.
22. Lincoln DW, Wakerley JB. Factors governing the periodic activation of supraoptic and paraventricular neurosecretory cells during suckling in the rat. J Physiol 1975;250:443–61.
23. Freund-Mercier MJ, Richard P. Electrophysiological evidence for facilitatory control of oxytocin neurones by oxytocin during suckling in the rat. J Physiol 1984;352:447–66.
24. Bicknell RJ, Flint APF, Leng G, Sheldrick EL. Phasic patterns of electrical stimulation enhances oxytocin secretion from the isolated neurohypophysis. Neurosci Lett 1982;30:47–50.

25. Robinson ICAF. The growth hormone secretory pattern: a response to neuroendocrine signals. Acta Paediatr Scand 1991;suppl 372:70–8.
26. Dickson SL, Leng G, Robinson ICAF. Stimulus patterning and GH secretion in male rats. J Physiol 1992;452:215P.
27. Dickson SL, Leng G, Robinson ICAF. Induction of fos-like immunoreactivity in the rat hypothalamus following systemic administration of growth hormone-releasing peptide (GHRP-6). J Physiol 1993.
28. Bowers CY, Momany F, Reynolds GA, Hong A. On the in vitro and in vivo activity of a new synthetic hexapeptide that acts on the pituitary to specifically release growth hormone. Endocrinology 1984;114:1537–45.
29. Bowers CY, Reynolds GA, Durham D, Barrera CM, Pezzoli SS, Thorner MO. Growth hormone (GH)-releasing peptide stimulates GH release in normal men and acts synergistically with GH-releasing hormone. J Clin Endocrinol Metab 1990;70:975–80.
30. Clark RG, Carlsson LMS, Trojnar J, Robinson ICAF. The effects of a growth hormone-releasing peptide and growth hormone-releasing factor in conscious and anaesthetized rats. J Neuroendocrinol 1989;1:249–55.
31. Barinaga M, Bilezikjian LM, Vale WW, Rosenfeld MG, Evans RM. Independent effects of growth hormone releasing factor on growth hormone release and gene transcription. Nature 1985;314:279–81.
32. Mayo KE, Hammer RE, Swanson LW, Brinster RL, Rosenfeld MG, Evans RM. Dramatic pituitary hyperplasia in transgenic mice expressing a human growth hormone-releasing factor gene. Mol Endocrinol 1988;2:606–12.
33. Stefaneanu L, Kovacs K, Horvath E, et al. Adenohypophysial changes in mice transgenic for human growth hormone-releasing factor—a histological, immunocytochemical, and electron microscopic investigation. Endocrinology 1989;125:2710–8.
34. Arsenijevic Y, Rivest R, Eshkol A, Sizonenko P, Aubert M. Plasma growth hormone (GH) responses to intravenous GH-releasing factor (GRF) in adult rats: evidence for transient pituitary desensitization after GRF stimulation. Endocrinology 1987;121:1487–96.
35. Bilezikjian LM, Vale WW. Chronic exposure of cultured rat anterior pituitary cells to GRF causes partial loss of responsiveness to GRF. Endocrinology 1984;115:2032–4.
36. Clark RG, Robinson ICAF. Growth induced by pulsatile infusion of an amidated fragment of human growth hormone releasing-factor in normal and GHRF-deficient rats. Nature 1985;314:281–3.
37. Vance ML, Kaiser DL, Evans WS, et al. Pulsatile growth hormone secretion in normal man during a continuous 24-hour infusion of human growth hormone releasing factor (1–40): evidence for intermittent somatostatin secretion. J Clin Invest 1985;75:1584–90.
38. Wehrenberg WB. Continuous infusion of growth hormone releasing factor: effects on pulsatile growth hormone secretion in normal rats. Neuroendocrinology 1986;43:391–4.
39. Brain CE, Hindmarsh PC, Brook CGD. Continuous subcutaneous GHRH (1–29)NH2 promotes growth over one year in short, slowly growing children. Clin Endocrinol 1990;32:153–63.
40. Burton FH, Hasel KW, Bloom FE, Sutcliffe JG. Pituitary hyperplasia and gigantism in mice caused by a cholera toxin transgene. Nature 1991;350:74–7.

41. Vallar L, Spada A, Giannattasio G. Altered Gs and adenylate cyclase activity in human GH-secreting pituitary adenomas. Nature 1987;330:566–8.
42. Struthers RS, Vale WW, Arias C, Sawchenko PE, Montminy MR. Somatotroph hypoplasia and dwarfism in transgenic mice expressing a non-phosphorylatable CREB mutant. Nature 1991;350:622–4.
43. Charlton HM, Clark RG, Robinson ICAF, et al. Growth hormone-deficient dwarfism in the rat: a new mutation. J Endocrinol 1988;119:51–8.
44. Downs T, Frohman L. Evidence for a defect in growth hormone-releasing factor signal transduction in the dwarf (dw/dw) rat pituitary. Endocrinology 1991;129:58–67.
45. Kineman RD, Chen TT, Frawley LS. A cellular basis for growth hormone deficiency in the dwarf rat—analysis of growth hormone and prolactin release by reverse hemolytic plaque assay. Endocrinology 1989;125:2035–40.
46. Takeuchi T, Suzuki H, Sakurai S, Nogami H, Okuma S, Ishikawa H. Molecular mechanism of growth hormone (GH) deficiency in the spontaneous dwarf rat—detection of abnormal splicing of GH messenger ribonucleic acid by the polymerase chain reaction. Endocrinology 1990;126:31–8.
47. Clark RG, Robinson ICAF. Effect of a fragment of human growth hormone releasing factor in normal and "little" mice. J Endocrinol 1985;106:1–5.
48. Jansson JO, Downs TR, Beamer WG, Frohman LA. Receptor associated resistance to growth hormone-releasing hormone in dwarf "little" mice. Science 1986;232:511–3.
49. Carmignac DF, Robinson ICAF. Growth hormone (GH) secretion in the dwarf rat: release, clearance and responsiveness to GH-releasing factor and somatostatin. J Endocrinol 1990;127:69–75.
50. Mayo K. Molecular cloning and expression of a pituitary-specific receptor for growth hormone-releasing hormone. Mol Endocrinol 1992;6:1734–44.
51. Mode A, Norstedt G, Simic B, Eneroth P, Gustafsson JA. Continuous infusion of growth hormone feminizes hepatic steroid metabolism in the rat. Endocrinology 1981;108:2103–8.
52. Jeffery S, Carter ND, Clark RG, Robinson ICAF. The episodic secretory pattern of growth hormone regulates liver carbonic anhydrase-III—studies in normal and mutant growth-hormone-deficient dwarf rats. Biochem J 1990; 266:69–74.
53. Waxman DJ, Morrissey JJ, Macleod JN, Shapiro BH. Depletion of serum growth hormone in adult female rats by neonatal monosodium glutamate treatment without loss of female-specific hepatic enzymes-P450 2D (IIC12) and steroid 5-alpha-reductase. Endocrinology 1990;126:712–20.
54. Baumann G, Amburn K, Shaw M. The circulating growth hormone (GH)-binding protein complex: a major constituent of plasma GH in man. Endocrinology 1988;122:976–84.
55. Leung DW, Spencer SA, Cachianes G, et al. Growth hormone receptor and serum binding protein: purification, cloning and expression. Nature 1987;330: 537–43.
56. Trivedi B, Daughaday WH. Release of growth hormone binding protein from IM-9 lymphocytes by endopeptidase is dependent on sulphydryl group inactivation. Endocrinology 1988;123:2201–6.
57. Baumbach WR, Horner DL, Logan JS. The growth hormone-binding protein in rat serum is an alternatively spliced form of the rat growth hormone receptor. Genes Dev 1989;3:1199–205.

58. Sadeghi H, Wang BS, Lumanglas AL, Logan JS, Baumbach WR. Identification of the origin of the growth hormone-binding protein in rat serum. Mol Endocrinol 1990;4:1799–805.
59. Carmignac D, Wells T, Carlsson L, Clark R, Robinson I. Growth hormone (GH)-binding protein in normal and GH-deficient dwarf rats. J Endocrinol 1992;135:447–57.
60. Bick T, Amit T, Barkey RJ, Hertz P, Youdim MBH, Hochberg Z. The interrelationship of growth hormone (GH), liver membrane GH receptor, serum GH-binding protein activity, and insulin-like growth factor I in the male rat. Endocrinology 1990;126:1914–20.
61. Fairhall KM, Carmignac DF, Robinson ICAF. Growth hormone (GH) binding protein and GH interactions in vivo in the guinea pig. Endocrinology 1992; 131:1963–9.
62. Massa G, Mulumba N, Ketelslegers JM, Maes M. Initial characterization and sexual dimorphism of serum growth hormone-binding protein in adult-rats. Endocrinology 1990;126:1976–80.
63. Robinson ICAF, Clark RG. The secretory pattern of GH and its significance for growth in the rat. In: Isaksson et al., eds. Growth hormone—basic and clinical aspects. Amsterdam: Elsevier: 109–27.

5

Regulation of Growth Hormone Secretion in Man

MARY LEE VANCE

Growth hormone (GH) secretion occurs in a pulsatile fashion because of its primary regulation by two hypothalamic hormones, *growth hormone releasing hormone* (GHRH) and *somatostatin* (SRIF). GHRH stimulates GH synthesis; it increases transcription of growth hormone messenger RNA within minutes by increasing cyclic AMP (1). SRIF is probably more important in determining the timing and amplitude of GH pulses, but has no effect on GH synthesis.

Figure 5.1 shows an overview of the regulation of GH secretion in schematic form. In addition to multiple hypothalamic influences, somatotroph function is regulated by negative feedback by circulating IGF-I at the pituitary and hypothalamic levels and by "short loop" feedback by GH itself on the hypothalamus.

Studies using frequent sampling and sensitive GH assays have modified our understanding of GH physiology. The pattern of GH secretion depends on a number of factors, including stage of development, age, gender, nutritional state, sleep stage, stress, and exercise. The metabolic clearance of GH in healthy subjects and in patients with various disorders must also be taken into account.

Over the past 10 years, attempts to delineate the pattern of hormone secretion and clearance have resulted in formulation of several computerized programs that chacterize the number of significant excursions or pulses in a hormone concentration profile over time (i.e., frequent hormone measurements over 24h). Some of the frequently used programs to identify hormone pulses include Pulsar, Detect, Cluster, and a modified program formulated by Santen and Bardin. More recently, a computer algorithm to estimate hormone secretion and clearance rates from a concentration profile simultaneously was developed by Veldhuis and Johnson. Using mathematical modeling, a hormone concentration profile is resolved into pure secretory events independent of hormone clearance. The half-life of the particular hormone is also calculated,

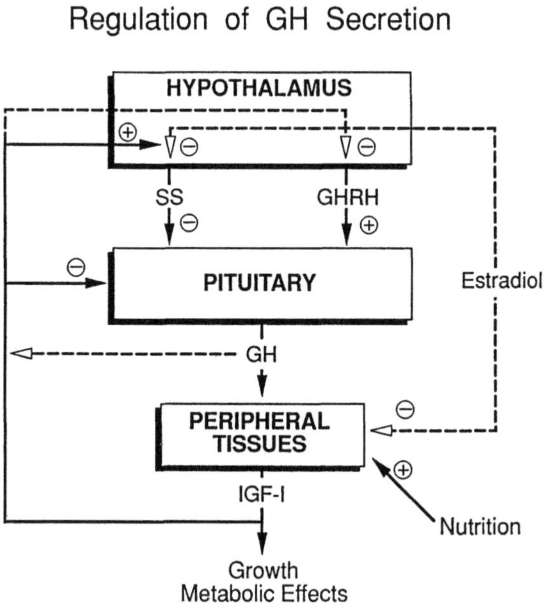

FIGURE 5.1. Schematic representation of the regulation of GH secretion in man. The primary regulators are the hypothalamic hormones, GHRH and SS. Some modulating factors include circulating gonadal steroids, GH and IGF-I, and the nutritional state. Reprinted with permission from M.O. Thorner, M.L. Vance, E. Horvath, and K. Kovacs, The Anterior Pituitary, in Wilson and Foster, eds., *Williams Textbook of Endocrinology*, 8th edition (Philadelphia: Saunders, 1992), p. 231.

and the results can be compared with independent studies of hormone degradation.

In several studies, the half-life of exogenously administered GH ranged from 8.9 to 27 min (2–5). In a more recent study, the half-life of endogenous GH in normal men was determined after sequential administration of intravenous boluses of GHRH and SRIF, followed by a continuous SRIF infusion; GHRH stimulated GH secretion, and additional secretion was inhibited by SRIF. The GH half-life was 18 ± 0.08 min (6, 7). Using deconvolution analysis, the monoexponential half-life of endogenous GH was 17 ± 1.7 min (7). This independent validation of the deconvolution analysis-calculated GH half-life permits application of the program to the study of GH secretion under various conditions.

Between bursts of GH release, serum GH concentrations are undetectable in conventional assays (<0.2 μg/l). In adults, as in early childhood and puberty, GH is secreted in volleys of multiple secretory bursts. These complex volleys likely arise from secretion of multiple GHRH secretory bursts into the hypophyseal portal blood during periods of reduced SRIF secretion (6). Regardless of nutritional state, the amplitude of GH secretory bursts is greatest during the night. The most consistent period of GH secretion in both children and young adults is with the onset of the first slow-wave sleep (stages III and IV), which usually occurs within the first hour of sleep. Delay in the onset of deep sleep usually delays the onset of the largest GH peak. GH levels are highest during slow-wave sleep and are at their lowest during *rapid-eye-movement* (REM) sleep. Plasma levels of glucose, fatty acids, and insulin do not change with the burst of GH secretion, although resistance to exogenous insulin and decreased glucose tolerance develop after the peak of GH secretion. REM sleep may be responsible for the termination of sleep-related GH secretion.

Similar observations were made with GHRH administration during different stages of sleep, at different times during the night, and with sleep deprivation (8). These studies support the hypothesis that decreased hypothalamic SRIF secretion initiates GH secretory bursts. Sleep studies suggest that there is a circadian rhythm for growth hormone in man and that variations of SRIF secretion may account for differences in sensitivity to GHRH during different phases of sleep. Thus, slow-wave sleep may be associated with low hypothalamic SRIF secretion, and REM sleep with high SRIF secretion.

GH secretion also varies with age and gender. Analysis of GH concentration profiles (measured every 20 min over 24 h) in young men, young women, older men, and older women revealed that serum estradiol, but not testosterone, correlated with the 24-h integrated GH concentration (9). When the effects of estradiol were removed from the stepwise regression analysis, neither age nor gender influenced the integrated GH concentration. This study and others characterizing the increased GH secretion during puberty (10) suggest that gonadal steroids, particularly estradiol, are important in the regulation of GH secretion. The site of estrogen action is unknown and may reside at the pituitary or hypothalamic levels and/or the periphery. Studies in rats suggest that gonadal steroids affect GHRH and SRIF gene expression (11, 12). Estrogens may also influence peripheral factors, such as synthesis of IGF-I and/or IGF-I and GH binding proteins.

Fasting increases both the number and the amplitude of GH secretory bursts. It has been proposed that nutritional deprivation is associated with low hypothalamic SRIF secretion. In contrast, GH secretion is decreased in obese subjects; the number of GH secretory bursts is reduced, but the amplitude of these secretory bursts is similar to those of age-matched normal-weight subjects. The metabolic clearance rate of GH is accelerated in obesity, which contributes to the hyposomatotropism (13). The diminished GH release in obesity may be associated with increased SRIF secretion. Studies of patients with GHRH-secreting tumors and normal adults given intravenous GHRH infusions support the concept that SRIF is responsible for the timing of GH secretory pulses.

The significance of pulsatile GH secretion is not established. However, it is likely that the pattern may determine tissue responses. Thus, pulsatile GH secretion may be particularly important for growth and its multiple effects on metabolism (14).

GH secretion is augmented or inhibited by a number of neurogenic, metabolic, and hormonal influences. These factors are often only effective in the acute situation. For example, hyperglycemia in normal subjects acutely suppresses GH secretion (15), while in poorly controlled diabetics, 24-h integrated GH concentrations are increased (16). Similarly, although acute increases in free fatty acid concentration inhibit GH release, fatty acid levels are markedly increased during starvation when GH secretion is augmented. It appears that GH secretion is inversely related to insulin concentrations. Thus, GH secretion is augmented in fasting, anorexia nervosa, insulin-dependent diabetes mellitus, and hepatic cirrhosis when circulating insulin levels are low. In contrast, in obesity hyperinsulinemia is present and GH secretion is suppressed.

Certain amino acids, particularly arginine and leucine, transiently stimulate GH secretion and are used clinically as tests of GH reserve. While exercise, stress, and some neurogenic factors stimulate GH secretion, emotional deprivation can inhibit its release in children and lead to diminished linear growth (17, 18). Central alpha adrenergic agonists (norepinephrine) stimulate GH release (19); clonidine, a central alpha adrenergic agonist, stimulates growth hormone in this manner and is also used as a test of GH reserve. Beta adrenergic antagonists augment the efficacy of various stimuli for GH secretion. Dopamine agonists stimulate GH secretion in normal subjects by stimulating dopamine receptors; conversely, these drugs may decrease circulating GH concentrations in patients with acromegaly. Acetylcholine agonists stimulate (20, 21), while agents that lower acetylcholine tone cause suppression of GH release. Stimulation of GH by dopamine agonists, alpha adrenergic agonists, glucagon, and amino acids is inhibited by pirenzapine, a cholinergic muscarinic blocking drug (22–25). In addition, sleep-induced GH secretion is blocked by another cholinergic blocking drug, methylscopolamine. These agents are considered to act at the level of hypothalamus or the median eminence.

GH secretory responses to a number of stimuli are augmented after pretreatment with estrogens (e.g., diethylstilbestrol 3 mg/day for 3 days). The role of serotonin in human GH secretion is unclear (19). Histamine does not appear to play a role. Endogenous opiates probably do not directly influence GH secretion, since spontaneous GH secretion is not modified by naloxone. Morphine and beta-endorphin do not affect GH secretion, while nalorphine and an enkephalin analog (DAMME, FK33824) stimulate GH secretion through naloxone-sensitive mechanisms (26).

The pharmacological agent GH releasing peptide (27) is a potent stimulator of GH release; this hexapeptide was developed from an enkephalin series of peptides. Its mechanism of action is unknown, but it does not act through characterized GHRH, SRIF, or enkephalin receptors. It is likely that GH releasing peptides will have a role in clinical medicine, possibly as a test of GH reserve or to augment GH release from a potentially functional pituitary gland, such as occurs in obesity, aging, or in some children with a GH neurosecretory defect.

All the factors influencing GH secretion are probably not yet known. It is anticipated that new compounds will be developed that stimulate GH release or diminish the inhibitory influence of SRIF. Since GH is an important hormone for maintenance of muscle mass, carbohydrate and lipid utilization, and for its overall anabolic effects, continuing investigation in this area will likely yield significant practical information that can be applied to clinical medicine.

References

1. Barinaga M, Yamonoto G, Rivier C, Vale W, Evans R, Rosenfeld MG. Transcriptional regulation of growth hormone gene expression by growth hormone-releasing factor. Nature 1983;306:84–5.
2. Hendricks CM, Eastman RC, Takeda S, Asakawa K, Gorden P. Plasma clearance of intravenously administered pituitary human growth hormone: gel filtration studies of heterogeneous components. J Clin Endocrinol Metab 1985;60:864–7.
3. Hindmarsh PC, Matthews DR, Brain CE, et al. The half-life of exogenous growth hormone after suppression of endogenous growth hormone secretion with somatostatin. Clin Endocrinol (Oxf) 1989;30:443–50.
4. Owens D, Srivastava MC, Tompkins CV, Nabarro JD, Sonksen PH. Studies on the metabolic clearance rate, apparent distribution space and plasma half-disappearance time of unlabelled human growth hormone in normal subjects and in patients with liver disease, renal disease, thyroid disease and diabetes mellitus. Eur J Clin Invest 1973;3:284–94.
5. Parker ML, Utiger RD, Daughaday WH. Studies on human growth hormone, II. The physiological disposition and metabolic rate of human growth hormone in man. J Clin Invest 1962;41:262–8.
6. Hartman ML, Faria ACS, Vance ML, Johnson ML, Thorner MO, Veldhuis JD. Temporal structure of in vivo growth hormone secretory events in man. Am J Physiol 1991;260:E101–10.
7. Faria AC, Veldhuis JD, Thorner MO, Vance ML. Half-time of endogenous growth hormone (GH) disappearance in normal man after stimulation of GH secretion by GH-releasing hormone and suppression with somatostatin. J Clin Endocrinol Metab 1989;68:535–41.
8. Van Cauter E, Kerkhofs M, Van Onderbergen A, Caufriez A, Copinschi G. Modulation of spontaneous and GHRH-stimulated GH secretion by sleep [Abstract]. Prog 71st meet Endocr Soc, 1989:no. 220.
9. Ho KY, Evans WS, Blizzard RM, et al. Effects of sex and age on the 24-hour profile of growth hormone secretion in man: importance of endogenous estradiol concentrations. J Clin Endocrinol Metab 1987;64:51–8.

10. Martha PM Jr, Rogol AD, Veldhuis JD, Kerrigan JR, Goodman DW, Blizzard RM. Alterations in the pulsatile properties of circulating growth hormone concentrations during puberty in boys. J Clin Endocrinol Metab 1989;69:563–70.
11. Zeitler P, Argente J, Chowen-Breede JA, Clifton DK, Steiner RA. Growth hormone releasing hormone messenger ribonucleic acid in the hypothalamus of the adult male rat is increased by testosterone. Endocrinology 1989;127: 362–8.
12. Werner H, Koch Y, Baldino F, Gozes I. Steroid regulation of somatostatin mRNA in the rat hypothalamus. J Biol Chem 1988;263:7666–71.
13. Veldhuis JD, Iranmanesh A, Ho KK, Waters MJ, Johnson ML, Lizarralde G. Dual defects in pulsatile growth hormone secretion and clearance subserve the hyposomatotropism of obesity in man. J Clin Endocrinol Metab 1991; 72(1):51–9.
14. Jansson JO, Isaksson OG, Eden S, Isgaard J, Carlsson L, Ekberg S. Effects of plasma GH pattern on growth factors and body growth. In: Frisch H, Thorner MO, eds. Hormonal regulation of growth. New York: Raven Press, 1988:185–99.
15. Stewart PM, Smith S, Seth J, Stewart SE, Cole D, Edwards CR. Normal growth hormone response to the 75 g oral glucose tolerance test measured by immunoradiometric assay. Ann Clin Biochem 1989;26:205–6.
16. Asplin CM, Faria AC, Carlsen EC, et al. Alterations in the pulsatile mode of growth hormone release in men and women with insulin-dependent diabetes mellitus. J Clin Endocrinol Metab 1989;69:239–45.
17. Silver HK, Finkelstein M. Deprivation dwarfism. J Pediatr 1967;70:317–24.
18. Powell GF, Brasel JA, Blizzard RM. Emotional deprivation and growth retardation simulating idiopathic hypopituitarism, I. Clinical evaluation of the syndrome. N Engl J Med 1967;276:1271–8.
19. Delitala G. Clinical neuropharmacology in the management of disorders of the pituitary and hypothalamus. In: DeGroot LJ, Besser GM, Cahill GF Jr, et al., eds. Endocrinology. Philadelphia: WB Saunders, 1989:454–73.
20. Soulairac A, Schaub C, Franchimont P, Aymard N, van Cauwenberge H. A study of the pharmacological activation of the central pole of the hypothalamo-hypophyseal axis. Ann Endocrinol (Paris) 1968;29(1):45–54.
21. Salvadorini F, Galeone F, Nicotere M. Clinical evaluation of CDP-choline: efficacy as antidepressant treatment. Curr Ther Res Clin Exp 1968;18:513–20.
22. Mendelson WB, Sitaram N, Wyatt RJ, Gillin JC, Jacobs LS. Methoscopolamine inhibition of sleep-related growth hormone secretion. Evidence for a cholinergic secretory mechanism. J Clin Invest 1978;61:1683–90.
23. Delitala G, Frulio T, Pacifico A, Maioli M. Participation of cholinergic muscarinic receptors in glucagon- and arginine-mediated growth hormone secretion in man. J Clin Endocrinol Metab 1982;55:1231–3.
24. Delitala G, Grossman A, Besser GM. Opiate peptides control growth hormone through a cholinergic mechanism in man. Clin Endocrinol (Oxf) 1982; 18:401–5.
25. Delitala G, Maioli M, Pacifico A, Brianda S, Palermo M, Mannelli M. Cholinergic receptor control mechanisms for L-dopa, apomorphine, and clonidine-induced growth hormone secretion in man. J Clin Endocrinol Metab 1983;57:1145–9.

26. Stubbs WA, Delitala G, Jones A, et al. Hormonal and metabolic responses to an enkephalin analogue in normal man. Lancet 1978;2:1225–7.
27. Bowers CY, Reynolds GA, Durham D, Barrera CM, Pezzoli SS, Thorner MO. Growth hormone (GH)-releasing peptide stimulates GH release in normal men and acts synergistically with GH-releasing hormone. J Clin Endocrinol Metab 1990;70:975–82.

6

Mechanisms for Insulin-Like Growth Factor I Regulation of Growth Hormone Secretion

DIANE PRAGER AND SHLOMO MELMED

Growth hormone releasing hormone (GHRH) and *somatostatin* (SRIF) are secreted by the hypothalamus. After traversing the hypophyseal portal system, these two peptides will bind to their respective receptors situated on the somatotroph cells. Nevertheless, regulation of pituitary *growth hormone* (GH) secretion is mediated by several hormonal and central nervous system signals. GH is secreted in a wave-like pulsatile fashion. The pattern of GH secretion is considered in more detail elsewhere in this volume. GHRH and SRIF act out of phase to stimulate or suppress GH secretion, respectively. Peripheral hormones, including steroid and thyroid hormones as well as *insulin-like growth factor I* (IGF-I) and *insulin-like growth factor II* (IGF-II), also participate in the control of GH secretion.

IGF-I mediates most of the growth-promoting actions of GH. IGF-I, a potent mitogenic growth factor, bears a striking homology to proinsulin. It binds to specific receptors that also bind insulin at lower affinity. The predominant site of GH-stimulated IGF-I is the liver, and several extra-hepatic tissues also synthesize IGF-I. IGF-I regulates GH gene expression and secretion by a negative feedback regulation, analogous to the inhibition by thyroid and adrenal hormones of their respective pituitary trophic hormones (1, 2).

Pituitary Action of IGF-I

Although several lines of evidence indicate that IGF-I has a specific inhibitory action on GH secretion both in vivo and in vitro, the pituitary itself has been shown to express IGF-I (3). It is therefore as yet unclear whether IGF-I behaves as a true endocrine hormone or whether its pituitary action is paracrine mediated. IGF-I inhibits GH gene transcription, mRNA levels, and secretion in rat and human pituitary cells. This action is selective and apparently specific for IGF-I (4, 5). IGF-I also suppresses expression of a transfected *human growth hormone* (hGH) gene (6). IGF-I does not regulate secretion of other pituitary hormones (7). The doses of IGF-I required to elicit this in vitro response (~2 nM) are similar to the kd for ligand binding to pituitary IGF-I receptors (8). IGF-I also inhibits GH secretion that has been induced in vitro by GHRH, hydrocortisone, triiodothyronine, cyclic AMP-mediated signals (forskolin and 8 bromo-cAMP), and 12-0-tetradecanoylphorbol-13-acetate (9). The IGF-I effects on GH are proportional to the number of pituitary IGF-I receptors (10). Interestingly, overexpression of the human IGF-I receptor in heterologous pituitary cells enhances the somatotroph response to IGF-I (11). Presumably, the increased mass of IGF-I receptors enhances the ligand-mediated signal to the somatotroph cell nucleus. Although IGF-II also suppresses GH secretion, its effect has been shown to be mediated by the pituitary IGF-I receptor (12). In vivo, centrally administered somatomedin C also inhibits rat GH secretion (13). The combined administration of both IGF-I and IGF-II is additive in suppressing GH secretion in vivo (13).

The studies indicating ligand-mediated binding and action of IGF-I in pituitary tumor cells led us to conclude that the action of IGF-I in pituitary cells correlated with the relative abundance and function of the pituitary IGF-I receptor (10). We therefore isolated stably transfected pituitary cell lines expressing either wild-type human IGF-I receptor cDNA or mutants of the human receptor. Using these stable transfectants, we demonstrated that ^{950}Tyrosine (^{950}Tyr) located in the submembrane domain of the human IGF-I receptor β-subunit is critical for transducing the ligand signal to the GH gene within the pituitary cell. Furthermore, exogenous mutant human IGF-I receptors form hybrids with endogenous rat pituitary receptors and act dominantly to inhibit pituitary ligand-mediated signal transduction.

Determination of Critical IGF-I Receptor Sites for IGF-I Signal Transduction

Mutations of the human IGF-I receptor cDNA were constructed as depicted in Figure 6.1, and GC rat pituitary cells secreting GH were stably transfected (Fig. 6.1) (14). ^{950}Tyr situated in the submembrane domain was replaced with either Cys, Ser, Ala, Leu, or Thr. IGFIR-CRH consisted of three additional amino acid (Cys-Arg-His) insertions at position 950. Twenty-two amino acids (944–965) were deleted to form IGFIR Δ22. A truncated human IGF-I receptor cDNA was isolated by inserting

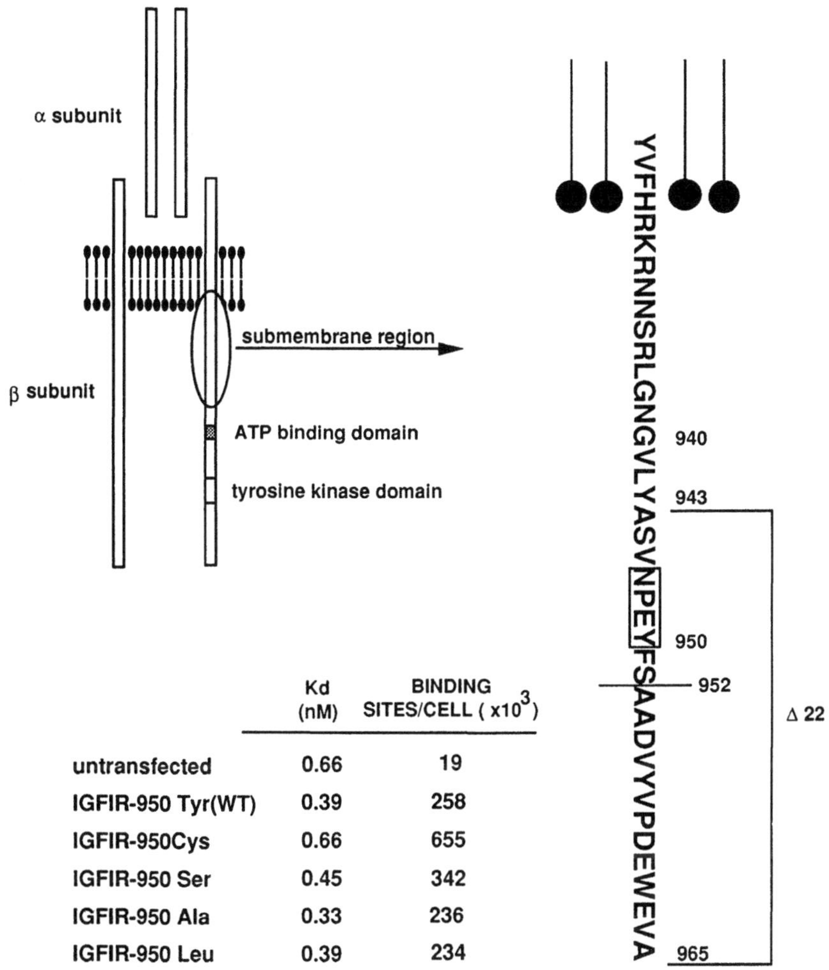

	Kd (nM)	BINDING SITES/CELL (x10^3)
untransfected	0.66	19
IGFIR-950 Tyr(WT)	0.39	258
IGFIR-950Cys	0.66	655
IGFIR-950 Ser	0.45	342
IGFIR-950 Ala	0.33	236
IGFIR-950 Leu	0.39	234
IGFIR-950 Thr	0.36	185
IGFIR-950 CRH	0.35	281
IGFIR-950 Δ22	0.25	133
IGFIR-952 STOP	0.92	964

FIGURE 6.1. IGF-I receptor cDNA constructs (β-subunit) are shown. Transfectants were subjected to Scatchard analysis, and derived kd and number of binding sites for untransfected and stable transfectants are depicted.

a synthetic stop codon at position 952 (^{952}STOP). Southern blot analysis confirmed the genomic integration of the exogenous human IGF-I receptor into rat GC cells. After neomycin selection, binding of [^{125}I] IGF-I was performed on all transfectants. Fifty percent displacement of maximum cell-associated binding was achieved by 0.6–2 nM IGF-I, and Scatchard analysis yielded linear slopes for all the mutant transfectants tested. These results suggest that a single class of high-affinity receptors is present on the transfectant cells. The derived kd for [^{125}I] IGF-I binding ranged from 0.25 to 0.92 nM, which is similar to WT untransfected cells. The number of mutant IGF-I receptors was markedly increased in transfected cells, with derived receptor mass showing up to a 50-fold induction.

Ligand-Induced Receptor Autophosphorylation and Phosphorylation of Endogenous Substrate

The transfected IGF-I receptor was tested for autophosphorylation, as well as phosphorylation of endogenous pp183, as cytoplasmic phosphorylation has been shown to be a critical step in the function of growth factor receptors. Cells overexpressing the WT IGF-I receptor contained

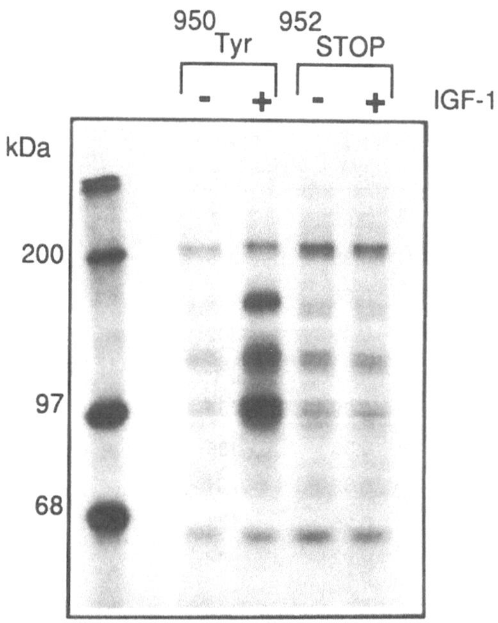

FIGURE 6.2. Phosphorylation of tyrosine residues in the IGF-I receptor and in cytoplasmic substrate. WT or ^{952}STOP IGF-I receptor transfectants were metabolically labeled with ^{35}S-methionine and then treated with IGF-I 6.5 nM for 1 min at 37°C. Cells were lysed, immunoprecipitated with a monoclonal anti-phosphotyrosine antibody as described and analyzed on 7.5% SDS-polyacrylamide gels. Reproduced from Prager, Yamasaki, Weber, Gebremedhin, and Melmed (18) by copyright permission of the American Society for Clinical Investigation.

phosphorylated tyrosine residues (97 kd) representing the β-subunit of the IGF-I receptor, as visualized by immunoprecipitation with monoclonal antiphosphotyrosine antibody. A larger phosphorylated protein corresponding to the predicted size of IRS-1 was also visualized. The 135-kd α-subunit of human IGF-I receptor was also immunoprecipitated, indicating the integrity of covalent αβ receptor subunit bonding. The site-directed mutant, ^{950}Ala, did not phosphorylate cytoplasmic substrate in response to ligand treatment, and the ^{952}STOP transfectants failed to undergo both ligand-mediated IGF-I receptor autophosphorylation as well as endogenous cytoplasmic protein phosphorylation (Fig. 6.2).

Exon 16 of the insulin receptor interacted with IRS-1, the putative cellular messenger for transduction of the insulin signal (15), and phosphorylation of IRS-1 was absent in ^{960}Tyr and ^{962}Ser insulin receptor mutants (14, 16). The results of the IGF-I receptor mutant phosphory-

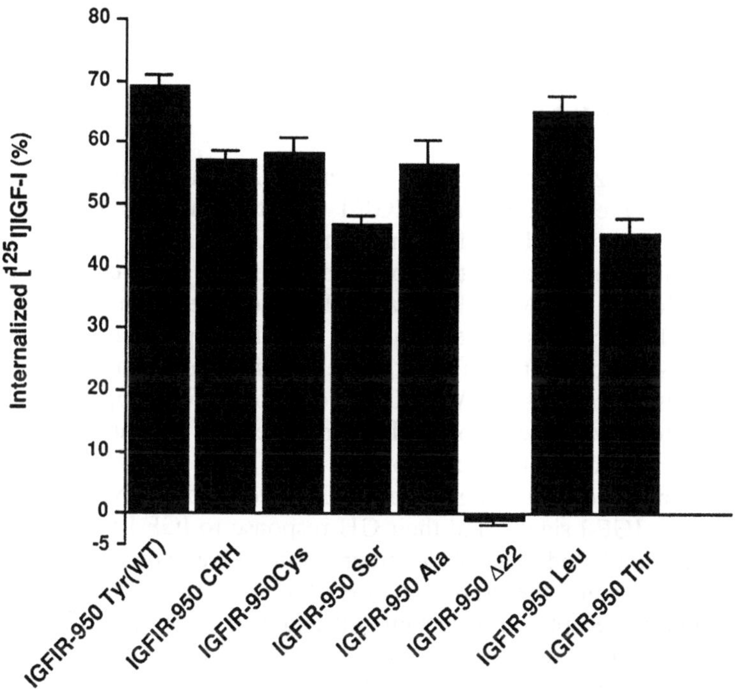

FIGURE 6.3. Internalization of [^{125}I] IGF-I in transfected cells. Transfectants were incubated with [^{125}I] IGF-I for 30 min and internalization determined by stripping ligand bound on the cell membrane by acidification of incubation buffer. Internalization (%) was calculated by dividing intracellular and degraded ligand by surface-associated intracellular and degraded radiolabel. Values for untransfected control cells were substracted. Each value represents mean ± SE of at least 3 separate experiments, each performed in duplicate. Reproduced with permission from Yamasaki, Prager, Gebremedhin, and Melmed (14).

lation experiments here suggest that pituitary IGF-I receptor autophosphorylation is not necessarily accompanied by phosphorylation of somatotroph cytoplasmic substrates (14).

Pituitary IGF-I Receptor Internalization

^{950}Tyr of the IGF-I receptor β-subunit is situated within an asparagine-proline-X-tyrosine (NPXY) motif that is necessary for internalization of the low-density lipoprotein receptor (17). As ligand-mediated growth factor receptor internalization may be important for postreceptor signaling, transfectants were tested for their ability to internalize radiolabeled IGF-I. Cells expressing either WT or mutant IGF-I receptors internalized 46%–55% of exogenous [^{125}I] IGF-I, and IGFIR Δ22 failed to internalize labeled IGF-I ligand (Fig. 6.3). It would therefore appear that ^{950}Tyr is not critical for IGF-I receptor internalization in these cells, and internalization may not be required for signal transduction (see below).

IGF-I Pituitary Signal Transduction

After characterizing the binding, internalization, and phosphorylation properties of the mutant human IGF-I receptor transfectants, receptor function was studied. Transduction of the IGF-I signal to the GH gene was the unique postreceptor function assessed in these cells. A concentration of 6.5 nM IGF-I maximally attenuates somatotroph GH secretion (11), and this concentration of ligand occupies more than 90% of the IGF-I binding sites on pituitary cells. After treatment with 6.5 nM IGF-I, cells overexpressing WT IGF-I receptor showed suppressed GH secretion to 60% of controls, while treatment of untransfected cells resulted in a modest suppression of GH ($P < 0.001$ vs. WT) (Fig. 6.4). Transfectants overexpressing mutant IGF-I receptor cDNAs failed to enhance the IGF-I signal, and their GH response to IGF-I was similar to that of untransfected cells (20% suppression). Overexpression of intact WT IGF-I receptor therefore enhanced IGF-I receptor function, while ^{950}Tyr mutant receptor transfectants all failed to alter the IGF-I signal. The presence of intact ^{950}Tyr in the pituitary IGF-I receptor is therefore critical for IGF-I signal transduction to the GH gene.

Pituitary IGF-I Receptor Function Is Suppressed by a Transdominant Human/Rat Hybrid Receptor Mutant

The deletion mutant (IGFIR Δ22) failed to suppress GH secretion even modestly in response to ligand, despite the presumed presence of intact endogenous rat receptors on these cells. These results indicated that

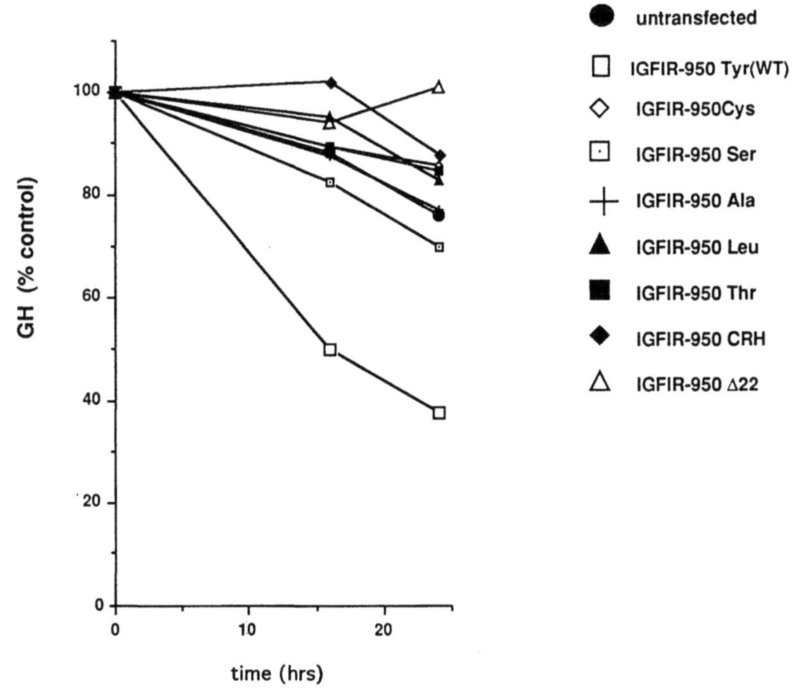

FIGURE 6.4. ^{950}Tyr is required for ligand-mediated GH suppression. Cells were treated with 6.5 nM IGF-I for the indicated time periods. GH (% of control) represents hormone levels from IGF-I-treated versus untreated transfectants. Each value represents the mean of 2 or 3 separate experiments, each with 6 determinations. All values for ^{950}Tyr (WT) transfectants were lower than those from mutant receptor transfectants and untransfected cells ($P < 0.001$). Reproduced with permission from Yamasaki, Prager, Gebremedhin, and Melmed (14).

nonfunctional hybrid exogenous/endogenous IGF-I receptors had been formed. We therefore tested whether or not mutant human IGF-I half-receptors had formed hybrids with endogenous rat IGF-I half-receptors and whether these hybrids possess a dominant negative effect on endogenous pituitary IGF-I receptor function (18).

IGF-I Receptor Immunoprecipitation Shows Presence of Human/Rat Hybrid IGF-I Receptors

Quantification of endogenous rat IGF-I receptors on the transfectants was assessed by [^{125}I] IGF-I binding in the presence of increasing amounts of the monoclonal antibody for the human IGF-I receptor α-IR3, which specifically inhibits [^{125}I] IGF-I binding to human IGF-I receptors, but

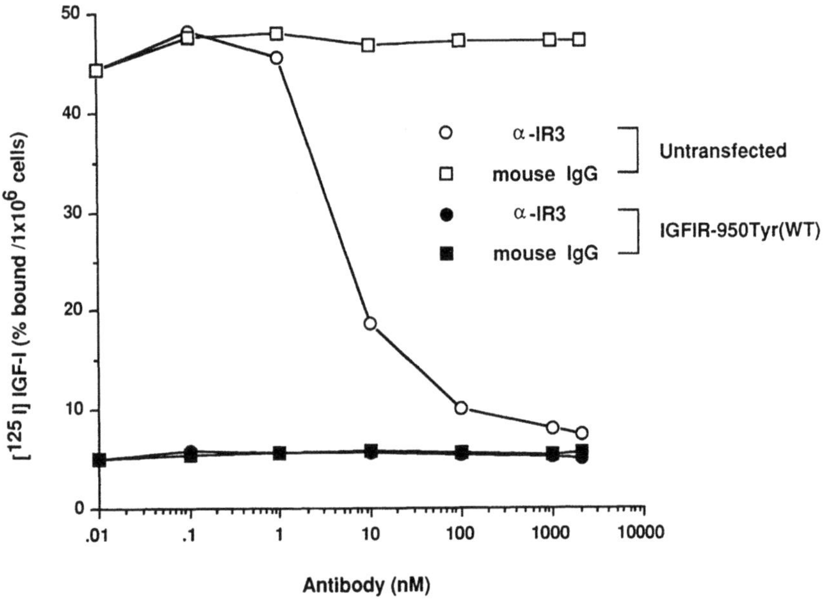

FIGURE 6.5. [^{125}I] IGF-I displacement by α-IR3 in untransfected cells and cells overexpressing ^{950}Tyr (WT) receptor. Cells were incubated with radiolabeled IGF-I and increasing amounts of either α-IR3 or mouse IgG. Each point represents the mean of duplicate values. Reproduced with permission from Yamasaki, Prager, Gebremedhin, and Melmed (14).

not to rat IGF-I receptors (19). A residual 7% nondisplaceable [^{125}I] IGF-I binding was still present on WT transfectants after displacement with very high concentrations of α-IR3. (Fig. 6.5). This failure to completely displace radiolabeled IGF-I from the transfectants suggested that despite overexpression of the human IGF-I receptor, rat/human hybrid receptors and/or endogenous rat/rat holotetrameric receptors were also present. In fact, subsequent experiments indicate that most of the endogenous rat IGF-I receptors present on transfected GC cells do in fact form hybrids with the overexpressed human IGF-I receptors.

Immunoprecipitation of receptors derived from cells overexpressing either WT or mutant IGF-I receptors with either α-IR3 antibody or a monoclonal antibody that recognizes both human and rat IGF-I receptor β-carboxyterminal subunits (Ab 1–2) was performed (20). The endogenous rat IGF-I receptors that are present in modest amounts on untransfected cells failed to immunoprecipitate with either α-IR3 or Ab 1–2, as expected. The 460-kd WT human IGF-I holotetramer, the 360-kd rat/mutant human hybrid, and 300-kd mutant human holotetramer bands were visualized in receptors derived from WT and ^{952}STOP cells, respectively (Fig. 6.6). α-IR3 also precipitated the truncated (^{952}STOP) human

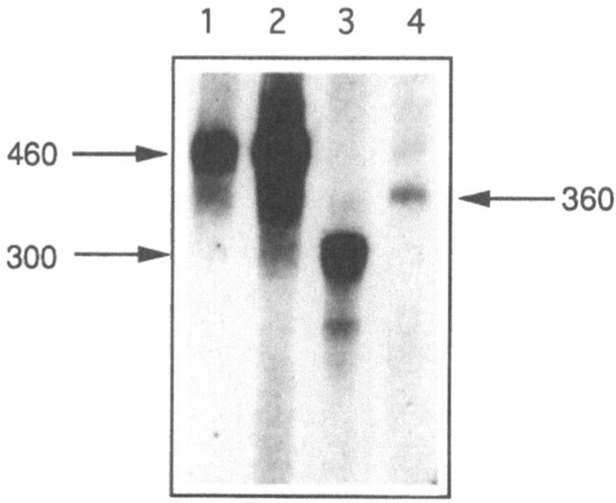

FIGURE 6.6. Immunoprecipitation of ^{35}S-methionine-labeled hybrid IGF-I receptors. ^{950}Tyr (WT) (lanes 1 and 2) or ^{952}STOP (lanes 3 and 4) cells were immunoprecipitated with either Ab 1–2 (lanes 2 and 4) or α-IR3 (lanes 1 and 3) and then electrophoresed on 4% nondenaturing gels. The 460-kd intact human or rat IGF-I receptor protein, 360-kd rat/mutant human IGF hybrid receptor protein, and 300-kd truncated human IGF-I receptor proteins are depicted. Reproduced from Prager, Yamasaki, Weber, Gebremedhin, and Melmed (18) by copyright permission of the American Society for Clinical Investigation.

receptor holotetramer (300 kd), and Ab 1–2 recognized the rat/mutant human hybrid receptor (360 kd). α-IR3 antibody recognized human IGF-I receptors, but not endogenous rat IGF-I receptors, and only bound weakly to the rat/human hybrid. The 360-kd IGF-I receptor protein species precipitated by Ab 1-2 receptor was absent in cells overexpressing WT receptors, indicating the formation of hybrids between endogenous rat αβ IGF-I half-receptor and exogenous human αβ IGF-I half-receptor. The ratio of hybrid receptor band intensity to endogenous rat receptor band intensity strongly suggests that most endogenous rat IGF-I receptors are in hybrid formation with the mutant human IGF-I receptors.

Half-life of WT human receptors in these cells was 2–6 h, with marked receptor degradation observed by 16 h. Newly synthesized mutant rat/human IGF-I hybrid receptors in ^{952}STOP cells have a prolonged half-life as compared to the WT human IGF-I receptor. Functional interactions between rat and defective human half-receptors may have an inhibiting influence on the rate of IGF-I receptor degradation.

^{952}STOP cells could not transduce ligand-mediated biological signaling to GH despite the presence of endogenous rat IGF-I receptors. Untransfected cells only suppressed GH by 40% in response to 36 h treatment with IGF-I, and WT transfectants exhibited a 60% suppression

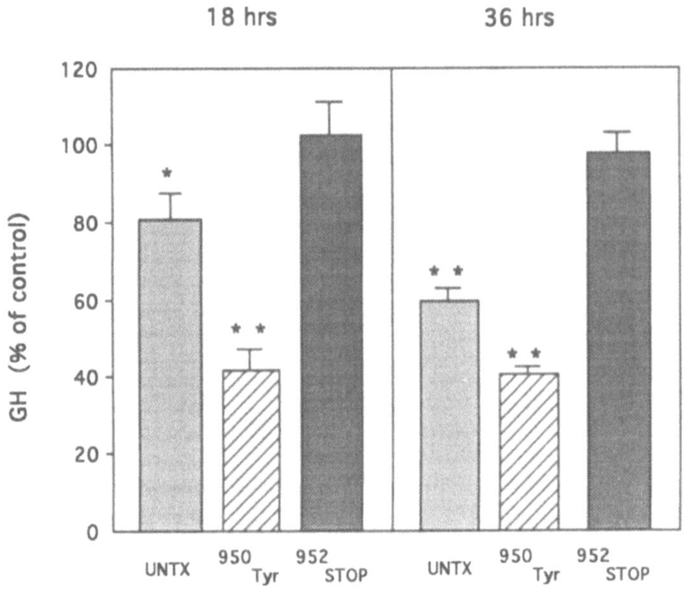

FIGURE 6.7. ^{952}STOP blocks IGF-I receptor signaling to GH. GH secretion in untransfected (UNTX), ^{950}Tyr (WT), and ^{952}STOP cells was measured after treating with IGF-I (6.5 nM). GH levels from untreated cells (4.2 ng/ml, 100% at 18 h; 12 ng/ml, 100% at 36 h) were compared to effects of IGF-I on GH secretion by the 3 different cell types. (* = $P < 0.05$ UNTX vs. ^{952}STOP, ** = $P < 0.001$ UNTX vs. ^{952}STOP, ** = $P < 0.001$ ^{950}Tyr vs. ^{952}STOP. Mean ± SEM of 3 separate experiments.) Reproduced from Prager, Yamasaki, Weber, Gebremedhin, and Melmed (18) by copyright permission of the American Society for Clinical Investigation.

of GH secretion as shown (11). ^{952}STOP cells consistently failed to respond to the IGF-I ligand, and GH secretion was higher in the ^{952}STOP cells than in nontransfected control cells despite the presence of endogenous rat IGF-I receptors in the transfectants ($P < 0.001$) (Fig. 6.7). These results indicate that ^{952}STOP half-receptors function as dominant negative inhibitors of endogenous rat IGF-I receptor function despite abundant IGF-I binding.

The results shown here indicate that ^{950}Tyr of the human IGF-I receptor β-subunit transmembrane domain is a critical requirement for pituitary IGF-I receptor function. IGF-I suppression of GH gene expression requires a 12- to 16-h lag period, unlike the early metabolic effects of IGF-I that manifest within minutes. This pituitary cell model will facilitate studies of mechanisms for the IGF-I signaling in a cell type expressing a well-defined and easily assayable polypeptide.

The physiologic relevance of the demonstrated effects of IGF-I on GH secretion studies has been underscored by several recent in vivo studies in man. Administered IGF-I causes a potent inhibition of GH secretion (21, 22) in normal human subjects. IGF-I receptor function may therefore play an important pathophysiologic role in several conditions. Fasting and malnutrition are associated with suppressed IGF-I and elevated GH levels. Lack of pituitary IGF-I feedback regulation may allow unrestrained GH secretion, thus counteracting the catabolic effects of malnutrition. This regulation of GH secretion may be determined by the relative abundance of pituitary IGF-I receptors. However, the relative contributions of either circulating or pituitary-derived IGF-I to the regulation of GH is presently unclear. Regardless of the source of pituitary IGF-I, the structure and function of the pituitary IGF-I receptor is a major determinant in the net effect of IGF-I on GH secretion.

References

1. Abe H, Molitch ME, Van Wyk JJ. Human growth hormone and somatomedin C suppress the spontaneous release of growth hormone in unanesthetized rats. Endocrinology 1983;113:1319.
2. Berelowitz M, Szabo M, Frohman LA. Somatomedin-C mediates GH negative feedback by effects on both the hypothalamus and the pituitary. Science 1981;212:1279.
3. Fagin J, Fernandez-Mejia C, Melmed S. Pituitary IGF-I gene expression: regulation of T3 and GH. Endocrinology 1989;125:2385–91.
4. Yamashita S, Melmed S. Insulin-like growth factor I regulation of growth hormone gene transcription in primary rat pituitary cells. J Clin Invest 1987;449–52.
5. Yamashita S, Weiss M, Melmed S. IGF-I regulates growth hormone secretion and mRNA levels in human pituitary tumor cells. J Clin Endocrinol Metab 1986;63:730–5.
6. Yamashita S, Ong J, Melmed S. Regulation of human growth hormone gene expresion by insulin-like growth factor I in transfected cells. J Biol Chem 1987;262:13254–7.
7. Yamashita S, Melmed S. IGF-I action on rat anterior pituitary cells: suppression of GH secretion and mRNA levels. Endocrinology 1986;118:176–82.
8. Rosenfeld RG, Ceda G, Cutler CW, Dollar LA, Hoffman AR. Insulin and insulin-like growth factor (somatomedin) receptors on cloned rat pituitary tumor cells. Endocrinology 1985;117:2208.
9. Morita S, Yamashita S, Melmed S. IGF-I action on rat anterior pituitary cells: effects of intracellular messengers on GH secretion and mRNA levels. Endocrinology 1987;121:2000.
10. Yamasaki H, Prager D, Gebremedhin S, Moise L, Melmed S. Binding and action of insulin-like growth factor I in pituitary tumor cells. Endocrinology 1991;128:857–62.
11. Yamasaki H, Prager D, Gebremedhin S, Moise L, Melmed S. IGF-I attenuation of growth hormone is enhanced by over-expression of pituitary cell IGF-I receptors. Mol Endocrinol 1991;5:890–6.

12. Weber M, Melmed S, Rosenbloom J, Yamasaki H, Prager D. Rat somatotroph IGF-II signalling: role of the IGF-I receptor. Endocrinology 1992;131:2147–53.
13. Harel Z, Tannenbaum GS. Synergistic interaction between insulin-like growth factors I and II in central regulation of pulsatile growth hormone secretion. Endocrinology 1992;131:758–64.
14. Yamasaki H, Prager D, Gebremedhin S, Melmed S. Human IGF-I receptor ^{950}Tyrosine is required for somatotroph growth factor signal transduction. J Biol Chem 1992;267:20953–8.
15. Sun XJ, Rothenburg P, Kahn CR, et al. Structure of the insulin receptor substrate IRS-1 defines a unique signal transduction protein. Nature 1991;352:73–7.
16. White MF, Livingston JN, Backer JM, et al. Mutation of the insulin receptor at tyrosine 960 inhibits signal transmission but does not affect tyrosine kinase activity. Cell 1988;54:641–9.
17. Chen WJ, Goldstein JL, Brown MS. NPXY, a sequence often found in cytoplasmic tails, is required for coated pit-mediated internalization of the low density lipoprotein receptor. J Biol Chem 1990;265:3116–23.
18. Prager D, Yamasaki H, Weber MM, Gebremedhin S, Melmed S. Human IGF-I receptor function in pituitary growth hormone-secreting cells is suppressed by a dominant negative mutant. J Clin Invest 1992;90:2117–22.
19. Gustafson TA, Rutter WJ. The cysteine-rich domains of the insulin and insulin-like growth factor I receptors are primary determinants of hormone binding specificity. J Biol Chem 1990;265:18663–7.
20. Soos MA, Siddle K. Immunological relationships between receptors for insulin and insulin-like growth factor I. Biochem J 1989;263:553–63.
21. Cutler HP, Schmid C, Zapf K, Froesch ER. Effects of recombinant insulin-like growth factor I on insulin secretion and renal function in normal human subjects. Proc Natl Acad Sci USA 1989;86:2868–72.
22. Walker JL, Baxter RC, Young SCJ, Pucilowska JB, Underwood LE. Effects of recombinant insulin-like growth factor I (IGF-I) on IGF binding proteins (IGF-BPS) and the acid labile subunit (ALS) in growth hormone insensitivity syndrome (GHIS) [Abstract]. 74th annu meet Endocr Soc, 1992;190.

Part II

The Immune System and Growth Hormone Interaction

7

Role of Growth Hormone and Insulin-Like Growth Factor I in Immunoregulation

KEITH W. KELLEY, SEAN ARKINS, AND YONG MING LI

It has long been recognized that the steroidal, adrenal-derived glucocorticoid hormones have potent anti-inflammatory properties and suppress effector functions of both T lymphocytes and macrophages. Until recently, the idea that protein hormones from the adenohypophysis other than *adrenocorticotropic hormone* (ACTH) could play a role in immunoregulation was little more than a curiosity. Instead, the most likely candidates for soluble communication molecules that regulate the immune response were a group of molecules synthesized primarily by leukocytes and given the generic name of interleukins. *Interleukins* are proteins that are synthesized and secreted by either resting or activated leukocytes and deliver signals to other cells that are either necessary for or amplify some activity of another type of leukocyte. Because these proteins were originally thought to serve as communication molecules between leukocytes, they were named interleukins. At least twelve of these interleukins have now been identified in leukocytes, their genes have been cloned, and the recombinant proteins have been expressed in bacteria or yeast. This list of interleukins is rapidly growing. However, as discussed below, it is becoming increasingly difficult to distinguish between an interleukin (e.g., IL-1) and a hormone (e.g., GH). The recent observations that certain hormones increase the secretion of some cytokines, and vice versa, also serve to blur the classical boundaries between interleukins and hormones.

The general view that cells of the immune system function autonomously is no longer tenable. Indeed, leukocytes are not only affected by neuroendocrine hormones, but they also synthesize and secrete hormones that were once thought to be synthesized only by the hypophysis. Furthermore, cytokines that are secreted by leukocytes have important effects on the synthesis and release of neuroendocrine hormones. Findings that have been reported within the past five years have now made it clear that cells of the immune system are inextricably linked with other physiological systems, such as the neuroendocrine, respiratory, reproductive, and central nervous systems. Changes in one system evoke changes in the other, and it is likely that feedback loops have evolved to coordinate and regulate functional activities of the immune system. The integrated view of immunophysiologists is that cells of the immune system interact with the entire body, rather than existing as a separate physiologic system that operates autonomously.

Like many other natural biologic substances, *growth hormone* (GH), *prolactin* (PRL), and *insulin-like growth factor I* (IGF-I) were named for their first recognized biologic activities, that is, somatic growth, milk production, and cartilage proliferation. Yet, neither GH nor IGF-I totally restores normal body growth in hypophysectomized rats. The function of PRL in males continues to puzzle endocrinologists. The clinical significance of hormonally related autoimmune diseases, such as diabetes, and the similarity of insulin and IGF-I continue to attract the research interests of both immunologists and endocrinologists.

The major target tissues for GH, IGF-I, insulin, and PRL have been previously defined as liver, muscle, adipocytes, and mammary epithelial cells. However, it now appears that cells of the immune system are also targets for these hormones, so cells that should now be added to this list are T and B lymphocytes, monocytes, natural killer cells, and all of the hematopoietic progenitors (reviewed by us in 1–6 and by others in 7–9). The conserved nature of cytokine, GH, and PRL receptors (10, 11) provides a biochemical basis for trying to understand how both GH and PRL affect the immune system. Evidence is accumulating within both the immune and central nervous systems that IGF-I is an important intermediate in mediating these effects (12, 13). It is the purpose of this chapter to highlight new findings that document a role for the somatolactogenic gene family, as well as IGF-I, in regulating developmental and functional activities of leukocytes.

Pituitary Gland Required for Host Resistance

Some of the first experiments on the potential role of GH in immunoregulation were published in the 1960s by Nicola Fabris, Ernst Sorkin, and their colleagues in Davos-Platz, Switzerland (reviewed in 1). Istvan Berczi and Eva Nagy significantly extended these findings by studying a number of immune responses in hypophysectomized rats that had been injected with either GH or PRL. Even though both of these hormones were shown to augment several components of both the humoral and cell-mediated immune systems, it had not been demonstrated that they could actually improve host resistance to an infectious pathogen.

To test this hypothesis, we infected three groups of rats with *Salmonella typhimurium* and evaluated mortality over the next 7 days (14). Injection of 1200×10^6 *colony-forming units* (CFU) of *S. typhimurium* caused around 50% of both the normal and sham-operated rats to die, whereas hypophysectomized rats injected with this dose of bacteria all died (Fig. 7.1). When the number of *S. typhimurium* was reduced 600-fold, nearly all of the intact and sham-operated control rats survived, compared to a survival rate of around 25% for rats without a pituitary gland. Although these data clearly established that the pituitary gland was necessary for rats to effectively resist an experimental infection with an infectious pathogen, both the pituitary-derived protective substance and its mechanism of action remained unknown.

The ability of GH to increase survival of hypophysectomized and pituitary-intact rats was tested by injecting GH 6 days before and 6 days after an experimental injection of *S. typhimurium*. Forty-four percent of the intact and 13% of the hypophysectomized, placebo-treated rats were alive 7 days after infection, whereas survival of GH-injected intact and

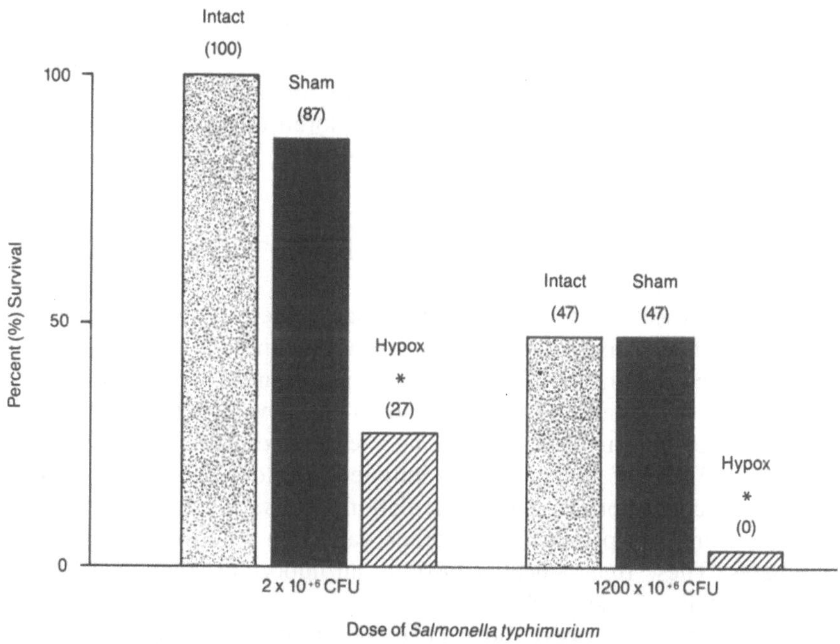

FIGURE 7.1. The pituitary gland is required for protection against the lethal effects of *S. typhimurium*. Rats (n = 15 per group) remained intact, were sham-operated, or were hypophysectomized (Hypox) and infected with one of two doses of *S. typhimurium*. Survival at 7 days after infection is shown on the ordinate, and the exact percentage survival for each treatment is shown in parentheses for each treatment. The asterisk indicates that survival of hypophysectomized rats was significantly less than either of the controls. Data from Edwards, Yunger, Lorence, Dantzer, and Kelley (14).

hypophysectomized rats was significantly greater at 75% and 53%, respectively (14). Furthermore, administration of an antiserum specific for GH, but not a control antiserum, abrogated the enhancement of host resistance in hypophysectomized rats. These data established that GH derived from the pituitary gland can provide significant protection when the host is infected with a bacterial pathogen. We have found similar results with recombinant GH (14, 15) and have found that the survival-promoting effects of recombinant GH can be improved significantly by incorporation of the hormone into liposomes (16).

While these experiments demonstrated that GH enhances the resistance of both intact and hypophysectomized rats to *S. typhimurium*, they did not establish the mechanism of action of GH. One could imagine several pathways by which GH could promote the survival of rats infected with a virulent bacterial pathogen. One example is that GH might simulate the synthesis of certain cytokines, as was recently reported for *tumor necrosis factor α* (TNF-α) in macrophages taken from hypophysectomized rats injected with GH (17), human myeloid cells incubated with GH in vitro (18), and with the in vitro secretion IL-1α and IL-2 following injection of GH into GH-deficient children (19). However, since macrophages are well known to be the primary effector cells for providing host defenses against *S. typhimurium*, we specifically explored the possibility that GH might affect functional activities of myeloid cells.

Growth Hormone and Prolactin Prime Macrophages

Macrophages isolated from infected animals secrete copious amounts of microbicidal oxygen metabolites, and this enhanced response of both macrophages and neutrophils is known as priming. One of the best-characterized substances that can lead to the development of activated macrophages in vitro is the T lymphocyte-derived cytokine, *interferon γ* (IFN-γ) (20). After phagocytosis is initiated, myeloid cells display a burst of oxygen consumption, glucose depletion, and *superoxide anion* (O_2^-) secretion and the production of other damaging reactive oxygen intermediates. These free radicals constitute an important component of the antimicrobial defense mechanisms of animals.

In the report described above (14), we cultured significantly fewer *S. typhimurium* from the spleen of both intact and hypophysectomized rats injected with GH than from placebo-treated rats. We then isolated resident peritoneal macrophages from uninfected GH-treated rats and cultured these cells with *S. typhimurium* in vitro. We found that GH significantly increased the killing of *S. typhimurium* by macrophages derived from either intact or hypophysectomized rats. Since the enhanced bacterial killing activity of macrophages from GH-treated rats could be abrogated by the inclusion in the culture medium of superoxide dismutase and catalase, which are scavengers of O_2^- and hydrogen peroxide, respectively, we tested the possibility that GH increased the capability of macrophages to secrete O_2^-. We have also recently reported that porcine alveolar macrophages cultured for 4 h with porcine GH killed significantly more *Pasteurella multocida*, an important respiratory bacterial pathogen in pigs, than did control macrophages (16).

The respiratory burst can be triggered by the binding of particulate stimuli, such as opsonized bacteria, to the membrane of phagocytic cells. In the absence of a triggering stimulus, pretreatment of macrophages with GH does not increase the secretion of O_2^- (21). However, if a secondary stimulus is added, such as opsonized zymosan, GH induces a significant enhancement in the secretion of O_2^-, and this in vitro priming can be abrogated by an antibody specific for GH (21). Similar results are obtained when macrophages are isolated from either normal (15) or hypophysectomized (15, 21) rats treated in vivo with pituitary-derived or recombinant GH (Table 7.1). Three daily injections of GH are required to obtain primed macrophages in hypophysectomized rats (15). Daily treatment with GH increases the secretion of hydrogen peroxide and the

TABLE 7.1. Pituitary-derived rat GH and PRL, as well as recombinant porcine GH and rat interferon γ (IFN-γ), augment the secretion of superoxide anion (O_2^-) by resident peritoneal macrophages from rats.

In vivo treatment	n	Superoxide anion (nmol O_2^-/mg protein/4 h) (mean ± SEM)
Hypophysectomized rats		
Placebo (Parlow's buffer)	6	-32.1 ± 21.9^{e}
Rat IFN-γ (500 U)	6	$280.1 \pm 25.9^{a,b}$
Rat GH (48 μg)	6	$274.8 \pm 24.3^{a,b}$
Rat prolactin (48 μg)	6	$276.4 \pm 32.5^{a,b}$
Recombinant GH (48 μg)	6	$208.2 \pm 32.9^{b,c}$
Pituitary-intact rats		
Placebo (Parlow's buffer)	6	$24.9 \pm 17.3^{d,e}$
Rat IFN-γ (500 U)	3	$122.3 \pm 29.8^{c,d}$
Rat IFN-γ (2500 U)	3	$231.3 \pm 80.3^{a,b}$
Rat GH (48 μg)	3	$9.2 \pm 19.3^{d,e}$
Rat GH (500 μg)	3	345.1 ± 42.2^{a}
Rat prolactin (48 μg)	3	$21.0 \pm 59.6^{d,e}$
Rat prolactin (500 μg)	3	$307.2 \pm 22.8^{a,b}$
Recombinant GH (48 μg)	3	$60.1 \pm 56.4^{d,e}$
Recombinant GH (500 μg)	3	$262.5 \pm 20.8^{a,b}$

Note: Animals were treated 9 days with the doses of each substance shown in parentheses. Treatments with different superscripts are significantly different.

Source: Data used with permission from Edwards, Ghiasuddin, Yunger, et al. (15).

phagocytosis of opsonized *Listeria monocytogenes* by resident peritoneal macrophages from both hypophysectomized and intact rats (15). Prolactin injections can also increase the respiratory burst of macrophages isolated from normal and hypophysectomized rats (15).

These data are consistent with new findings that show that PRL increases both O_2^- and hydrogen peroxide secretion by macrophages from intact mice (22). However, approximately 10-fold higher concentrations of both GH and PRL are needed to prime macrophages isolated from hypophysectomized rats as compared to intact rats (Table 7.1). The response of macrophages to both of these pituitary hormones is as great as that from the classic macrophage-activating factor, IFN-γ. However, it appears that IFN-γ is not required in vitro for GH to augment the respiratory burst of macrophages because an antibody to IFN-γ does not block the GH-induced priming of macrophages (16).

Priming of Human Neutrophils by Somatolactogenic Hormones and IGF-I

All of the experiments described thus far used resident peritoneal macrophages from rodents or pulmonary alveolar macrophages from pigs. However, other myeloid cells, such as polymorphonuclear leukocytes, also secrete O_2^- in response to both particulate and pharmacological stimuli. We used opsonized zymosan as well as the protein kinase C activator, *phorbol myristate acetate* (PMA), to demonstrate that recombinant human GH and PRL at minimal concentrations of 100 ng/ml and 50 ng/ml, respectively, augment the secretion of O_2^- by human *polymorphonuclear neutrophils* (PMN) (23, 24). Since at least some of the effects of GH on hemopoiesis are mediated by an increase in the secretion of macrophage-derived IGF-I (see below), we tested the hypothesis that GH primes human PMN by increasing their ability to secrete IGF-I. Several lines of evidence supported this hypopthesis (23): (i) de novo protein synthesis was required as assessed by experiments in which cycloheximide completely blocked the enhanced secretion of O_2^- that was caused by GH; (ii) recombinant human IGF-I caused a significant 4-fold increase in O_2^- secretion in vitro at a concentration as low as 12 ng/ml; and (iii) a monoclonal antibody directed against the human IGF-I receptor abrogated the priming that was induced by IGF-I. However, the most compelling argument against this hypothesis is that the antibody directed against the human IGF-I receptor did not affect the GH-induced priming of human PMN. These data suggested that the effect of GH on priming human PMN for O_2^- is not mediated by inducing the extracellular release of IGF-I. Based on the effects of GH being mediated by IGF-I on human myeloid (25) and erythroid (26) cell differentiation, as well as the clonogenesis of human T lymphocytes (27), it is likely that the GH-induced augmentation of O_2^- by human PMN is a special case. However, it was important to define this mechanism, so we continued to search for another mechanism of action of human GH on human PMN.

It is well accepted that human GH, but not GH from other species, has a pleiotropic receptor-binding activity (28). Many of the biological effects of human GH are shared with prolactin since human GH can bind and activate both somatogenic and lactogenic receptors. However, human PRL binds only to the lactogenic receptor and does not bind to the somatogenic receptor. We first observed that recombinant human GH and recombinant human PRL, but not GH of bovine or porcine origin, augment O_2^- secretion of human PMN (24). A monoclonal antibody directed against the human PRL receptor completely abrogated O_2^- secretion by PMN incubated with either human GH or PRL, whereas a monoclonal antibody to the human GH receptor had no effect. These results suggested that human GH probably binds to the PRL receptor on PMN to enhance their secretion of O_2^-.

In order to further explore the mechanism by which GH primes human PMN, we collaborated with Jim Wells and his colleagues at Genentech, Inc., who had created a panel of human GH variants by site-directed mutagenesis to aid in defining the contact residues that are responsible for this molecule to bind to somatogenic and lactogenic receptors (e.g., 29). One of these human GH variants, which has reduced affinity for both the human GH binding protein and the human PRL binding protein and designated K172A/F176A, was unable to prime human PMN (24) (Fig. 7.2). This finding indicated that priming is initiated by a ligand-receptor interaction, the affinity of which is near that defined for the receptors for human PRL and GH. Another GH variant, K168A/E174A, which has relatively low affinity for the human PRL binding protein but slightly increased affinity for the human GH binding protein, had a much-reduced ability to prime human PMN. In contrast, human GH variant E56D/R64M, which has a similar affinity as wild-type human GH for the human PRL binding protein but a lower affinity for the human GH binding protein, primed PMN as effectively as the wild-type human GH. We also found that the priming of human PMN by human GH required the presence of zinc, which is consistent with the fact that binding of human GH to the human PRL binding protein, but not to the human GH binding protein, is zinc dependent.

Collectively, these data indicate that human GH utilizes the PRL receptor of human PMN for augmenting the secretion of O_2^-, thereby providing direct evidence for coupling the binding of human GH to the human PRL receptor with one aspect of functional activation of human leukocytic target cells. All of these data are consistent with those of others who have found that human GH primes human myeloid cells for an enhanced respiratory burst (30, 31, 32), that prolactin acts similarly

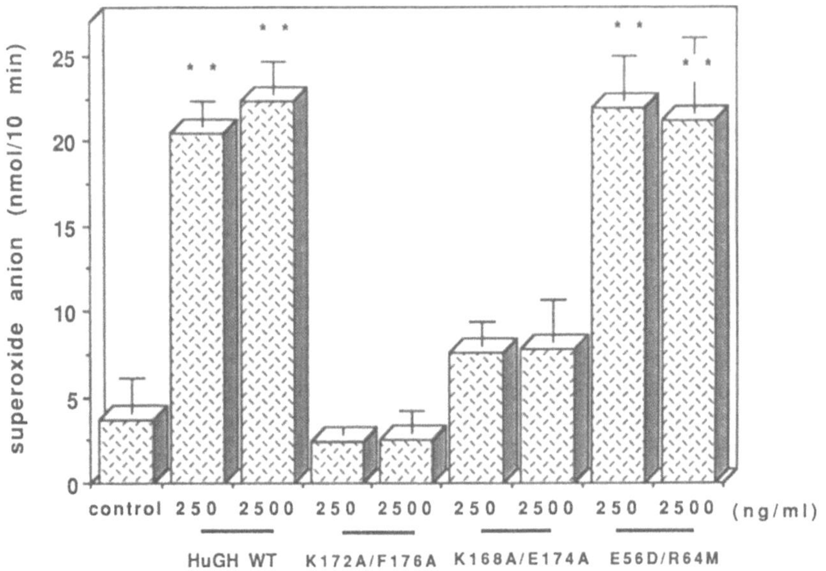

FIGURE 7.2. Evidence that human GH primes human PMN for binding to the PRL receptor. Significant enhancement (indicated by asterisks) was observed in human PMN treated with wild-type human GH (HuGH WT) and the variant E56D/R64M, which binds to the human PRL receptor. The human GH variant K172A/F176A, which does not effectively bind to either the human GH or PRL receptor, and variant K168A/E174A, which binds effectively only to the human GH receptor, did not significantly increase O_2^- secretion. Reproduced from Fu, Arkins, Fuh, et al. (24) by copyright permission of the American Society for Clinical Investigation.

on rat macrophages (15, 22), that human leukocytes (33) and thymic epithelial cells (34) express receptors for prolactin, and that monocytes taken from patients with either acromegaly or prolactinomas are more effective in killing *Mycobacterium avium* in vitro (35).

Growth Hormone, Immunity, and Aging

Several years ago we demonstrated that implantation of GH3 pituitary cells, which can secrete high quantities of GH in vivo, into syngeneic aged rats reversed the involution of their thymus gland (36). These results demonstated that the well-known thymic atrophy of aging is not totally irreversible and is at least in part dependent on the neuroendocrine status of the aged animal. We recently found that thymic reconstitution is accomplished by reversing age-associated changes in specific, immature T lymphocyte subsets in the thymus gland (37). Implantation of GH3 pituitary cells also restores the responsiveness of peritoneal macrophages from aged rats to IFN-γ (38). A number of other laboratories have reported that direct injections of GH have a number of effects on the thymus gland of both aged and grafted immunodeficient mice (reviewed in 4). The role of GH and IGF-I in regulating both metabolic and immune events in aged animals and humans is quite important, but is beyond the scope of this chapter.

IGF-I and Leukocytes

While IGF-I was originally identified as an intermediary growth factor induced by the actions of GH on skeletal tissues and mediating this hormone's anabolic actions, concerted efforts over the past thirty years have considerably expanded the knowledge of this growth factor's modus operandi. Consequently, it is no longer tenable to view IGF-I simply as the mediator of the anabolic actions of GH. It is now evident that IGF-I is a pleiotropic growth factor that functions in endocrine, paracrine, and autocrine pathways and is capable of eliciting a range of responses, including growth, maturation, or differentiation, depending on the phenotype of the responding cell. Such a pleiotropic role for IGF-I in tissues of the immune system is evidenced by the demonstrations that IGF-I restores lymphoid organ size and cellularity in hypophysectomized (39, 40) and diabetic (41) rats, stimulates the proliferation of leukemic blasts (42, 43) and T lymphocytes (44–46), and promotes the growth and differentiation of erythroid (26, 47–49) and myeloid (25, 50) cells. The generalized role for IGF-I in maintaining commitment to proliferation by preventing apoptosis of differentiating progenitors (13) may also pertain to hematopoietic precursors, since Rodriguez-Tarduchy et al. (12) showed that the addition of IGF-I in vitro prevented apoptosis in IL-3-derived bone-marrow lineages upon withdrawal of this cytokine. Indeed, it could be argued that the pleiotropic circle has been closed by the demonstration that IGF-I also modulates effector functions of mature myeloid cell types by priming macrophages and PMN to produce O_2^- and by the recent finding that IGF-I increases the uptake and degradation of low-density lipoprotein by macrophages (51).

Synthesis of IGF-I by Leukocytes

The evidence described above demonstrates that IGF-I influences a wide range of cell types and events in the immune system. Evidence exists that indicates that IGF-I is actually synthesized by leukocytes (52–54), so we sought to identify and characterize the production of IGF-I mRNA transcripts by lymphocytes. Transcripts encoding the entire prepro-peptide

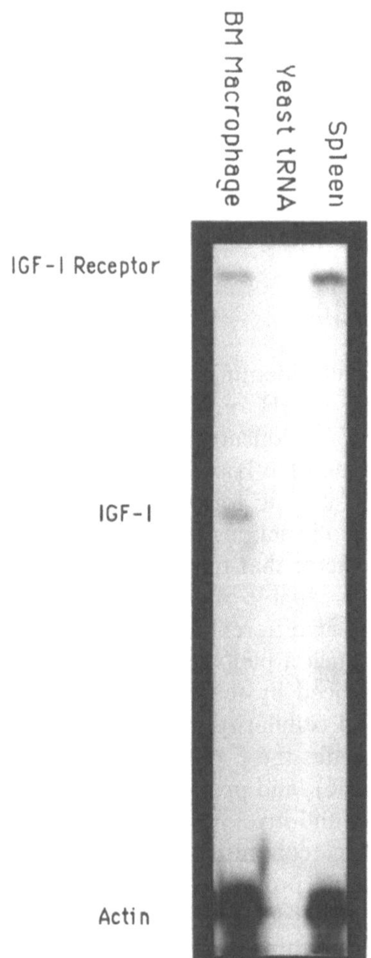

FIGURE 7.3. Rat bone marrow (BM)-derived macrophages express abundant steady state mRNA for IGF-I as well as the type I IGF-I receptor, whereas rat spleen cells express fewer transcripts for IGF-I and abundant transcripts for the IGF-I receptor, as assessed by solution hybridization, RNase protection assays. Total cellular RNA from rat BM-derived macrophages (10 μg), rat spleen (15 μg), or yeast tRNA (10 μg) was cohybridized with 265-nucleotide antisense transcripts to the rat IGF-I receptor (60) and with antisense transcripts from an expression vector cassette containing the entire coding region for IGF-I exon 4 (182 bp; 61) and flanking intronic sequences and a 122-bp fragment of rat β-actin (Arkins et al., in preparation, Biragyn et al., in preparation).

can be detected by PCR amplification in murine leukocytes of both the lymphoid and myeloid lineages (Arkins et al., in preparation). On a quantitative basis, however, IGF-I steady state mRNA levels are negligible or nondetectable in T and B cell lines when assessed by Northern blotting or ribonuclease protection assays, whereas macrophages express abundant IGF-I mRNA transcripts (Fig. 7.3) that are translated into a large molecular weight (26 kd) prepro-IGF-I peptide. Macrophage IGF-I mRNA transcripts initiate exclusively within exon 1, characteristic of extrahepatic IGF-I mRNA. Our demonstration that IGF-I mRNA transcripts are specifically induced during the in vitro differentiation of macrophage progenitors strongly suggests an autocrine role for macrophage-derived IGF-I in the growth and differentiation of hemopoietic cells. The precise role of IGF-I production by mature macrophages remains conjectural. However, we have advanced a working hypothesis (5) that proposes that this growth factor plays important autocrine and paracrine roles in such events as macrophage differentiation, the expansion of responsive T and B cell clones following antigenic stimulation, and in wound healing (see below).

Growth Hormone and IGF-I in Immunoregulation

It appears that classic pituitary hormones, such as GH (55, 56) and PRL (33, 57), as well as IGF-I (52–54), are synthesized by leukocytes. Furthermore, all three of these molecules affect developmental and functional activities of lymphoid and myeloid cells, and in some cases, IGF-I may mediate the effects of GH on leukocytes. Accordingly, we recently proposed a hypothesis to integrate existing knowledge on somatolactogens, somatomedins, and immune events into an immunoregulatory model (5) (Fig. 7.4). This model proposes that IGF-I modulates immune events in at least three distinct immunological compartments.

First, within the bone marrow, myeloid, erythroid, and lymphoid progenitors express receptors for IGF-I (e.g., 58). Localized IGF-I synthesis from stromal cells and maturing macrophages and endocrine IGF-I from the circulation synergizes with colony-stimulating factors to prevent apoptosis of differentiating progenitor cells (12) during hemopoiesis and to promote their proliferation (e.g., 48, 50) and possibly maturation.

Second, in peripheral lymphoid tissues, the sequestration of an inflammatory stimulus induces IGF-I mRNA transcription in macrophages. Secreted IGF-I acts on a small subset of antigen-responsive T and B cells, possibly in concert with certain interleukins, acting initially as a chemoattractant (44) and later helping to promote proliferation and clonal expansion (e.g., 43–46). Autocrine IGF-I also acts as a maturational influence on tissue macrophages, influencing effector functions such as the genera-

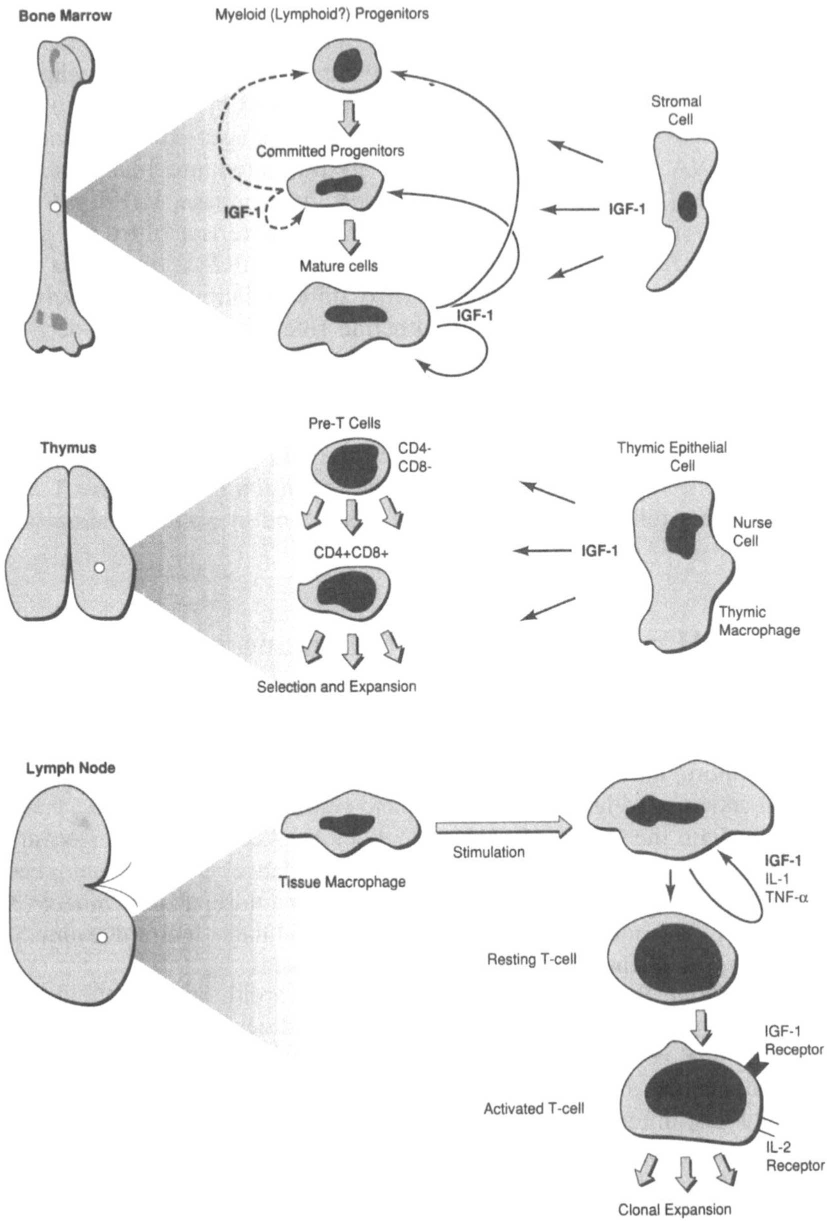

FIGURE 7.4. A hypothetical model for summarizing the role of IGF-I in immunoregulation. Reprinted with permission from Kelley, Arkins, Li, and Biragyn (5).

tion of O_2^-, degradation of low-density lipoprotein, and possibly the secretion of IL-1 and the ability to process and present antigen.

Third, within the thymus, thymic progenitors are influenced by circulatory IGF-I and by IGF-I produced by thymic epithelial cells (59). IGF-I exerts chemoattractant effects, promoting the migration of T cell progenitors to the thymus gland (44). Intrathymic T cell progenitors express IGF-I receptors and proliferate in response to this growth factor before demonstrating responsiveness to any known cytokine (58), suggesting that IGF-I may function within the thymus both by preventing apoptosis and by enhancing progression through the cell cycle. While much of this model remains speculative, it provides a rational framework for trying to understand the mechanisms by which GH, PRL, and IGF-I affect the variety of progenitor and mature cells that compose the immune system.

References

1. Kelley KW. Growth hormone, lymphocytes and macrophages. Biochem Pharmacol 1989;38:705–13.
2. Kelley KW. Growth hormone in immunobiology. In: Ader R, Felten DL, Cohen N, eds. Psychoneuroimmunology. 2nd ed. New York: Academic Press, 1991:377–402.
3. Kelley KW, Dantzer R. Growth hormone and prolactin as natural antagonists of glucocorticoids in immunoregulation. In: Plotnikoff N, Murgo A, Faith R, Wybran J, eds. Stress and immunity. Boca Raton, FL: CRC Press, 1991: 433–52.
4. Kelley KW, Arkins S, Li YM. Growth hormone, prolactin and insulin-like growth factor: new jobs for old players. Brain Behav Immun 1992;6 (in press).
5. Kelley KW, Arkins S, Li YM, Biragyn A. Growth hormone, insulin-like growth factor-I and immune function. In: Muller EE, Cocchi D, Locatelli V, eds. Growth hormone and somatomedins during lifespan. Heidelberg: Springer-Verlag, 1993.
6. Arkins S, Dantzer R, Kelley KW. Somatolactogens, somatomedins and immunity. J Dairy Sci 1993.
7. Weigent DA, Blalock JE. Growth hormone and the immune system. Prog Neuroendocrinimmunology 1990;3:231–41.
8. Berczi I, Nagy E. Effects of hypophysectomy on immune function. In: Ader R, Felten DL, Cohen N, eds. Psychoneuroimmunology. 2nd ed. New York: Academic Press, 1991:339–75.
9. Gala RR. Prolactin and growth hormone in the regulation of the immune system. Proc Soc Exp Biol Med 1991;198:513–27.
10. Bazan JF. Structural design and molecular evolution of a cytokine receptor superfamily. Proc Natl Acad Sci USA 1990;87:6934–8.
11. Cosman D, Layman SD, Idzerda RL, et al. A new cytokine receptor superfamily. Trends Biochem Sci 1990;15:265–70.
12. Rodriguez-Tarduchy G, Collins MKL, Garcia I, Lopez-Rivas A. Insulin-like growth factor I inhibits apoptosis in IL-3 dependent hemopoietic cells. J Immunol 1992;149:535–40.

13. Barres BA, Hart IK, Coles HSR, et al. Cell death and control of cell survival in the oligodendrocyte lineage. Cell 1992;70:31–46.
14. Edwards CK III, Yunger LM, Lorence RM, Dantzer R, Kelley KW. The pituitary gland is required for protection against lethal effects of *Salmonella typhimurium*. Proc Natl Acad Sci USA 1991;88:2274–7.
15. Edwards CK III, Ghiasuddin SM, Yunger LM, et al. In vivo administration of recombinant growth hormone or interferon-γ activates macrophages: enhanced resistance to experimental *Salmonella typhimurium* infection is correlated with the generation of reactive oxygen intermediates. Infect Immun 1992;60:2514–21.
16. Edwards CK III, Arkins S, Yunger LM, Blum A, Dantzer R, Kelley KW. The macrophage-activating properties of growth hormone. Cell Mol Neurobiol 1992;12:499–510.
17. Edwards CK III, Lorence RM, Dunham DM, et al. Hypophysectomy inhibits the synthesis of tumor necrosis factor-α by rat macrophages: partial restoration by exogenous growth hormone or interferon-γ. Endocrinology 1991;128:989–96.
18. Laurence J, Grimison B, Gonenne A. Effect of recombinant human growth hormone on acute and chronic human immunodeficiency virus infection in vitro. Blood 1992;79:467–72.
19. Casanova S, Repellin AM, Schimpff RM. Production of interleukin-1α and interleukin-2 by mononuclear cells from children with growth delay in relation to the degree of growth hormone deficiency: effects of substitutive treatment. Horm Res 1990;34:209–14.
20. Adams DO, Hamilton TA. Molecular transductional mechanisms by which IFNγ and other signals regulate macrophage development. Immunol Rev 1987;97:5–27.
21. Edwards CK III, Ghiasuddin SM, Schepper JM, Yunger LM, Kelley KW. A newly defined property of somatotropin: priming of macrophages for production of superoxide anion. Science 1988;239:769–71.
22. Chen Y, Johnson AG. In vivo activation of macrophages by prolactin from young and aging mice. Int J Immunopharmacol 1992.
23. Fu YK, Arkins S, Wang BS, Kelley KW. A novel role of growth hormone and insulin-like growth factor-I: priming neutrophils for superoxide anion secretion. J Immunol 1991;146:1602–8.
24. Fu YK, Arkins S, Fuh G, et al. Growth hormone augments superoxide anion secretion of human neutrophils by binding to the prolactin receptor. J Clin Invest 1992;89:451–7.
25. Merchav S, Tatarsky I, Hochberg Z. Enhancement of human granulopoiesis in vitro by biosynthetic insulin-like growth factor-I/somatomedin C and human growth hormone. J Clin Invest 1988;81:791–7.
26. Merchav S, Tatarsky I, Hochberg Z. Enhancement of erythropoiesis in vitro by human growth hormone is mediated by insulin-like growth factor I. Br J Haematol 1988;70:267–71.
27. Geffner ME, Bersch N, Lippe BM, Rosenfeld RG, Hintz RL, Golde DW. Growth hormone mediates the growth of T-lymphoblast cell lines via locally generated insulin-like growth factor-I. J Clin Endocrinol Metab 1990;71:464–9.

28. Hughes JP, Friesen HG. The nature and regulation of the receptors for pituitary growth hormone. Annu Rev Physiol 1985;47:469–82.
29. Cunningham B, Wells JA. Rational design of receptor-specific variants of human growth hormone. Proc Natl Acad Sci USA 1991;88:3407–11.
30. Wiedermann CJ, Niedermuhlbichler M, Geissler D, Beimpold H, Braunsteiner H. Priming of normal human neutrophils by recombinant human growth hormone. Br J Haematol 1991;78:19–22.
31. Wiedermann CJ, Reinisch N, Kahler C, et al. In vivo activation of circulating monocytes by exogenous growth hormone in man. Brain Behav Immun 1992;6.
32. Spadoni GL, Spagnoli A, Cianfarani S, et al. Enhancement by growth hormone of phorbol diester-stimulated respiratory burst in human polymorphonuclear leukocytes. Acta Endocrinol (Copenh) 1991;124:589–94.
33. Pellegrini I, Lebrun JJ, Ali S, Kelly PA. Expression of prolactin and its receptor in human lymphoid cells. Mol Endocrinol 1992;6:1023–31.
34. Dardenne M, Kelly PA, Bach JF, Savino W. Identification and functional activity of prolactin receptors in thymic epithelial cells. Proc Natl Acad Sci USA 88:9700–4.
35. Sabharwal P, Zwilling B, Glaser R, Malarkey WB. Cellular immunity in patients with acromegaly and prolactinomas. Prog Neuroendocrinimmunology 192;5:120–5.
36. Kelley KW, Brief S, Westly HJ, et al. GH_3 pituitary adenoma implants can reverse thymic aging. Proc Natl Acad Sci USA 1986;83:5663–7.
37. Li YM, Brunke DL, Dantzer R, Kelley KW. Pituitary epithelial cell implants reverse the accumulation of $CD4^-CD8^-$ lymphocytes in thymus glands of aged rats. Endocrinology 1992;130:2703–9.
38. Davila DR, Edwards CK III, Arkins S, Simon J, Kelley KW. Interferon-γ-induced priming for secretion of superoxide anion and tumor necrosis factor-α declines in macrophages from aged rats. FASEB J 1990;4:2906–11.
39. Guler HP, Zapf J, Scheiwiller E, Froesch ER. Recombinant human insulin-like growth factor I stimulates growth and has distinct effects on organ size in hypophysectomized rats. Proc Natl Acad Sci USA 1988;85:4889–93.
40. Thissen JP, Underwood LE, Maiter D, Maes M, Clemmons DR, Ketelslegers JM. Failure of insulin-like growth factor-I (IGF-I) infusion to promote growth in protein-restricted rats despite normalization of serum IGF-I concentrations. Endocrinology 1991;128:885–90.
41. Binz K, Joller P, Froesch P, Binz H, Zapf J, Froesch ER. Repopulation of the atrophied thymus in diabetic rats by insulin-like growth factor I. Proc Natl Acad Sci USA 1990;87:3690–4.
42. Pepe MG, Ginztion NH, Lee PDK, Hintz RL, Greenberg PL. Receptor binding and mitogenic effects of insulin and insulin-like growth factors I and II for human myeloid leukemia cells. J Cell Physiol 1987;133:219–27.
43. Geffner ME, Bersch N, Golde DW. Insulin and IGF-I stimulate normal and virally-transformed T-lymphocyte cell growth in vitro. Brain Behav Immun 1992;6.
44. Tapson VF, Boni-Schnetzler M, Pilch PF, Center DM, Berman JS. Structural and functional characterization of the human T lymphocyte receptor for insulin-like growth factor I in vitro. J Clin Invest 1988;82:950–7.

45. Kooijman R, Willems M, Rijkers GT, et al. Effects of insulin-like growth factors and growth hormone on the in vitro proliferation of T lymphocytes. J Neuroimmunol 1992;38:95–104.
46. Johnson EW, Jones LA, Kozak RW. Expression and function of insulin-like growth factor receptors on anti-CD3-activated human T lymphocytes. J Immunol 1992;148:63–71.
47. Kurtz A, Zapf J, Eckardt KU, Clemons G, Froesch ER, Bauer C. Insulin-like growth factor I stimulates erythropoiesis in hypophysectomized rats. Proc Natl Acad Sci USA 1988;85:7825–9.
48. Merchav S, Silvian-Drachsler I, Tatarsky I, Lake M, Skottner A. Comparative studies of the erythroid-potentiating effects of biosynthetic human insulin-like growth factors-I and -II. J Clin Endocrinol Metab 1992;73:447–52.
49. Werther GA, Haynes K, Johnson GR. Insulin-like growth factors promote DNA synthesis and support cell viability in fetal hemopoietic tissue by paracrine mechanisms. Growth Factors 1990;3:171–9.
50. Scheven BAA, Hamilton NJ. Stimulation of macrophage growth and multinucleated cell formation in rat bone marrow cultures by insulin-like growth factor I. Biochem Biophys Res Commun 1991;174:647–53.
51. Hochberg Z, Hertz P, Maor G, Oiknine J, Aviram M. Growth hormone and insulin-like growth factor-I increase macrophage uptake and degradation of low density lipoprotein. Endocrinology 1992;131:430–5.
52. Rappolee DA, Mark D, Banda MJ, Werb Z. Wound macrophages express TGF-α and other growth factors in vivo: analysis by mRNA phenotyping. Science 1988;241:708–12.
53. Nagaoka I, Trapnell BC, Crystal RG. Regulation of insulin-like growth factor I gene expression in the human macrophage-like cell line U937. J Clin Invest 1990;85:448–55.
54. Baxter JB, Blalock JE, Weigent DA. Characterization of immunoreactive insulin-like growth factor-I from leukocytes and its regulation by growth hormone. Endocrinology 1991;129:1727–34.
55. Weigent DA, Blalock JE, LeBoeuf RD. An antisense oligodeoxynucleotide to growth hormone messenger ribonucleic acid inhibits lymphocyte proliferation. Endocrinology 1991;128:2053–7.
56. Weigent DA, Baxter JB, Blalock JE. The production of growth hormone and insulin-like growth factor-I by the same subpopulation of rat mononuclear leukocytes. Brain Behav Immun 1992;6.
57. Sabharwal P, Glaser R, Lafuse W, et al. Prolactin synthesized and secreted by human peripheral blood mononuclear cells: an autocrine growth factor for lymphoproliferation. Proc Natl Acad Sci USA 1992;89:7713–6.
58. Gjerset RA, Yeatgin J, Volkman SK, Vila V, Arya J, Haas M. Insulin-like growth factor-I supports proliferation of autocrine thymic lymphoma cells with a pre-T cell phenotype. J Immunol 1990;145:3497–501.
59. Timsit J, Savino W, Safieh B, et al. Growth hormone and insulin-like growth factor-I stimulate hormonal function and proliferation of thymic epithelial cells. J Clin Endocrinol Metab 1992;74.
60. Werner H, Woloschak M, Adamo M, Shen-Orr Z, Roberts CT Jr, LeRoith D. Developmental regulation of the rat insulin-like growth factor I receptor gene. Proc Natl Acad Sci USA 1989;86:7451–5.

61. Tollefesen SE, Lajaras R, McCusker RH, Clemmons DR, Rotwein P. Insulin-like growth factors (IGF) in muscle development. Expression of IGF-I, the IGF-I receptor, and IGF binding protein during myoblast differentiation. J Biol Chem 1989;264:138810–7.

8

Production and Function of Growth Hormone in the Immune System

Douglas A. Weigent and J. Edwin Blalock

An important and relatively new finding is the ability of cells of the immune system to produce and secrete neuroendocrine hormones (reviewed in 1). Currently, the list includes *adrenocorticotropin* (ACTH) and endorphins, thyrotropin, luteinizing hormone, and *growth hormone* (GH). In addition, the synthesis of corticotropin releasing hormone (2) and *growth hormone releasing hormone* (GHRH) (3) has also been documented. Further, functional receptors for the neuroendocrine hormones have also been observed in the immune system (4). Thus, the evidence taken together supports the existence of the same signal molecules and receptors that potentially mediate interactions both between and within the neuroendocrine and immune systems.

For several years now, we have investigated the importance of neuroendocrine hormone production by leukocytes, using the GH axis as our model. The studies have involved: (i) examining the structure of lymphocyte-derived GH, (ii) identifying factors involved in the synthesis of GH by leukocytes, and (iii) identifying immune functions where the endogenous production of this hormone may play an important role. The results discussed below show that a GH molecule structurally similar to pituitary GH can be produced by cells of the immune system. The GH produced by the immune system plays a role in lymphocyte proliferation and is regulated, in some respects, similarly to pituitary GH, whereas clear differences also exist.

Materials and Methods

Cell Preparations

Adult (150–200 g) male Sprague-Dawley rats were obtained from Harlan (Prattville, AL). Following sacrifice, lymphoid tissues including spleens, thymus, PBL, and bone marrow were immediately collected and teased with sterile ground-glass-edged slides to obtain cell suspensions. Greater than 95% cell viability by trypan blue exclusion was observed in all cell fractions. For proliferation studies, cell suspensions (1×10^6/ml) of fractionated lymphoid cells were prepared in RPMI-1640 supplemented with 10% fetal calf serum containing 100 U/ml of penicillin and streptomycin (GIBCO Laboratories, Grand Island, NY). These suspensions were incubated in the presence of ^{3}H-thymidine (24 h; 2 μC/ml, New England Nuclear, Boston, MA) in plastic tissue culture dishes with and without various concentrations of ODNs as indicated. The cultures were harvested using a Whittaker Mini-Mash Harvester and the glass fiber filters counted in a Tracor Analytic Model 6892 liquid scintillation system (Elk Village, IL).

RIA of GH and IGF

GH was measured by RIA using ^{125}I-GH as radioligand and a polyclonal monkey anti-rat GH antibody. This antiserum (NIDDK-anti-rGH-S-5) was a generous gift from the hormone distribution program of the NIADDK and provided by Dr. S. Raiti. This antibody is highly specific with low crossreactivity with other highly purified rat pituitary hormones (rat prolactin <0.0001%; rat follicle stimulating hormone, 0.0009%; rat thyrotropin, 0.008%) and has a sensitivity of 500 pg/ml. The reagents, including buffer, cold standard or unknown, ^{125}I-GH, and antiserum were incubated at room temperature for 24 h prior to the addition of insoluble protein A (8 mg/tube) as a precipitating agent. The immune complexes were pelleted by centrifugation and counted in a gamma counter. Concentrated fluids were measured for their IGF-I content by a standard RIA procedure using a commercial kit available from Incstar Corporation (Stillwater, MN). The RIA kit includes an *octadecasilyl* (ODS)-silica extraction procedure to separate IGF-I from binding proteins. This involves incubating the controls and unknown samples with 0.5 N HCl and then applying the samples to ODS-silica/C18 columns. The columns are then washed 2 times with 4% acetic acid, and the samples are eluted with methanol, evaporated to dryness, and reconstituted in a BSA-borate buffer provided in the kit. According to the manufacturer of the RIA kit, the rabbit antisomatomedin C antibody used in this assay has less than 1% crossreactivity with IGF-II, human GH, fibroblast growth factor, transforming growth factor, and platelet-derived growth factor. The assay has a sensitivity of not less than 15 ng/ml.

Synthesis and Purification of Oligodeoxynucleotide (ODN)

Unmodified 20-base deoxyribonucleotides were synthesized on an automated solid-phase synthesizer (Genesis 2000, DuPont) and purified by passage over Nensorb-20 columns by standard procedures. The sequences of synthesized oligomers based on the rat GH mRNA sequence are: sense 1–7 amino acids: (30 bases) exon II 5′-TTACCTGCCATGCCC-TTGTCCAGTGTCTTT-3′; antisense 1–7 amino acids: (30 bases) exon II 3′-AATGGACGGTACGGGAACAGGTCACAGAAA-5; ODN H (31 bases) (GH gene 5′ promoter region) 5′-TCTCCCCTCAGGACA-CATTGTGCCCAAAGGG-3′.

Nucleic Acid Analysis

Total cytoplasmic RNA was isolated by the proteinase K method (5). Preparation of cDNA molecules, labeling, and probing of membranes were done as previously described (6). After washing, the nitrocellulose papers were exposed to X-ray film at −70°C with DuPont (Wilmington, DE) Cronex Lightning-Plus intensifying screens for 2–3 days. The autoradiographs were analyzed using a GS 300 densitometer (Hoefer Scientific Instruments, San Francisco, CA).

Protein Analysis

GH in cells was detected by immunofluorescence as previously described (7). Culture supernatant fluids were analyzed by immunoaffinity chromatography as previously described (7).

Materials

Monkey anti-rat GH antiserum was generously donated by Dr. S. Raiti of The National Pituitary Agency (NIDDK, NIH, Bethesda, MD). [^{3}H]TdR was purchased from New England Nuclear/DuPont (Wilmington, DE). The IGF-I antiserum (UBK487) was generously donated by Drs. L.E. Underwood and J.J. Van Wyk of the University of North Carolina at Chapel Hill through the National Hormone and Pituitary Program. The antiserum is highly specific, shows a 0.5% crossreactivity with IGF-II, and crossreacts minimally with insulin at 10^{-6} M, according to the suppliers. The recombinant human IGF-I used in these studies was obtained from Chemicon (El Segundo, CA). Additional reagents of the highest purity available were purchased from GIBCO (Grand Island, NY) and Sigma (St. Louis, MO) unless otherwise stated.

Structure of Lymphocyte-Derived GH

The strategy employed by ourselves and others to identify and establish the nature of GH produced by cells of the immune system is outlined in Table 8.1. The list includes analysis of the transcriptional products as

TABLE 8.1. Experimental criteria used to establish the structural similarity of lymphocyte-derived GH to pituitary-derived GH.

1. Detection of GH production in leukocytes by immunofluorescence with FITC-conjugated specific antibodies to pituitary GH.
2. De novo synthesis of GH by intrinsically labeling the protein in culture and purification on specific antibody affinity columns.
3. Demonstration of similar molecular weight by gel filtration and HPLC.
4. Lymphocyte-derived GH can compete with iodinated pituitary GH for binding to rat lymphocyte GH receptors.
5. The replication of NB2 rat node lymphoma cells is stimulated by both affinity-purified lymphocyte GH and pituitary GH. The growth stimulation is blocked by specific antibodies to pituitary GH.
6. Northern gel analysis demonstrated that the specific leukocyte GH RNA was polyadenylated and had a molecular mass of 1.0 kb similar to pituitary GH RNA.
7. Synthesis of similar cDNA molecules by reverse transcription and PCR and restriction enzyme analysis from the lymphocyte and pituitary RNA molecules.
8. Preliminary nucleotide sequence analysis suggesting almost complete identity between the cDNA product from lymphocytes and the pituitary.

well as the protein product. The first evidence that immune cells could produce GH was obtained by stimulating cells with mitogen or antigen and at various times removing cells to glass slides and staining for hormone production with fluorescent-labeled specific antibodies to the hormone (7). The results showed that approximately 10% of the nontreated control cells were positive for GH, whereas mitogen treatment enhanced the level twofold. GH secreted in supernatant fluids was shown to be de novo synthesized by incorporating radioactive amino acids into protein and immunoaffinity purifying the fluids on columns coupled with specific antibodies to the hormone. The eluted material was chromatographed on gel filtration columns to ascertain the molecular weight and extent of incorporated label. On occasion, several peaks were observed; however, a major peak was always found in the expected place for bona fide hormone.

This material was collected and further tested for specificity. Bioassays for growth with NB2 cells were performed, as well as the ability of the purified material to block the binding of iodinated authentic GH to its receptor. These data confirmed the similarity of the lymphocyte product to the pituitary product. Finally, the presence of direct transcription products by Northern analysis and/or conversion of RNA to cDNA for restriction enzyme analysis and cloning and sequencing has been performed (7, 8). The studies utilizing reverse transcription and the polymerase chain reaction to generate cDNA molecules resulted in a single major DNA band corresponding in length to the distance between the 5′ ends of the two GH primers and identical to the product obtained when pituitary RNA was used as the template. The DNA band specifically hybridized to a GH-specific probe after Southern transfer (8). The similarity of the samples amplified from leukocytes and the pituitary was confirmed by restriction enzyme analysis. The cloning and sequencing of leukocyte GH cDNA showed that there were only four nucleotide differences throughout the molecule potentially encoding two different amino acids, one near the amino terminal and the other near the carboxy-terminal end of the protein (Rohn, LeBoeuf, Blalock, Weigent, in preparation). Overall, the studies suggest that an almost identical GH molecule is produced by leukocytes, while the significance of the amino acid differences and whether they are allelic differences or sequencing errors is unclear. The production of GH by leukocytes has been confirmed by two independent groups using normal and transformed cell lines (9, 10).

Production of GH by Cells of the Pituitary and the Immune System

The synthesis and secretion of GH by the pituitary somatotroph is under the regulation of numerous factors, as listed briefly in Table 8.2 (11). Two hypothalamic neuropeptides, GHRH and SRIF, act primarily on the pituitary somatotroph to regulate GH secretion. IGF-I, the anabolic effector agent of GH, may inhibit the synthesis and secretion of GH by acting at the pituitary and hypothalamic levels. In vitro studies show that glucocorticoids also enhance basal GH secretion and GHRH-induced GH release from dispersed rat pituitary cells (12).

Numerous studies have been conducted by ourselves and others to identify whether molecules involved in pituitary GH regulation can be produced by cells of the immune system and play a role in the regulation of lymphocyte GH. Recently, we have shown the production of irGHRH and IGF-I molecules by cells of the immune system and their involvement

TABLE 8.2. Comparison of the response of the rat pituitary and lymphocytes to major regulators of GH secretion.

	Response	
Stimuli (concentration)	Pituitary	Lymphocyte
GHRH (10^{-6}–10^{-8})	+	+
SRIF	−	0
GH	−	+, 0*
IGF-I	−	−
Dexamethasone	+	−, 0*
TRH	+	+, 0*

+ = enhancement, − = suppression, 0 = no effect, * = conflicting reports.

in the regulation of synthesis of leukocyte-derived GH (3, 6). The GH gene appears to be under tonic suppression in lymphoid tissue in vivo because removal of spleen or thymic tissues results in rapid, spontaneous gene transcription; maximum levels of GH mRNA are reached after only 8 h without any obvious stimulation (7). The nature of the molecule and mechanism of the negative signal in vivo remains unclear. RIA of GH in conditioned media showed that 10^6 IM-9 cells secrete 1–10 pg GH/ml/day. The IM-9 cells can be stimulated to secrete additional GH by treatment with GHRH and TRH (9). In a study using a sensitive enzyme immunoassay, small amounts of GH were reported from normal and transformed cell lines (10). Addition of GH during the culture augmented endogenous GH secretion from nonstimulated and *phytohemagglutinin* (PHA)-stimulated peripheral blood mononuclear cells. GHRH and SRIF did not affect GH secretion.

Our own unpublished observations suggest that synthesis of GH mRNA by lymphocytes may be similar to (GHRH stimulation) and different from (no effect with GH) that by the pituitary (13). Mitogen treatment stimulated an increase in GH mRNA synthesis and GH secretion in humans and in rats (14). It is of interest that the cytokine regulation of pituitary GH by the actions of IL-1, IL-2, TNF, and IFN-γ has been reported (reviewed in 15). Although no data have yet been published, it is quite possible that GH made by leukocytes will be regulated by cytokine(s).

Role of GH in the Immune Response

The concept that GH affects cells of the immune system is now well established by the results of both in vivo and in vitro studies (reviewed in 16) (Table 8.3). Proliferative responses of both transformed and normal lymphoid cells are generally greater when treated with GH in vitro. The functional activity of cytotoxic T lymphocytes and NK cells is influenced by GH, and the hormone was necessary for T lymphocytes to develop

TABLE 8.3. Role of GH in the immune response.

Augments antibody synthesis
Increases lectin-induced T cell proliferation and IL-2 synthesis in vivo
Stimulates proliferation of human lymphoblastoid cells
Augments basal lymphocyte proliferation in vitro
Increases activity of cytotoxic T lymphocytes in vitro
Augments activity of natural killer cells in vivo
Primes macrophages for superoxide anion release in vitro and in vivo
Primes macrophages for TNF-α release in vivo
Augments neutrophil differentiation in vitro
Augments erythropoiesis

cytolytic activity against an allogeneic stimulus in serum-free media. The cytolytic activity of NK cells is reduced by hypophysectomy, an effect partially reversed by administration of GH in vivo. In addition, the decline of NK activity related to aging can be prevented in part by injection of GH. GH is involved in the normal differentiation of stem cells into erythrocytes and granulocytes. In the presence of granulocyte macrophage colony-stimulating factor, GH or IGF-I doubles the number of granulocyte colonies. This enhancement can be blocked by antibodies to the IGF-I receptor, suggesting that GH induces IGF-I, which then stimulates the granulopoiesis. Physiologic concentrations of GH augment the number of erythroid colonies that develop from bone marrow cells incubated with erythropoietin. GH can prime macrophages in both in vitro and in vivo systems for the production of superoxide anion. Hypothalamic releasing hormones have been shown to modulate the immune system. It has been shown that GHRH stimulates lymphocyte proliferation and inhibits both NK activity and the chemotactic response (17).

Most of the studies done to date have examined the effect of exogenously added GH on selected immune responses. Overall, it appears that GH stimulates all the major cell types in the immune system. Thus, GH augments T cell proliferation, cytotoxic T cells, NK cells, macrophages, and B cells. The presence of lymphocyte-derived GH in the system makes it more difficult to study. Nevertheless, we have employed antisense ODNs and antibodies to block the endogenous activity of GH. The results of these studies are described below (Tables 8.4–8.7) and clearly support the presence and importance of lymphocyte-derived GH in the immune system.

Antisense ODN to GH mRNA Inhibits Lymphocyte Proliferation

Although numerous leukocyte functions appear to be modulated by GH, for the most part, very little has been done to address the specific activities of leukocyte-derived GH. In part, this has proven difficult to study since leukocytes are already spontaneously producing GH once they have been removed from animals, and very potent inhibitors of this synthesis have not yet been identified. To study this, we have used a GH antisense ODN that encompassed the translation initiation codon to investigate the relationship between GH and lymphocyte proliferation (18) (Table 8.4).

Treatment of rat lymphocytes with GH antisense ODN decreased the amount of synthesized and secreted irGH (Table 8.4). Antisense ODN (10 μM) reduced the amount of irGH secreted by 61% compared to

TABLE 8.4. Effect of GH antisense ODN on splenocyte [^{3}H]TdR incorporation.

ODN treatment	% inhibition of [^{3}H]TdR incorporation[a]	irGH protein (pg/ml–10^7 cells)[b]	% inhibition of control GH
1. Sense 1–7 amino acids (30 bases; exon II)	10 ± 3	135 ± 18	9
2. Antisense 1–7 amino acids (30 bases; exon II)	87 ± 6	58 ± 9	61
3. ODN H (31 bases; GH gene 5′ promoter region)	8 ± 3	143 ± 17	5

[a] Rat lymphocytes (1 × 10^6/ml) were cultured with ^{3}H-TdR (2 µc/ml) and with 10 µM of one of the indicated ODNs. Cells were harvested 24 h later. The cell associated ^{3}H-radioactivity was determined by liquid scintillation counting. Results are from replicate cultures for each treatment from a representative (n = 6) experiment performed 3 times. Inhibition of [^{3}H]-TdR incorporation relative to the incorporation by control cultures is expressed as the average cpm ± SEM. ODNs were synthesized on a Coder 300 DNA synthesizer (DuPont, Wilmington, DE) and purified prior to the addition using Nensorb cartridges (DuPont) according to the manufacturer's protocol.

[b] Lymphocytes were obtained from adult male Sprague-Dawley rats (150–200 g) and cultured in RPMI-1640 medium containing 10% fetal calf serum (FCS) supplemented with penicillin (100 U/ml), streptomycin (100 µg/ml), and mycostatin (100 U/ml). Rat splenocytes (1 × 10^7/ml) were incubated with 10 µM of the indicated ODN for 24 h in culture medium. Control cultures were treated as above except ODNs were not added to the medium. Supernates from control and ODN-treated cultures were collected, cells removed by centrifugation, and fluids assayed for irGH protein by RIA as described in the methods. The cell number at the end of the experiment was 10^7/ml. Each value represents the mean ± SEM of duplicate samples assayed from 3 different experiments.

Source: Reprinted with permission from Weigent, LeBoeuf, and Blalock (18), © by The Endocrine Society, 1991.

untreated control cells, cells treated with a sense GH ODN, or an unrelated control antisense ODN (Table 8.4). The difference in the amount of irGH secreted into the culture supernates of sense and antisense GH-treated lymphocytes was not due to the differential uptake of the two ODNs. Cells were treated with ^{32}P-labeled sense or antisense ODNs (10^8 cpm per µg) and the amount of radioactivity in the cytoplasm determined at various times after the addition of ODN. The results of this experiment showed that the amount of cell-associated radioactivity in lymphocytes treated with either ^{32}P-labeled sense or antisense ODN for 16 h was essentially the same.

It has been previously shown that lymphocyte proliferation in vitro is increased by treatment with GH (16). The fact that lymphocytes produce irGH and that the normal physiological function of GH is to stimulate growth prompted us to investigate whether GH produced by lymphocytes affects DNA synthesis. Lymphocytes were treated with either sense or antisense GH ODN for 24 h, and changes in DNA synthesis were determined by measuring [^{3}H]TdR incorporation (Table 8.4). The results of

TABLE 8.5. Reversal of antisense GH ODN-mediated growth inhibition by the addition of exogenous GH.

Culture conditions	[^{3}H]TdR incorporation	% of control
Control, no additions	100,842 ± 7,404	100.0
Antisense oligo (1 μM)	65,428 ± 4,281	64.9
GH (1 μg/ml) + antisense oligo (1 μM)	100,019 ± 9,301	99.2

Note: Rat spleen cells (4 × 10^6/ml) were cultured with the indicated GH ODN concentration, [^{3}H]TdR (5 μCi/ml), and with or without rat GH (1 μg/ml) for 24 h. After 24 h, the cultures were harvested on glass fiber filters, and the amount of [^{3}H]TdR incorporated was determined. Each value represents the mean ± SEM of 6 replicate cultures.
Source: Reprinted with permission from Weigent, LeBoeuf, and Blalock (18), © by The Endocrine Society, 1991.

these experiments showed that GH stimulates lymphocyte proliferation since antisense-treated lymphocytes contained 87% less [^{3}H]TdR than did either control or sense-treated cells. Thus, a decrease in the synthesis and secretion of GH resulted in a reduction in cell replication. ODN treatment did not affect the total number of cells recovered or the cell viability. Incubation of lymphocytes with the same concentrations of sense and antisense ODNs blocked the antisense effect on cell proliferation. The results of these experiments showed that growth inhibition by GH antisense ODN could be prevented by the complementary GH sense ODN. In a similar series of experiments, we determined whether the antisense GH ODN-mediated growth inhibition could be reversed by the exogenous addition of rat GH. The data in Table 8.5 show that the growth inhibition by GH antisense ODN could be prevented by the addition of exogenous GH.

Regulation of Leukocyte-Derived IGF-I Synthesis by GH

In the neuroendocrine system, the level of IGF-I gene expression is thought to be principally regulated by GH (11). In view of this and with the aim to study the role(s) of lymphocyte-derived GH, we determined the effect of GH on the production of IGF-I by leukocytes (6). Spleens were removed from rats and the cells dispersed and washed in RPMI-1640 with antibiotics and without serum. The cells (50 × 10^6 cells/ml) were then cultured in 3 ml of RPMI-1640 and treated with 500 ng/ml of rat GH for either 4 or 24 h. The supernatant fluids were collected for analysis of irIGF-I content by RIA as described in "Methods." The data in Table 8.6 show an increased amount (1.38 pg/ml vs. 2.74 pg/ml) of irIGF-I in the supernatant fluids from leukocytes cultured for 24 h in the presence of 500 ng/ml GH. In order to confirm our results from the RIA,

TABLE 8.6. An increase in irIGF-I production from leukocytes after treatment with exogenous GH.

Treatment[a]	Time (h)	irIGF-I (pg/ml)[b]
No addition	4	0.86 ± 0.03
	24	1.38 ± 0.06
500 ng/ml GH	4	0.74 ± 0.03
	24	2.74 ± 0.15

[a] Spleens were removed from animals (3 rats per group) and immediately dispersed. The cells were washed in RPMI-1640 without serum. The cells were then incubated in 3 ml RPMI-1640 (50×10^6 cells/ml). Half of the cells were incubated in the presence of 500 ng/ml rat GH. The supernatant fluids were collected at 4 h and 24 h after the start of incubation.
[b] The supernatant fluid was lyophilized and then resuspended in 300 µl distilled water. The fluid was then analyzed (50 µl/tube) by RIA as described in "Materials and Methods." Data are presented as the average amount of irIGF-I ± SEM.
Source: Reprinted with permission from Baxter, Blalock, and Weigent (6), © by The Endocrine Society, 1991.

we determined whether there was also an increase in the presence of cytoplasmic irIGF-I by direct immunofluorescence. Spleen cells cultured for 24 h in the presence of GH were stained with FITC-conjugated antibodies to IGF-I. A twofold increase in the number of cells positive for IGF-I was seen in cells treated with GH (data not shown). The cultured spleen cells were also examined for IGF-I RNA by slot blot analysis, and

TABLE 8.7. Detection of a decrease in irIGF-I production from leukocytes treated with antibodies to GH by direct immunofluorescence.[a]

Dilution of antibody	% positive by immunofluorescence[b]
No antibody	6.6 ± 1.0
1:50	3.0 ± 0.7
1:500	3.9 ± 0.9
1:5000	5.6 ± 1.2

[a] Freshly dispersed spleen cells were cultured in RPMI-1640 without serum at 37°C, 5% CO_2 with the above final dilution of antibodies to rat GH (NIDDK anti-rGH-S-5, 0.4 ml). After 24 h, the cells were washed 3 times in PBS, fixed onto glass slides, and stained with FITC-conjugated antibodies to IGF-I as described in "Materials and Methods."
[b] At least 500 cells from each group were counted from randomly chosen fields. Data are presented as the percent of positive cells ± SEM.
Source: Reprinted with permission from Baxter, Blalock, and Weigent (6), © by The Endocrine Society, 1991.

no significant difference was seen in the amount of cytoplasmic IGF-I-related RNA in GH-treated cells compared to nontreated cells (data not shown). Taken together, it appears that the stimulatory effect of GH on irIGF-I production may be at the level of translation.

Since we have shown that the addition of exogenous GH can increase the production of irIGF-I, we next determined whether irGH can influence irIGF-I production in leukocytes by determining the effect of adding antibody to GH to leukocyte cultures (Table 8.7). Freshly dispersed spleen cells were cultured in the presence of GH antibodies. After 24h, the cells were washed three times with PBS, air dried onto glass slides, and fixed in 95% ethanol. The spleen cells were then stained with FITC-conjugated antibodies to IGF-I. The data show that at the highest concentration of antibodies to GH, the number of cells positive for irIGF-I decreased twofold. Taken together, the data suggest that the production of irIGF-I by leukocytes can be mediated by GH.

Summary

GH has been shown to be synthesized by the pituitary and, more recently, by cells of the immune system. The data gathered to date suggest that the molecule synthesized by cells of the immune system is structurally identical to the gene product of the pituitary. The regulation of pituitary GH secretion involves the integration of numerous factors acting at the hypothalamic and pituitary levels. The regulation of leukocyte GH synthesis and secretion is just beginning to be investigated and exhibits both similarities to and differences from pituitary GH regulation. The important signal molecules—GH, GHRH, SRIF, and IGF—and their receptors are present in both the immune and neuroendocrine systems. The available data suggest that these molecules from lymphocytes are involved in one another's synthesis, in some instances similar to and in others different from what has been observed for pituitary GH. Recent evidence also suggests that GH modulates cells involved in the immune response. The activity of all major immune cell types, including T cells, B cells, *natural killer* (NK) cells, and macrophages, can be altered by GH. We have shown that lymphocyte-derived GH is involved in lymphocyte proliferation and can induce IGF-I production by cells of the immune system. Taken together, these observations have begun to provide a biochemical basis for one line of communication between the immune and neuroendocrine systems through the actions of GH.

References

1. Blalock JE. A molecular basis for bidirectional communication between the immune and neuroendocrine systems. Physiol Rev 1989;69:1–27.

2. Stephanou A, Jessop DS, Knight RA, Lightman SL. Corticotropin-releasing factor-like immunoreactivity and mRNA in human leukocytes. Brain Behav Immun 1990;4:67–73.
3. Weigent DA, Blalock JE. Immunoreactive growth hormone-releasing hormone in rat leukocytes. J Neuroimmunol 1990;29:1–13.
4. Weigent DA, Blalock JE. Structural and functional relationships between the immune and neuroendocrine systems. Bull Inst Pasteur 1989;87:61–92.
5. Maniatis T, Fritsch EF, Sambrook J. In: Molecular cloning: a laboratory manual. Cold Spring Harbor, NY: Cold Spring Harbor Laboratory, 1982.
6. Baxter JB, Blalock JE, Weigent DA. Characterization of immunoreactive insulin-like growth factor-I from leukocytes and its regulation by growth hormone. Endocrinology 1991;129:1727–34.
7. Weigent DA, Baxter JB, Wear WE, Smith LR, Bost KL, Blalock JE. Production of immunoreactive growth hormone by mononuclear leukocytes. FASEB J 1988;2:2812–8.
8. Weigent DA, Riley JE, Galin FS, LeBoeuf RD, Blalock JE. Detection of GH and GHRH-related messenger RNA in rat lymphocytes by the polymerase chain reaction. Proc Soc Exp Biol Med 1991;198:643–8.
9. Kao T-L, Harbour DV, Smith EM, Meyer WJ. Immunoreactive growth hormone production by cultured lymphocytes [Abstract]. Prog 71st annu meet Endocr Soc, Seattle, 1989:108.
10. Hattori N, Shimatsu A, Sugita M, Kumagi S, Imura H. Immunoreactive growth hormone (GH) secretion by human lymphocytes: augmented release by exogenous GH. Biochem Biophys Res Commun 1990;168:396–401.
11. Root AW. Neurophysiological regulation of the secretion of growth hormone. J Endocrinol Invest 1989;12:3–17.
12. Vale W, Vaughan J, Yamamoto G, Spiess J, Rivier J. Effects of synthetic human pancreative (tumor) GH releasing factor and somatostatin, triiodothyronine and dexamethasone on GH secretion in vitro. Endocrinology 1983;112:1553–8.
13. Weigent DA, Blalock JE. Growth hormone and the immune system. Prog Neuro Endocrin Immunology 1990;3:231–41.
14. Weigent DA, Blalock JE. Expression of growth hormone in leukocytes. Int Rev Immunol 1989;4:195–213.
15. Scarborough DE. Cytokine modulation of pituitary hormone secretion. Ann NY Acad Sci 1990;594:169–87.
16. Kelley KW. Growth hormone lymphocytes and macrophages. Biochem Pharmacol 1989;35:705–13.
17. Zelazowski P, Dohler KD, Stepien H, Pawlikowski M. Effect of growth hormone releasing hormone on human peripheral blood leukocyte chemotaxis and migration in normal subjects. Neuroendocrinology 1989;50:236–9.
18. Weigent DA, LeBoeuf RD, Blalock JE. An antisense oligonucleotide to growth hormone mRNA inhibits lymphocyte proliferation. Endocrinology 1991;128:2053–7.

9

Immunomodulating Role of Growth Hormone

NICOLA FABRIS AND EUGENIO MOCCHEGIANI

It is generally accepted that the immune system, although regulated to a large extent by intrinsic cellular and humoral events, is sensitive to signals generated out of the immune network, namely, in the nervous and endocrine systems. This assumption is supported by two lines of experimental evidence: first, that spontaneous or induced alterations in the neuroendocrine system may cause functional modifications of immune reactivity (1) and, second, that receptor sites for many neuropeptides and protein hormones are present in lymphoid cells (2). On the other hand, the neuroendocrine system seems to act not only as a modulator of the immune network, but as a target for signals generated within the immune system. Examples of such interactions are the alterations that can be induced in the neuroendocrine balance either by removal of relevant lymphoid organs, such as the thymus (3), or by the functioning of the immune system itself, such as the production of lymphokines and cytokines with neuroendocrine-modulating effects (4, 5).

Although the whole picture of neuroendocrine-immune interactions is still not fully discovered, the network seems already quite complex, including interactions that appear to be recruited in different physiological situations and, in particular, either in conditions that are essential for baseline functioning or in conditions characterized by emergency events. This consideration has led us to distinguish at least two levels of neuroendocrine-immune interrelationships (6, 7).

The first level is based on the interactions between the neuroendocrine system and the thymus (strategic circuit). Neuroendocrine-thymus interactions are primarily involved in maturative steps of both immune and neuroendocrine systems, which occur, in absence of pathological events, independently from the degree of antigenic stimulation.

The second level of interaction is at the periphery, between neuroendocrine signals and the humoral products that are secreted by immune cells during specific reactions to various stimuli (4). At this level both the presence of fully differentiated immune cells and the occurrence of a specific antigenic or stress-mediated hormonal stimulus are required. The main role played by these peripheral interactions appears to be that of recovering back to normal the neuroendocrine or immune balance when suddenly altered by a stressful cognitive or noncognitive event (1, 4, 7).

Growth hormone plays a relevant role in the context of these neuro-endocrine-immune interactions, likely acting both in the strategic and in the emergency circuits. This chapter aims to summarize the data available supporting a precise role of GH in the strategic circuit and the impact that this hormone may have both for developmental stages and for aging processes.

Pituitary-Thymus Interactions

The neuroendocrine-thymus interactions are sustained by a high number of neuropeptides and hormones capable of acting on various thymic functions (8). The actions may be positive—that is, may increase thymic functions, as for thyroid hormones, growth hormone, melatonin, beta-adrenergics, luteotropic hormone releasing hormone, and endogenous opioids—or negative, as for corticosteroids or gonadal hormones (Fig. 9.1). The role played by prolactin is controversial and will be discussed later.

The neuroendocrine-immune analysis of the strategic circuit should take into account two major aspects of thymic physiology that have been discovered in the last decade. First, in addition to T cells at different maturation steps from stem cells to the mature forms, epithelial cells do exist in the thymus capable of producing thymic peptides (9) with differentiative properties on T cell maturation and, in given circumstances, to synthesize and secrete neuropeptides (10), whose significance is still to be defined. These epithelial cells, identified by a monoclonal antibody (A2B5) (11) show at intracellular level the presence of various peptides, such as thymopoietin, thymosin alpha-1, alpha-7, and beta-4, and thymulin (9). This last peptide requires binding to zinc ions in order to be biologically active (12).

The second aspect is related to the relevance that a trace element, namely zinc, has for thymic functioning. Zinc is in fact required by thymulin in order to be biologically active. The zinc-unbound peptide, in addition to being inactive, prevents the active form from exerting its action on the specific targets, thus suggesting that zinc bioavailability may be of relevance for thymic hormone function (13).

More generally, the relevance of zinc for thymus functioning is supported by the observation that in the course of severe zinc deficiencies, atrophy of the thymus does occur, with reduction of the cortical area,

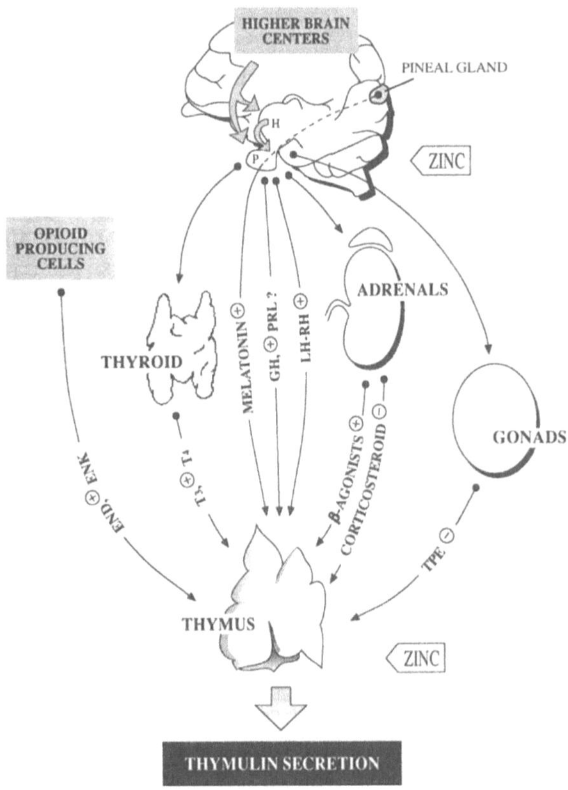

FIGURE 9.1. Proposed neuroendocrine pathways controlling thymulin secretion by the thymus.

decreased thymic endocrine activity, and alterations of various peripheral T-dependent immune functions (14). Furthermore, a marginal zinc deficiency, as it may occur in physiological aging or in Down's syndrome, is associated with thymulin hypofunctioning, and zinc supplementation is able to correct such a defect (15, 16). In old mice zinc supplementation induces regrowth of the thymus, with recovery of both thymic hormone production and thymocyte proliferation rate (16, 17).

In addition, one should not underestimate the observation that zinc turnover may also affect neuroendocrine function: Zinc deficiency causes reduced synthesis and release of developmental hormones, including GH, by the pituitary, as well as alterations in many peripheral endocrine functions (18). Thus, zinc may modulate thymic functions through its capacity to influence neuroendocrine profiles relevant for thymic activity (15).

The existence of an endocrine activity of the thymus and the role played by zinc for such a function are relevant in the context of the role of GH on the thymus because, on the one hand, GH may modulate thymic endocrine activity and, on the other hand, zinc can directly or indirectly interfere with GH-thymus interactions.

Historical Findings

One of the first experimental evidences of the existence of thymus-pituitary interactions was based on the discovery that, on the one hand, congenital mutation affecting the pituitary dwarf mice caused concomitant alterations in the thymus and in the thymus-dependent system (19, 20) and, on the other hand, that removal of the thymus in neonatal age was followed by a wasting syndrome, one of its features being a pituitary dysfunction (3) involving growth hormone-producing cells (Fig. 9.2).

Congenital hypopituitarism, which occurs as a recessive autosomal character in Snell dwarf (dw) mice, is characterized by nearly absent production of growth hormone and diminished synthesis of thyroid stimulating hormone, adrenocorticotropic hormone, and prolactin. Pituitary dwarf mice show a normal development of the lymphoid system during the first 2 weeks of life, whereas in the following periods they suffer from a developmental arrest followed by a progressive involution. Such an immunodeficiency affects primarily the thymus-dependent system, as indicated by the prolonged allogenic skin-graft survival, the depressed capability of spleen cells to induce graft vs. host reactions, or to react to T mitogens, and the reduced humoral antibody response to thymus-dependent antigens, while the synthesis of immunoglobulins is within normal range (Fig. 9.2) (19). The relevance of the pituitary-thymus interaction for the immunodeficiency state of the dwarf mouse is further supported by the findings that the thymus underdevelopment and the immunological deficiencies may be completely corrected by treating dwarf mice daily for 30 days with growth hormone and thyroxine, provided, however, the thymus is not previously removed (20).

The existence of thymus-pituitary interactions has originated from the observation that post-thymectomy wasting disease, in addition to the obvious immunological disturbances, is characterized by pathological signs (Fig. 9.2), which can hardly be linked to the direct effect of immune deficiency itself. Mice thymectomized at birth show, in fact, a progressive impairment of body growth with reduced length of ears and tail, microsplancnia, microsomia, thinness of the skin and lack of subcutaneous fat, bone alterations, particularly evident in the vertebrae with consequent hunched posture, and hypotrophy of various tissues including submaxillary gland, hair follicles, and bone marrow.

At pituitary level it was demonstrated that thymectomized mice show a progressive degranulation of pituitary acidophilic growth hormone-producing cells, whereas other cell lineages in the hypophysis are left unmodified (3, 21). Quite similar findings have been obtained in athymic

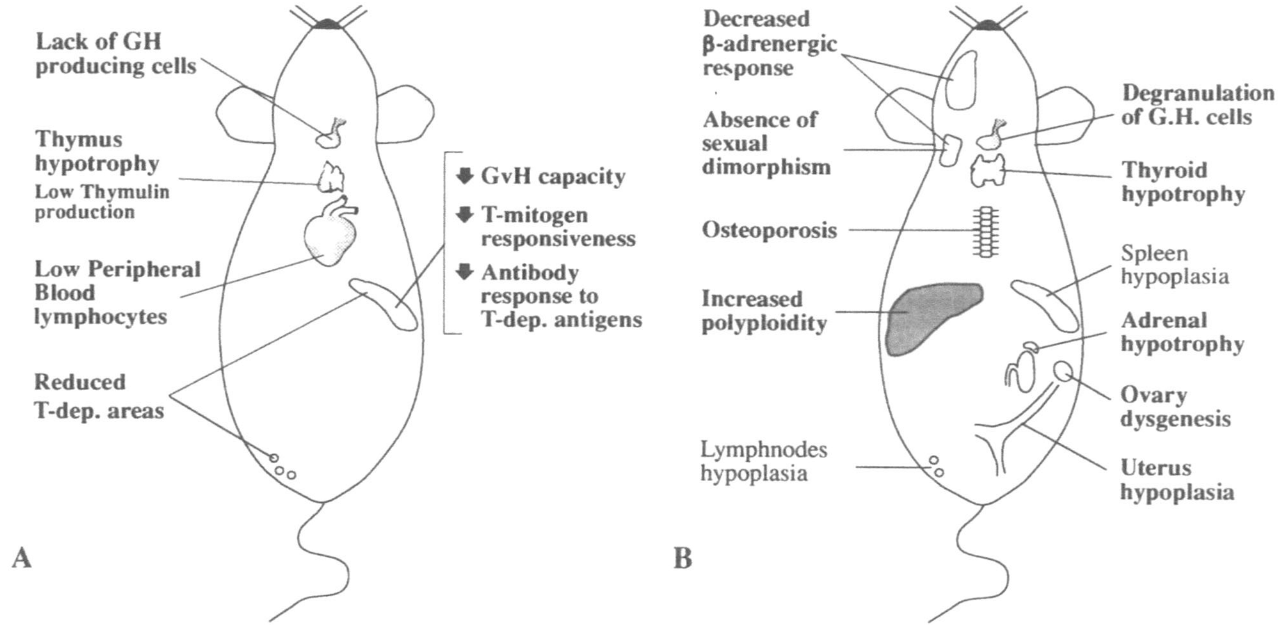

FIGURE 9.2. Summary of the major immunological alterations associated with GH deficiency (*A:* dwarf mouse model) and of the neuroendocrine modifications associated with thymus absence (*B:* nude mouse model). Data derived from Fabris (8).

nude mice (22). These data suggested that the thymus might, directly or indirectly, modulate the rate of synthesis and/or secretion of pituitary hormones relevant for its own efficiency in the feedback fashion other endocrine glands do and prompted a number of studies aimed to define the physiological extent and the mechanism of action of such a thymus-hypophysis axis and, more generally, of the neuroendocrine-immune network.

Further Evidence

With regard to the influence of GH on thymic function, the findings obtained in dwarf mice have gained further support from similar observations made in hypopituitary Weimaraner dogs (23) and in hypophysectomized rats (24, 25), both these conditions being characterized by hypotrophy of the thymus gland and by peripheral alterations of T-dependent immune function. Appropriate GH treatment was able to correct such defects. The opposite experimental conditions—that is, the hypersecretion of GH such as that observed in Buffalo rats—are associated with enlarged thymus and hyperimmune profile at the periphery (26).

In the context of GH effect on the thymus a great step ahead has been achieved following the discovery that the thymus produces thymic peptides with differentiative properties on the T cell lineage and that some of these peptides are present in the circulation where they can be measured, thus giving a marker of the functional activity of the thymus. The plasma determination of one of the best-known thymic peptides—that is, the *facteur thymique serique* (FTS), more recently called *thymulin* (ZnFTS) in this zinc-bound biologically active form (12)—has demonstrated that both pituitary dwarf mice (27) and dogs (23) show low plasma thymulin levels and that GH therapy is able in both conditions to restore them to normal values (23, 28) (Fig. 9.3).

With regard to humans, although a number of immune events, such as the percentage of B-cell and T-cell subsets (29) and lymphocyte proliferative response to mitogens (30), are not defective in growth hormone-deficient children, several other alterations in the immune systems of these subjects have been described. These changes include thymic hypoplasia (31) and low plasma levels of thymulin when compared to values of young healthy subjects (32). Recent evidence has shown that treatment with human growth hormone in growth hormone-deficient children is able to increase the reduced plasma circulating level of thymulin to the values present in age-matched healthy subjects (32) (Fig. 9.3). The effect of GH seems to be mediated by the increment of the reduced serum IGF-I concentrations, since a positive correlation between serum level of IGF-I, but not of GH, and thymulin activity has been documented (32).

In humans suffering from acromegaly, high circulating levels of both PRL and GH are present. The determination of plasma thymulin level

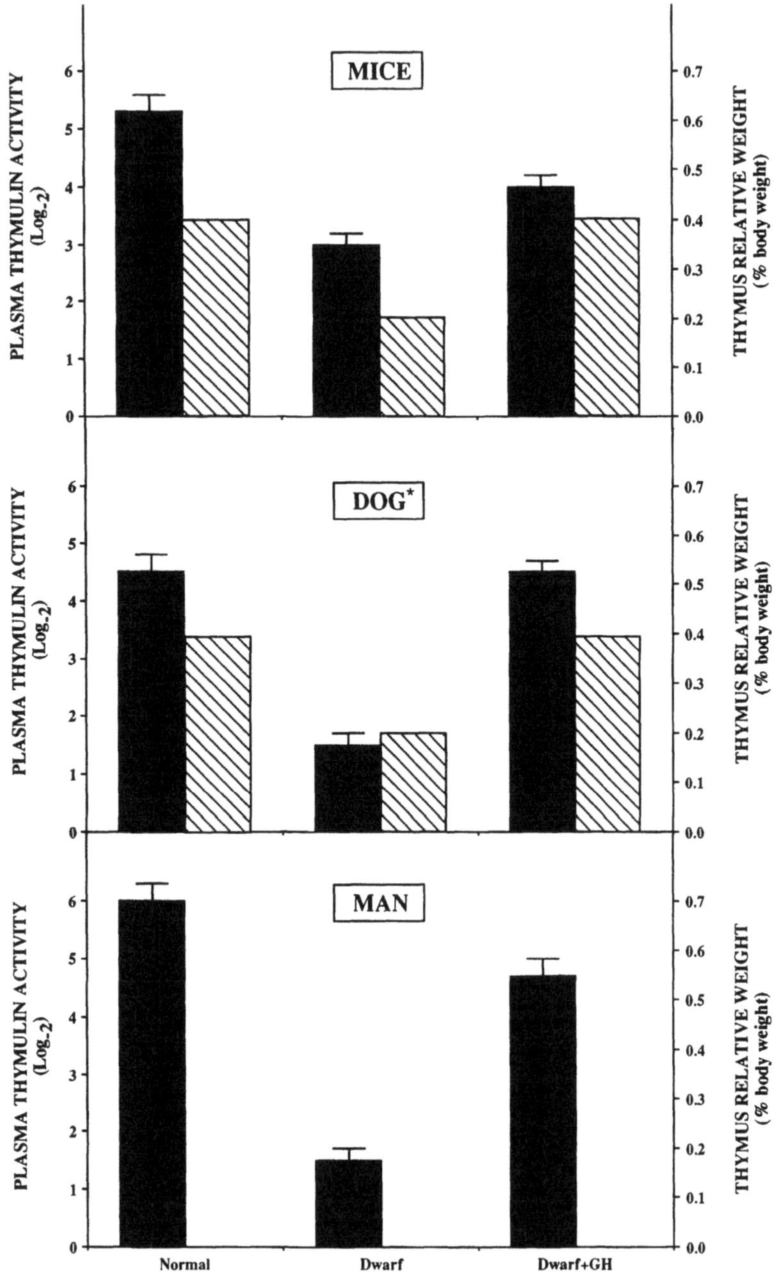

FIGURE 9.3. Reduced thymus weight and plasma thymulin concentration in GH-deficient mice and dogs and recovery following substitutive therapy. In humans only data related to thymulin concentration are available. Solid bars represent plasma thymulin levels. Hatched bars refer to thymus weight. Data for dogs redrawn from Roth, Kaeberle, Grier, Hopper, Spiegel, and McAllister (23).

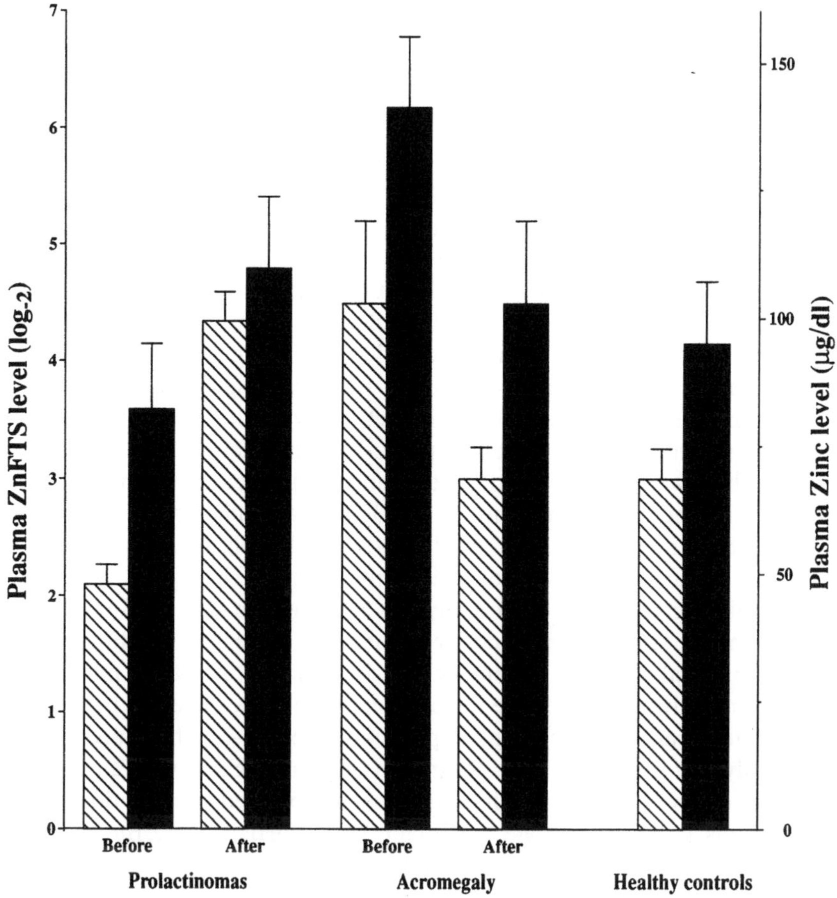

FIGURE 9.4. Plasma zinc and thymulin (ZnFTS) concentrations are reduced in patients with prolactinomas and increased in acromegalic subjects. Specific therapy corrects the alteration in both cases. Hatched bars represent thymulin concentrations. Solid bars refer to zinc levels. Data derived from Travaglini, Mocchegiani, De Min, Re, and Fabris (35).

has shown the presence of high thymic endocrine activity even in old acromegalic patients when compared to thymulin values of age-matched controls (33). Such an increment of thymic endocrine activity has already disappeared 6 months after surgical or pharmacological treatment (33). On the contrary, conditions characterized by pure hypersecretion of PRL (prolactinomas) are associated with reduced thymulin levels, and the appropriate therapy restores normal thymulin concentrations (34, 35) (Fig. 9.4).

Implication for Aging

Growth hormone secretion is highest around puberty and then declines with age both in humans (36, 37) and in rats (38). Although basal levels of growth hormone do not show dramatic changes, the age-associated reduction in amplitude of the secretory spikes of growth hormone is relevant, with differences exceeding 200 μg/ml of plasma in pulsating releases of growth hormone between old and young rats. As might be expected, the concentration of IGF-I also declines with age (39).

Such age-associated hormonal alterations may have a great impact in explaining at least some of the age-related alterations of the immune system and, particularly, thymus involution, including the failure of thymic endocrine activity, which is a common feature of aging both in animals and in man. Recent findings have, in fact, shown that thymic involution is not an intrinsic and irreversible process, but is largely dependent on the decline with age of various neuroendocrine factors. With regard to growth hormone, the implant of GH3 pituitary adenoma cells, which secrete high quantities of growth hormone and prolactin, in aged rats is able to induce a regrowth of the thymus with recovery of young-like histologic picture (40). Such a thymus regrowth in old GH3-implanted rats is not a sterile event since it induces an increased functionality of the peripheral immune system as assessed by the increased number of peripheral T lymphocytes, the restoration of the responsiveness to mitogen stimulation, and the augmented production of IL-2, which are all abnormally low in old rats (41). Regrowth of thymus gland with augmented thymic endocrine function has also been obtained in old dogs after exogenous growth hormone treatment (42). Furthermore, Buffalo rats, which are characterized by increased GH secretion, show persistence of the thymus with advancing age associated with hyperplasia of cortical epithelial cells and increased number of proliferating lymphocytes (26).

The implant under the kidney capsule of pituitary glands, which produce only prolactin in these conditions, is, on the contrary, ineffective on age-related immune deficiency (43). No experimental or clinical trials have been pursued with exogenous administration of prolactin. In this context it has to be emphasized, however, that aging is generally associated with higher prolactin secretion than young-adult age, a fact that tends to rule out a role of prolactin in immunological aging (44).

Mechanisms of Action

All these data clearly support the view that GH acts on the thymus and particularly on its endocrine component, represented by the thymic peptide-secreting epithelial cells. Studies performed in young mice, transgenic for the rat metallothionein growth hormone gene with high level of growth hormone, have in fact shown that in these animals the thymus gland contained more epithelial cells and Hassal's corpuscles when compared with normal, nontransgenic littermates (41). Furthermore, preliminary results using displacement and autoradiographic procedures showed the presence of a GH receptor in thymic epithelial cells (45).

It cannot be excluded, however, that GH may also have a direct action on thymocytes, at least during some steps of their differentiation. It has, in fact, been demonstrated that hypophysectomized rats show a reduced DNA and RNA synthesis in the thymus and spleen and that these effects can be reversed with *bovine growth hormone* (bGH) treatment, the stimulation of DNA synthesis in the thymus occurring earlier than in the spleen and lymph nodes (46). The presence of a specific receptor for growth hormone on thymocytes has also been documented (47).

Although the presence of GH receptors both on thymocytes and on thymic epithelial cells may support a direct effect of the hormone on these targets, on a functional basis it seems that the majority of GH actions are mediated through IGF-I. This assumption seems to apply not only to thymic function (32), but also to other immunomodulatory actions of GH (48).

An intriguing aspect of GH action is related to the role played by *prolactin* (PRL). Conflicting data are available in the literature on the effect of PRL on various immune functions. Receptors for PRL have been demonstrated on thymic epithelial cells, and in vitro administration of PRL on thymic epithelial cell cultures has been shown to be capable of increasing thymulin production in the culture supernatants (49).

In this context, a number of considerations should be made on the basis of available literature. First of all, it seems that increased PRL secretion is associated with reduced bioavailability of the trace element, though the causes of such a phenomenon are still unknown (50). This fact may, in part, explain the conflicting data between in vitro (49) and in vivo (35) effects of PRL: PRL increases thymulin production in vitro (49), whereas patients affected by prolactinomas show reduced plasma levels of thymulin (35). Although clear-cut data are not yet available, GH, on the contrary, seems to increase zinc bioavailability: Acromegalic patients show, in fact, increased plasma zinc level. Appropriate treatments of both prolactinoma and GH-producing tumor-affected patients correct both the zinc turnover and the thymulin alterations (33–35). GH and PRL might, therefore, modulate thymic functions through their effect on zinc bioavailability, which is of extremely high relevance for thymic function.

A further consideration in this context is the observation that zinc may play a role in GH/PRL modulation: GH, when bound to zinc ions, may utilize PRL receptors (50). This is quite intriguing information because, together with the previously reported data, it would suggest that prolactin may be acting positively on thymic function, but due to its effect on zinc turnover, such an effect is abortive. On the other hand, GH, in addition to acting directly on the thymus, may also exert indirect effects through both its positive action on zinc turnover and its capacity to utilize PRL receptors through binding to zinc ions. It is clear that something is contradictory in this context and that further work should be aimed at elucidating this point.

With regard to the counterregulatory branch of the feedback—that is, the possibility that thymic factors may modulate GH secretion or turnover—some evidence has been recently gathered. In particular, it has been experimentally demonstrated that thymic hormone injection into thymectomized mice is able to prevent some neuroendocrine abnormalities, such as the degranulation of GH-producing cells in the pituitary (51). More recently, in vitro experiments have shown that TF-5, a partially purified bovine thymic extract, stimulated both PRL and GH release from normal and neoplastic pituitary cells (52). In humans, the injection of thymic peptides is capable of inducing an active release of hormones, such as ACTH, prolactin, and GH (53).

Since these hormones have a modulatory action on the immune system—immunostimulatory for PRL and GH and, to a large extent, immunodepressive for ACTH—it may be supposed that thymic peptides, in addition to having a direct effect on thymocyte differentiation, can modulate peripheral immune function via PRL or GH. In all cases, their action on pituitary cells should be viewed as a feedback mechanism, allowing a precise regulation of the thymus-hypophysis axis.

Conclusions

There are several major deductions that can be derived from the experimental and clinical data reported above. First, GH plays a key role in modulating the function of the thymus as an organ deputized either to synthesize and release thymic hormones or to export mature T cells. Second, the immunomodulatory action of GH is exerted throughout life, its requirement being eventually more critical during ontogenesis because of the higher demand due to immune development or during aging because of the need to recover age-related losses. Third, the effectiveness of GH in restoring age-related immune deterioration at thymic and at peripheral levels, on the one hand, confirms that the majority of age-related alterations are not intrinsic and irreversible and, on the other hand, offers a new therapeutical approach to overcome the immunodeficiencies of elderly people. In this context it has to be noted, however, that the utilization of GH as an immunopotentiating drug should be carefully addressed and studied further, since we do not know the effect of GH treatment on pathological phenomena, such as autoimmune reactions or neoplastic transformation and growth, which are both frequent alterations in elderly and which might even be favored by GH (54).

References

1. Fabris N, Mocchegiani E, Muzzioli M, Provinciali M. Neuroendocrine-thymus interactions: perspectives for intervention in aging. Ann NY Acad Sci 1988;521:72–87.
2. Fabris N, Provinciali M. Hormones. In: Nelson D, ed. Natural immunity. Academic Press, 1990:306–47.
3. Pierpaoli W, Fabris N, Sorkin E. Developmental hormones and immunological maturation. In: Wolstenholme GE, Knight J, eds. Hormones and the immune response. Ciba Study Group no. 36. London: Churchill, 1970:126–43.
4. Weigent DA, Carr DJ, Blalock JE. Bidirectional communication between the neuroendocrine and immune system. Common hormones and hormone receptors. Ann NY Acad Sci 1990;579:17–27.
5. Besedovsky HO, Del Rey A. Immune-neuroendocrine circuits: integrative role of cytokines. In: Frontiers in neuroendocrinology. New York: Raven Press, 1992;13(1):61–94.
6. Fabris N. Pathways of neuroendocrine-immune interactions and their impact with aging process. In: Facchini A, Haaijman JJ, Labò G, eds. Immunoregulation in aging. Rijswijk, The Netherlands: Eurage, 1986:127–30.
7. Fabris N. Biomarkers of aging in the neuroendocrine-immune domain: time for a new theory of aging. In: Aging and cellular defence mechanisms. Ann NY Acad Sci 1992.
8. Fabris N. Neuroendocrine-thymus interactions during development and aging. In: Grossman CJ, ed. Hormones and immunity-bilateral communication between the endocrine and immune systems. New York: Springer-Verlag, 1992.
9. Savino W, Dardenne M. Thymic hormone containing cells, VI. Immunohistologic evidence for the simultaneous presence of thymulin, thymopoietin and thymosin α-1 in normal and pathological human thymuses. Eur J Immunol 1984;14:987–91.
10. Robert F, Geenen V. Thymic neuropeptides and T-lymphocytes development. Ann NY Acad Sci USA 1992;650:99–104.
11. Haynes BF, Shimizu K, Eisembarth GS. Identification of human and rodent thymus endothelium using tetanus toxin and monoclonal antibody A2B5. J Clin Invest 1983;71:9–14.
12. Dardenne M, Pleau JM, Nabama B, et al. Contribution of zinc and other metals to the biological activity of the serum thymic factors. Proc Natl Acad Sci USA 1982;79:5370–4.
13. Fabris N, Mocchegiani E, Amadio L, Zannotti M, Licastro F, Franceschi C. Thymic hormone deficiency in normal aging and Down's syndrome: is there a primary failure of the thymus? Lancet 1984;1:983–6.
14. Fraker PJ, Gershwin ME, Good MA, Prasad AS. Interrelationship between zinc and immune function. Fed Proc 1985;45:1474–9.
15. Fabris N, Mocchegiani E, Muzzioli M, Provinciali M. Role of zinc in neuroendocrine-immune interactions during aging. Ann NY Acad Sci 1991; 621:314–26.

16. Travaglini P, Moriondo P, Togni E, et al. Effect of oral zinc administration on prolactin and thymulin circulating levels in uremic patients. J Clin Endocrinol Metab 1989;68:186–90.
17. Fabris N, Mocchegiani E, Provinciali M. Zinc, immunity, and aging. In: Goldstein AL, ed. Zinc, immunity and aging. Biochemical advances in aging. New York: Plenum Press, 1990:271–81.
18. Prasad AS. Clinical, endocrinological and biochemical effects of zinc-deficiency. J Clin Endocrinol Metab 1985;14:567–89.
19. Fabris N, Pierpaoli W, Sorkin E. Hormones and the immunological capacity, III. The immunodeficiency diseases of the hypopituitary Snell-Bagg dwarf mouse. Clin Exp Immunol 1971a;9:209–25.
20. Fabris N, Pierpaoli W, Sorkin E. Hormones and the immunological capacity, IV. Restorative effects of developmental hormones of lymphocytes on the immunodeficiency syndrome of the dwarf mouse. Clin Exp Immunol 1971b; 9:227–40.
21. Bianchi E, Pierpaoli W, Sorkin E. Cytological changes in the mouse anterior pituitary after neonatal thymectomy: a light and electron microscopical study. J Endocrinol 1971;51:1–6.
22. Pierpaoli W, Besedovsky HO. Role of the thymus in programming of neuroendocrine function. Clin Exp Immunol 1975;20:325–30.
23. Roth JA, Kaeberle ML, Grier RL, Hopper JG, Spiegel HE, McAllister HA. Improvement in clinical condition and thymus morphological features associated with growth treatment of immunodeficient dwarf dogs. Am J Vet Res 1984;45:1151–5.
24. Nagy E, Berczi I, Friesen NG. Regulation of immunity in rats by lactogenic and growth hormones. Acta Endocrinol (Copenh) 1983;102:351–6.
25. Berczi I. The effects of growth hormone and related hormones on the immune system. In: Berczi I, ed. Pituitary function and immunity. Boca Raton, FL: CRC Press, 1983:133–59.
26. Hirokawa K, Utsuyama M, Kasai M, Konno A, Kurashima C, Morizumi E. Age-related hypoplasia of the thymus and T-cell system in the Buffalo rat. Virchows Arch [B] 1990;59:38–47.
27. Pelletier M, Montplaisir S, Dardenne M, Bach JF. Thymic hormone activity and spontaneous autoimmunity in dwarf mice and their littermates. Immunology 1976;30:783–8.
28. Fabris N, Mocchegiani E, Muzzioli M, Imberti R. Thymus-neuroendocrine network. In: Fabris N, Garaci E, Hadden J, Mitchison NA, eds. Immunoregulation. New York: Plenum Press, 1983:323–34.
29. Abbassi V, Bellanti JA. Humoral and cell-mediated immunity in growth hormone-deficient children: effect of therapy with human growth hormone. Pediatr Res 1985;19:299–301.
30. Rapaport R, Oleske J, Ahdieh H, et al. Effects of human growth hormone on immune functions: in vitro studies on cells of normal and growth hormone-deficient children. Life Sci 1987;41:2319–24.
31. Sipponen P, Similia S, Collan Y, Autere T, Herva R. Familial syndrome with panhypopituitarism, hypoplasia of hypophysis and poorly developed sella turgica. Arch Dis Child 1978;53:664–70.
32. Mocchegiani E, Paolucci P, Balsamo A, Cacciari E, Fabris N. Influence of growth hormone on thymic endocrine activity in humans. Horm Res 1990; 33:248–55.

33. Travaglini P, Mocchegiani E, Togni E, et al. Thymulin and zinc circulating level in patient with GH and PRL secreting pituitary adenomas. Int J Neurosci 1990;51:269–71.
34. Travaglini P, Mocchegiani E, De Min C, Re T, Fabris N, Faglia G. Zinc and bromocriptine long-term administration in patients with prolactinomas: effects on prolactin and thymulin circulating levels. Int J Neorosci 1991; 59:119–25.
35. Travaglini P, Mocchegiani E, De Min C, Re T, Fabris N. Modifications of thymulin titers in patients affected with prolonged low or high zinc circulating levels are independent of patients' age. Arch Gerontol Geriatr 1992; suppl 3:349–66.
36. Finkelstein JW, Roffwarg HP, Boyar RM, Kream J, Hellman L. Age-related change in the twenty-four spontaneous secretions of growth hormone. J Clin Endocrinol Metab 1972;35:665–70.
37. Gil-Ad I, Gurewitz R, Marcovivi O, Rosenfeld J, Laron Z. Effect of aging on human plasma growth hormone response to clonidine. Mech Ageing Dev 1984;27:97–100.
38. Sontag WE, Gough MA. Growth hormone releasing hormone induced release of growth hormone in aging male rats: dependence on pharmacological manipulation and endogenous somatostatin release. Neuroendocrinology 1988;47:482–8.
39. Rudman D, Nagraj HS, Jackson DL, Rudman IW, Boswell J, Pucci DC. Hyposomatomedinemia in the men of a veteran administration nursing home: prevalence and correlates. Gerontology 1987;33:307–14.
40. Kelley KW, Brief S, Westly HJ, et al. GH3 pituitary adenoma cells can reverse thymic aging in rats. Proc Natl Acad Sci USA 1986;83:5663–7.
41. Davila DR, Brief S, Simon J, Hammer RE, Brinster RL, Kelley KW. Role of growth hormone in regulating T-dependent immune events in aged, nude and transgenic rodents. J Neurosci Res 1987;18:108–16.
42. Goff BL, Roth JA, Arp LH, Incefy GS. Growth hormone treatment stimulates thymulin production in aged dogs. Clin Exp Immunol 1987;68:580–7.
43. Cross RJ, Campbell JL, Markzsbery WR, Roszman TL. Transplantation of pituitary grafts fail to restore immune function and to reconstitute the thymus glands of aged mice. Mech Ageing Dev 1990;56:11–22.
44. Weizman A, Weizman R, Hart J, Maoz B, Wijsenbeck H, David MB. The correlation of increased serum prolactin levels with decreased sexual desire and activity in elderly men. J Am Geriatr Soc 1983;31:485–8.
45. Ban E, Gagnerault MC, Jammes H, Postel Vinay MC, Haour F, Dardenne M. Specific binding sites for growth hormone in cultured mouse thymic epithelial cells. Life Sci 1991;48:2141–8.
46. Pandian MR, Talwar GP. Effect of growth hormone on the metabolism of thymus and on the immune response against sheep erythrocytes. J Exp Med 1971;134:1095–113.
47. Arrenbrecht S. Specific binding of growth hormone to thymocytes. Nature 1974;252:255–7.
48. Geffner ME, Bersch N, Lippe BM, Rosenfeld RG, Hintz RL, Golde DW. Growth hormone mediates the growth of T-lymphoblast cell lines via locally generated insulin-like growth factor I. J Clin Endocrinol Metab 1990;71:464–9.
49. Dardenne M, Savino W, Gagnerault MC, Itoh T, Bach JF. Neuroendocrine control of thymic hormonal production, I. Prolactin stimulates in vivo and in

vitro the production of thymulin by human and murine thymic epithelial cells. Endocrinology 1989;125:3–12.
50. Cunningham BC, Bass S, Fuh G, Wells JA. Zinc mediation of the binding of human growth hormone to the human prolactin receptor. Science 1990; 250:1709–12.
51. Deschaux P, Massengo B, Fontanges R. Endocrine interaction of the thymus with hypophysis, adrenals and testes: effects of two thymic extracts. Thymus 1979;1:95–100.
52. Spangelo BL, Judd AM, Ross PC, et al. Thymosin fraction 5 stimulates prolactin and growth hormone release from anterior pituitary cells in vitro. Endocrinology 1987;121:2035–43.
53. Hall HRS, O'Grady MP, Farrah JM Jr. The hypothalamic-pituitary adrenal axis by thymic peptides. In: Hadden JW, Massek K, Nisticò G, eds. Interactions among central nervous system, neuroendocrine and immune system. Rome: Pytagora Press, 1989:113–25.
54. Lippman M. Interactions of physic and endocrine factors with progression with neoplastic disease. In: Levy SM, ed. Biological mediators of behaviour and disease: neoplasia. Amsterdam: Elsevier, 1982:55–73.

Part III

Novel Aspects of Growth Hormone Research I

10

Further Observations on the Effects of Human Growth Hormone in Elderly Hyposomatomedinemic Men

LESTER COHN, AXEL G. FELLER, INGE W. RUDMAN, AND DANIEL RUDMAN

In 1990 the authors described the effects of *human growth hormone* (hGH) in 12 hyposomatomedinemic healthy men over 60 years of age (1). A significant increase in lean body mass and a decrease in adipose tissue mass were observed during a 6-month treatment period. The hGH doses were chosen in order to elevate the plasma level of *insulin-like growth factor I* (IGF-I, also termed somatomedin C) to a youthful range of 0.5–1.5 units/ml. Early adverse side effects were not observed. The present chapter extends the earlier report by describing a larger group of hyposomatomedinemic men over 60 years old, initially numbering 42, who were treated for up to 12 months. Complications of hGH treatment were encountered and a relation to the plasma level of IGF-I was identified.

Methods

Subjects and Protocol

The study population consisted of 81 community-based men aged 60 and over, who met the inclusion-exclusion criteria shown in Table 10.1. They were recruited by advertisements in local newspapers.

Figure 10.1 and Table 10.2 describe the protocol. Each participant was interviewed and examined monthly. The subjects were randomized in a ratio of 3 to 1 into group I or group II. The first 6 months served as a baseline period. The clinical characteristics of the 2 groups are summarized in Table 10.3. Hormone treatment was started in group I at the beginning of month 7 and continued until the end of month 18 or

TABLE 10.1. Criteria for inclusion and exclusion of study subjects.

Inclusion criteria
- a. Male gender, age 55 years or older
- b. Body weight 80%–120% of standard for age
- c. Provision of written informed consent

Exclusion criteria
- a. Symptomatic congestive heart failure, angina, or cardiac arrhythmia
- b. Impaired liver function
- c. Serum creatinine concentration greater than 133 μmol/l (2.5 mg %)
- d. Clinically evident pulmonary disease
- e. Sitting systolic blood pressure >180 mm Hg or diastolic blood pressure >100 mm Hg
- f. History of carpal tunnel syndrome
- g. Diabetes mellitus (i.e., fasting plasma sugar >6.6 mmol/l) (130 mg %)
- h. Diagnosis of malignant neoplastic disease within the past 5 years
- i. Any active chronic infection (e.g., tuberculosis, bronchiectasis, urinary tract infection, osteomyelitis) or any serious acute infection
- j. Active connective tissue disease
- k. Current systemic treatment with glucocorticoids
- l. Current thyroid dysfunction
- m. Significant proteinuria (i.e., >1+)
- n. Hematocrit less than 32%
- o. Consumption of excessive alcohol or drug abuse
- p. History of mental illness such that, in the opinion of the investigator, therapy with an investigational drug would be unwise
- q. Inability to comply with the examination schedule required by the protocol

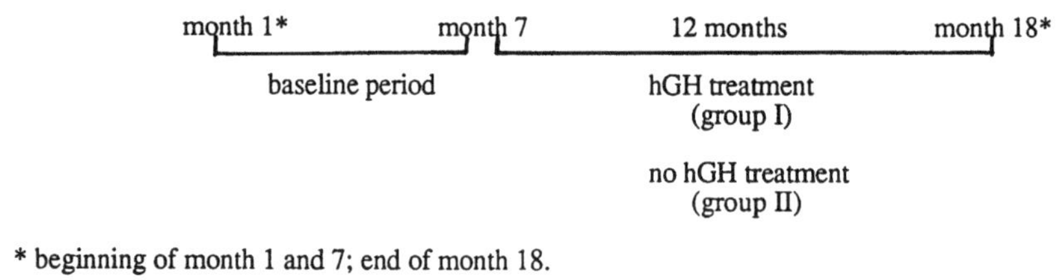

FIGURE 10.1. Description of the protocol.

TABLE 10.2. Schedule of tests.

Test (ref. for method)	Time point
History and physical examination	Monthly
CBC	Monthly
SMA-12	Monthly
Bone densities (26)	1, 6, 12, 18 months
Urinalysis	Monthly
Serum IGF-I (2)	Monthly
Lean body mass (6)	1, 6, 12, 18 months
Adipose mass (6)	1, 6, 12, 18 months

TABLE 10.3. Summary of the subjects.

	Group I (n = 61)	Group II (n = 20)
Age (years)	69.4 ± 1.60	69.10 ± 1.00
Body mass index (kg/M^2)	27.30 ± 0.72	27.43 ± 0.98
Initial plasma IGF-I (units/ml)	0.25 ± 0.019	0.27 ± 0.02
Continuing medical diagnoses*		
Degenerative joint disease	17	4
Hypertension	15	6
Arteriosclerotic heart disease	8	3
Benign prostatic hypertrophy	7	2
Hyperlipemia	3	1
Gout		1
Diverticulosis	1	
Psoriasis		1
Peripheral vascular disease	4	2
Chronic obstructive pulmonary disease	6	3
Continuing medications*		
Nonsteroidal anti-inflammatory drugs	15	6
Antihypertensives	20	6
Probenecid		1
Bronchodilators	3	1
Cholestyramine	1	
Allopurinol		1
Nitroglycerine	3	1

* Occurring in 10% or more of the group

until participants dropped out for personal or medical reasons. The biosynthetic hGH, 2.6 units/mg, was furnished by Eli Lilly Co. The initial dose was 0.03 mg/kg body weight s.c. (8 A.M.) 3 times a week (Monday, Wednesday, and Friday). Venous plasma was obtained 24 h after an hGH injection once a month and was analyzed for IGF-I by the method of Furlanetto et al. (2). If the IGF-I level was <0.50 units/ml, the hormone dose was increased by 25%; if the IGF-I level was >1.5 units/ml, the dose was reduced by 25%. Group II, the controls, received no hormone treatment. The schedule of tests was the same for both groups, as shown in Table 10.3, where references for the methods are also provided. The protocol was approved by the Human Research Committee of the Medical College of Wisconsin.

Special Tests

Neurologic Examination

At each monthly visit, subjects were asked whether they had experienced numbness or tingling in one or more fingers, discomfort or pain in the wrist, or pain in the lower arm or shoulder. Tinel's test (3) was also performed. Patients with one or more of the above symptoms in the upper extremities, or a positive Tinel's sign, underwent a nerve conduction study. The *electromyography* (EMG) nerve conduction examination was done with DISA 2000 by Dantec. The motor conduction study was performed using surface electrodes (4). Sensory latency was obtained antidromically by using standard techniques with ring electrodes on the second and fifth fingers for the median and ulnar nerves, respectively. The criteria used for the diagnosis of carpal tunnel syndrome were as follows: motor median latency more than 4.6 msec; motor median-ulnar difference more than 1.8 msec; or sensory palmar latency difference more than 0.5 msec with the same distance.

Assays for IGF-I

The work described was completed during 1987–1992. When the project was started, the only commercially available IGF-I assay was that of Furlanetto et al. (2), which was therefore adopted for the entire study. The intra-assay *coefficient of variation* (CV) was 12%, the sensitivity was 0.05 units/ml, and the crossreactivity with IGF-II was <2%. In 1990, the Daughaday acid ethanol extraction method became commercially available (5). For this method, the intra-assay CV was 3%, the sensitivity was 0.1 ng/ml, and the crossreactivity with IGF-II was <2%. For comparative purposes, since 1991 the last 105 plasma samples analyzed in the study for IGF-I were assayed not only by the Furlanetto method (2), but by the Daughaday method (5) as well. The results of the former test were expressed as units/ml and the results of the latter as ng/ml.

Body Composition by *Dual Energy X-Ray Absorptiometry* (DEXA)

As described in Table 10.3, lean body mass and adipose mass were calculated by counting of total body potassium 40 (6). For comparison, in 12 subjects these body composition masses were also measured by DEXA (7), using the XR-26 unit manufactured by Norland Inc., Fort Atkinson, WI. The correlation coefficient for lean body mass measured by the two methods was 0.92, and for adipose mass it was 0.96.

Results

Of the 81 men who entered the protocol at the beginning of the first month, 13 dropped out during the 6-month baseline period, including patients with carpal tunnel syndrome and hyperglycemia. At the beginning of month 7, 68 men entered the randomized portion of the protocol with 49 in group I and 19 in group II. The men in group I received hGH treatment. Nineteen of these individuals completed 1 year of hormone treatment without experiencing carpal tunnel syndrome, gynecomastia, or hyperglycemia. Sixteen men in group I had possible hGH-related complications of their therapy between months 7 and 18. A total of 28 men dropped out of group I after therapy had begun, including 9 with carpal tunnel syndrome, 3 with gynecomastia, and 3 with hyperglycemia. One of these patients had both carpal tunnel syndrome and gynecomastia. Other patients dropped out of group I and 2 patients dropped out of group II during this period for a variety of hGH-unrelated reasons.

All 10 cases of carpal tunnel syndrome that occurred during hGH treatment were bilateral and were manifested by a positive Tinel's sign in 9 patients and an abnormal EMG-nerve conduction study in all cases. Only 6 of the 10 individuals experienced symptoms. The time of onset of upper-extremity signs or symptoms varied from 2 to 11 months after hGH treatment was started. The hormone was discontinued when carpal tunnel syndrome was diagnosed. Symptoms subsided in 3 symptomatic cases within 5 months after hormone cessation. Tinel's sign became negative in 5 cases within 6 months. In 5 patients, EMGs and nerve conduction studies returned to normal at intervals ranging from 16 to 28 months after cessation of hGH. EMGs or nerve conduction studies continued to be abnormal in 4 patients 6–29 months after hGH treatment was discontinued. One patient was lost to follow-up.

Four men developed bilateral tender breast enlargement 3 to 11 months after hGH started. In one case, the gynecomastia was associated with a carpal tunnel syndrome. Mammograms, taken in 2 of the affected men, confirmed the condition. Hormone treatment was stopped when the gynecomastia was recognized, and the condition disappeared without exception within 4 months.

Three men developed fasting hyperglycemia after 2–9 months of hGH treatment. After oral glucose tolerance tests, 2 cases were diagnosed by the criteria of the American Diabetes Association (8) as glucose intolerance and one was diagnosed as diabetes mellitus. Hormone was stopped after recognition of the hyperglycemia, which then disappeared within 3 months.

Correlates of the Adverse hGH Reactions in Group I

The cases with adverse hGH reactions were divided into 3 subgroups—carpal tunnel syndrome ($n = 10$), gynecomastia ($n = 4$), and glucose intolerance ($n = 3$) (total of 16 patients because one patient had two reactions)—and compared with the 21 cases of no adverse hGH reaction with regard to age, initial body mass index, initial plasma IGF-I level, diagnoses, medications, mean IGF-I during treatment, and peak IGF-I during treatment. It was found that the carpal tunnel syndrome subgroup differed significantly from the no adverse hGH reaction subgroup with regard to two of the above variables: (i) mean IGF-I and (ii) peak IGF-I. The relationships are shown in Figure 10.2 and Table 10.4. First, in the untreated men of group II, the 95% confidence limits for mean plasma

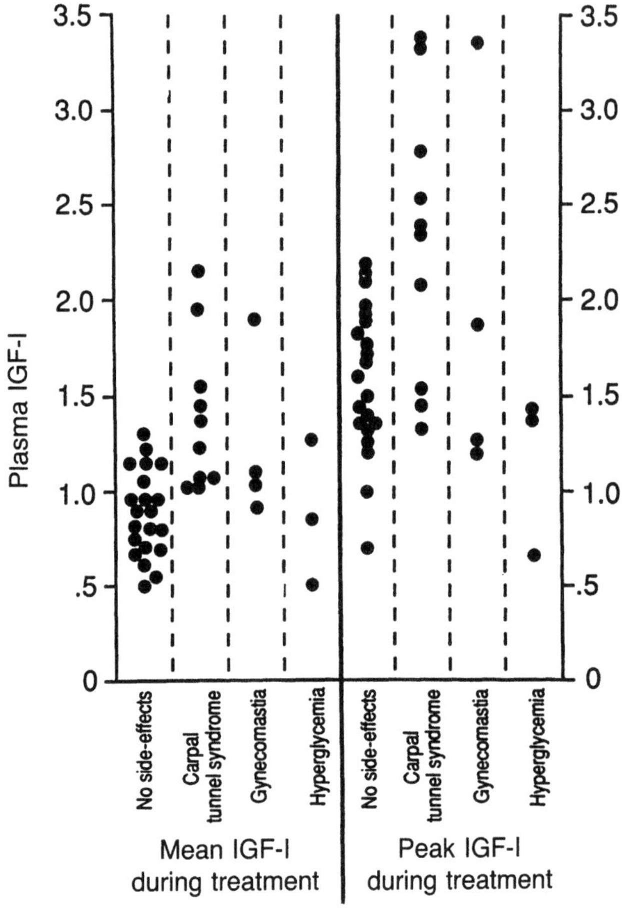

FIGURE 10.2. Mean and peak plasma IGF-I levels (units/ml) during hGH treatment in subgroups of group I.

IGF-I during months 6–18 were 0.1–0.3 units/ml. Among the 19 men who received hGH for 12 months without hGH-related side effects, the range for mean IGF-I during treatment was 0.5–1.3 units/ml. All 10 cases who developed carpal tunnel syndrome during hGH treatment had mean intra-treatment IGF-I levels above 1.0 units/ml. The mean IGF-I during treatment in the 10 men who developed carpal tunnel syndrome (group average, 1.4 units/ml) was significantly higher ($P < 0.01$) than in the 21 men who did not have an adverse reaction (group average, 0.9 units/ml). Second, the highest (peak) monthly IGF-I level during hGH treatment in group I men without side effects ranged from 0.6 to 2.2 units/ml, avg 1.6

TABLE 10.4. Mean and peak intra-treatment IGF-I levels (units/ml) in subgroups of group I and in group II.

	Mean IGF-I		Peak IGF-I	
	Avg	SE	Avg	SE
Group I, no hGH-related effects (n = 19)	0.88	0.03	1.56	0.10
Group I, carpal tunnel syndrome (n = 10)	1.38*	0.12	2.30*	0.23
Group I, gynecomastia (n = 4)	1.23†	0.23	1.93	0.44
Group II (n = 18)	0.26*	0.02	0.34*	0.02

* $P < 0.01$ vs. group I, no adverse effects.
† $P < 0.05$ vs. group I, no adverse effects.

units/ml. In the carpal tunnel syndrome cases, the corresponding range was 1.3 to 3.4 units/ml and the corresponding mean was 2.3 units/ml ($P < 0.01$ vs. the men with no adverse reactions). In the 4 cases with gynecomastia, average intra-treatment mean and peak IGF-I levels were 1.2 to 1.4 times higher than in the 19 men without adverse reactions. The difference in mean IGF-I value during treatment was significant ($P < 0.05$).

Body Composition Responses

These are summarized in Figure 10.3 and Table 10.5. The men in group II, who received no hGH, showed a significant ($P < 0.05$) progressive decline in lean body mass from the initial baseline amounting to −3.4% at month 18. Adipose mass increased by +5% over baseline ($P > 0.05$). The effect of hGH in group I was to cause a significant increase in lean body mass (+5.8% after 6 months of treatment, +5.9% after 12 months) and a significant decline in adipose mass (−17% after 6 months of treatment, −16.3% after 12 months). The increment in lean body mass and the decrement in adipose mass were significantly correlated (r = 0.68, $P < 0.01$) (Fig. 10.4).

Group I was subdivided according to mean IGF-I level during treatment and the occurrence of adverse hormone reactions (Table 10.5) (Fig. 10.5). In the men without side effects, those whose mean intra-treatment IGF-I was in the range of 0.5 to 1.0 units/ml exhibited lean body mass and adipose mass responses 1.5–3.0 times greater than in the men with mean intra-treatment IGF-I in the range of 1.0 to 1.5 units/ml. These differences were significant ($P < 0.05$) after 6 months of treatment, but no longer after 12 months. The 6 carpal tunnel syndrome men who reached month 12 (i.e., who received hGH for at least 6 months) had mean intra-treatment IGF-I levels in the range of 1.0–1.3 units/ml, and their body composition responses resembled the men without side effects whose mean IGF-I levels were in this range.

TABLE 10.5. Body composition responses of: subgroups of group I formed according to mean intra-treatment IGF-I level; group I carpal tunnel cases; and group II.

Subjects (*n*)	Mean IGF-I (units/ml), range during months 7–18	Month 12		Month 18	
		Lean body mass*	Adipose mass*	Lean body mass*	Adipose mass*
Group I, no adverse effects (19)	0.5–1.5	105.9 ± 1.5†	85.1 ± 2.9†	106.9 ± 1.8†	83.3 ± 3.4†
Group I, no adverse effects (14)	0.5–1.0	107.7 ± 1.6	80.4 ± 3.0	107.0 ± 2.4	81.2 ± 3.9
Group I, no adverse effects (5)	1.0–1.5	101.7 ± 2.6††	96.8 ± 4.9††	102.9 ± 1.8	90.0 ± 5.6
Group I, carpal tunnel syndrome** (6)	1.0–1.28	103.6 ± 3.2	95.1 ± 6.4		
Group II (18)	0.2–0.3	97.7 ± 1.4	107.3 ± 5.0	96.7 ± 0.8	105.1 ± 3.6

* Expressed as % of initial baseline, avg ± SEM.
** While 10 group I men developed carpal tunnel syndrome, only 6 are represented in this table because 4 dropped out before the body composition measurement at month 12.
† $P < 0.05$ for comparison with group II.
†† $P < 0.05$ for comparison with group I, no adverse effects, mean intra-treatment IGF-I 0.5 to 1.0 units/ml.

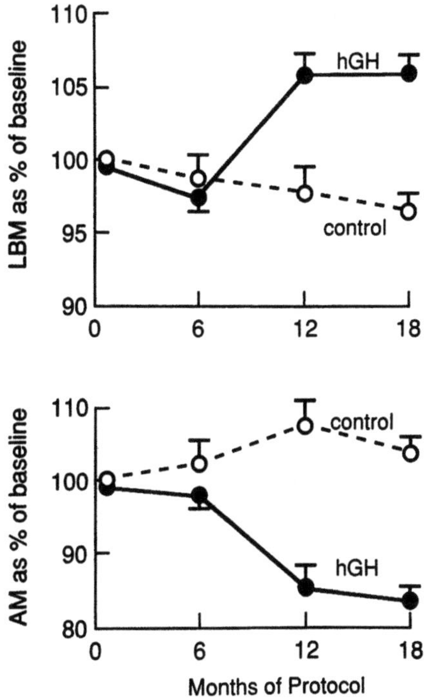

FIGURE 10.3. Lean body mass (LBM) and adipose mass (AM) as % of the initial baseline in the men of group I (n = 21) and of group II (n = 16) who completed the 18-month protocol. Each point is mean ± SEM.

Comparison of IGF-I Furlanetto and Daughaday Methods

All IGF-I data presented up to this point were obtained by the Furlanetto method (2). In the last 3 months of the study, 105 samples were assayed by the Daughaday (5) as well as by the Furlanetto assay. The results are shown in Figure 10.6. The correlation coefficient between the 2 assays was 0.55 ($P < 0.01$). The regression equation was IGF-I (ng/ml, Daughaday assay) = 240 IGF-I (units/ml, Furlanetto assay) + 105.

Other Laboratory Tests

For the men in both group I and group II who completed the 18-month protocol (n = 19 and n = 17, respectively), the authors compared the initial baseline and the 18-month values of the hematologic, urinalysis, blood chemistry, and bone density variables listed in Table 10.2. In neither group I nor group II did any of these variables change significantly

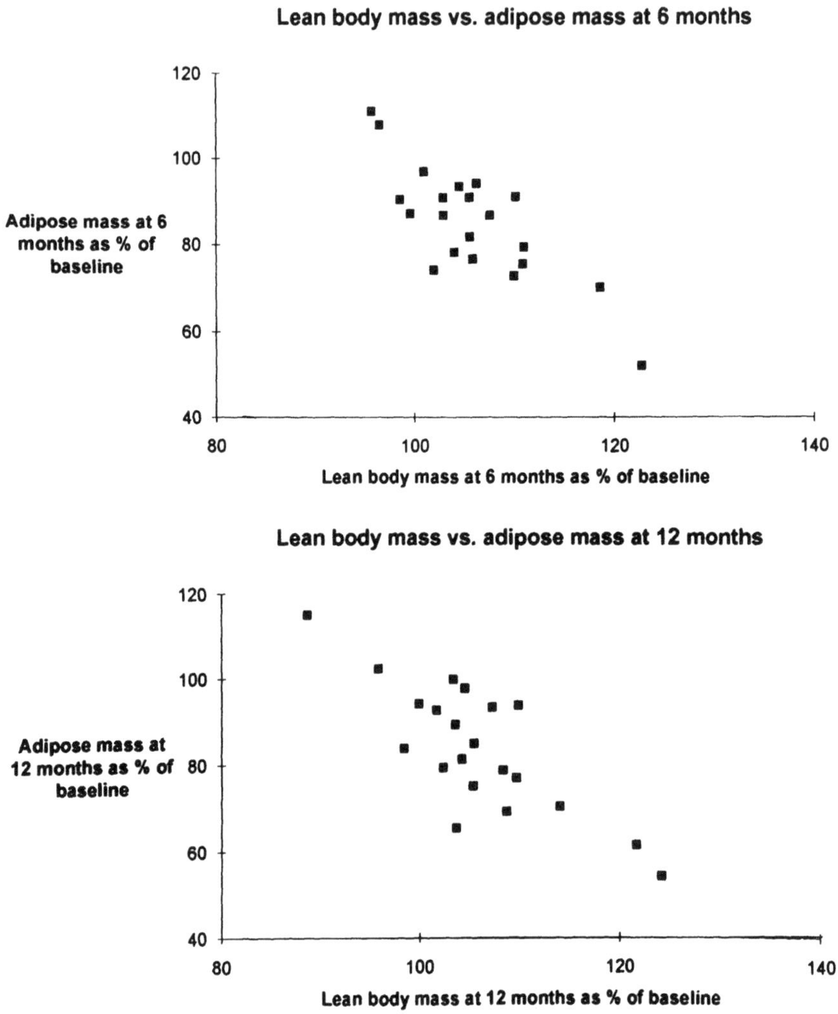

FIGURE 10.4. Relation of lean body mass and adipose mass (expressed as % of the initial baseline value) in the men of group I who completed the 18-month protocol.

($P < 0.05$) when the data were compared by the paired t-test between the beginning and the end of the protocol.

Discussion

Many subjects not receiving hGH (in either group during the lead-in period or in group II during the treatment period) dropped out of the study for various medical reasons. In these subjects, there were 23 medical

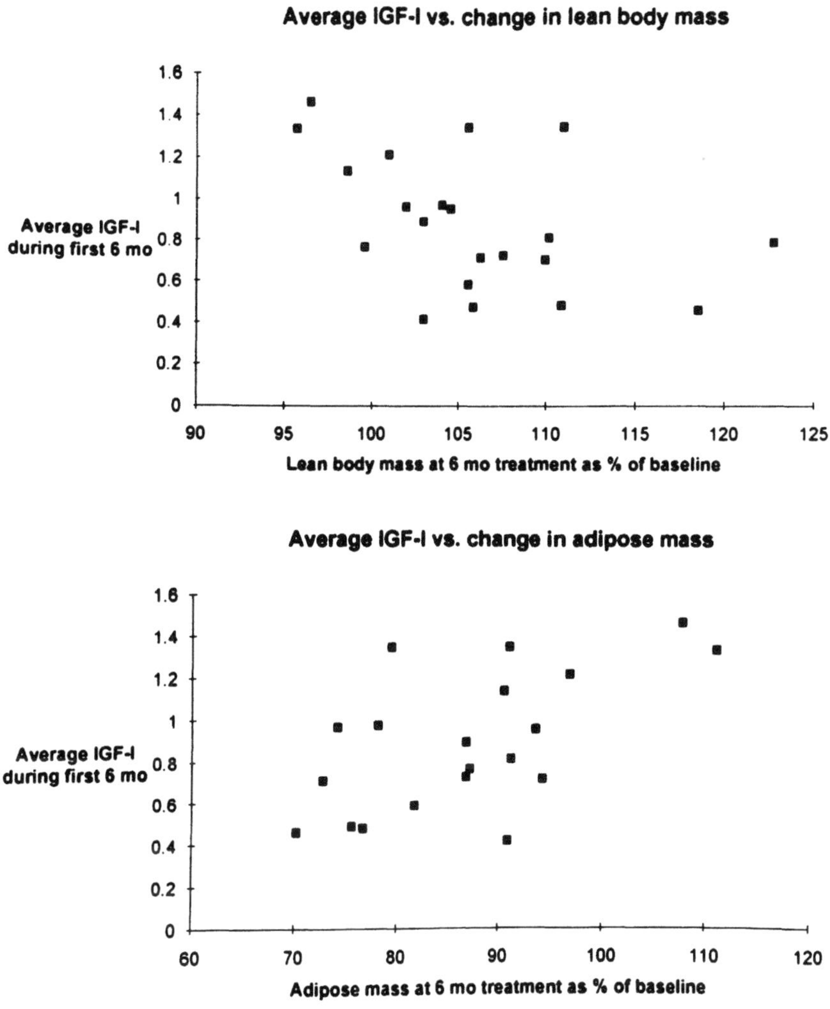

FIGURE 10.5. Relation of lean body mass and adipose mass (expressed as % of the initial baseline value) to mean intra-treatment plasma IGF-I level (units/ml) in the men of group I who completed the 18-month protocol.

dropouts per 100 subject years. Among the causes for these nonhormone-related medical dropouts, 5 might have been attributed to hGH (cancer of the prostate, leukemia, carpal tunnel syndrome, myocardial infarction, and diabetes mellitus) had they occurred during the hormone treatment. These observations show that when clinical trials are conducted with elderly subjects, even though they are initially asymptomatic, intercurrent medical events will occur at an appreciable rate and cannot be automatically classed as adverse drug reactions.

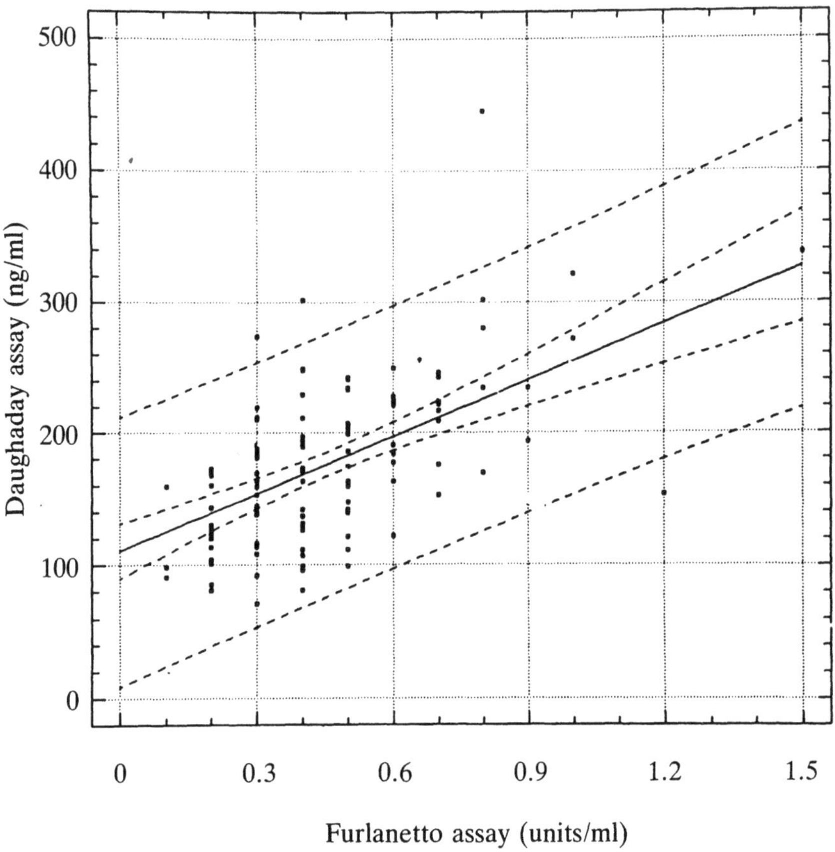

FIGURE 10.6. Relation of IGF-I values determined by the Furlanetto and Daughaday assays performed on the same sample of serum. One hundred and five samples are depicted.

Nevertheless, the high frequency of carpal tunnel syndrome, gynecomastia, and glucose intolerance in group I compared to group II, and the known relationships between these conditions and GH action and hypersomatotropic pathology, suggest that these events might have been induced by the hGH treatment. This suggestion was supported by the fact that the adverse event usually improved or disappeared after the hormone was stopped. Further supportive evidence was the bilaterality of the carpal tunnel syndrome and the gynecomastia, consistent with a systemic etiology.

In acromegaly, Clemmons et al. (9) found that the serum IGF-I level was elevated into the range of 2.6–21.7 units/ml (avg 6.8 units/ml).

In acromegaly of 3–30 years' duration, the prevalence of carpal tunnel syndrome has been reported to be 35%–43% (10), and of glucose intolerance to be 25%–50% (11). Gynecomastia has not been observed in acromegalics (10, 11). In the present study, it was somewhat unexpected to find that elevating the average plasma IGF-I level into the range 0.8–1.7 units/ml (peak values 1.0–2.9 units/ml) in the elderly men for only 2–11 months would be associated with carpal tunnel syndrome in 24% and gynecomastia in 9%. For comparison, the range of IGF-I in healthy young men is 0.5–1.5 units/ml (12). The observations suggest that in elderly men, structures such as connective tissue or cartilage (for carpal tunnel syndrome), breast (for gynecomastia), and liver, muscle, or adipose tissue (for glucose intolerance) may be more sensitive to hGH or IGF-I than in younger men. An alternative possibility was a reduced activity of the IGF-I binding proteins in old men, leading to a higher level of free plasma IGF-I at any specific level of total IGF-I. The system of IGF binding proteins (13) has recently been studied by ligand blotting and immunoprecipitation in 8 of the subjects in this study during their baseline period (14). The levels of binding proteins 2, 3, and 4 did not differ from those of healthy young men, but the level of IGF binding protein 1 was significantly lower. Other workers have also detected changes in IGF binding proteins in elderly women (13). Future studies need to correlate the occurrence of adverse hGH reactions in elderly people not only with the total serum IGF-I level, but also with the concentrations of the binding proteins and with estimates of IGF-I availability.

Both carpal tunnel syndrome and gynecomastia were significantly more common when the initial (<0.35 units/ml) IGF-I level was raised into the range of 1.0–1.5 units/ml than when it was raised into the range of 0.5–1.0 units/ml. In contrast, the anabolic and lipolytic effects on body composition tended to be greater when the intra-treatment IGF-I level did not exceed 1.0 units/ml. The implication was that the body composition responses may be achieved, while the carpal tunnel syndrome/gynecomastia reactions are largely avoided, by adjusting the hGH dose to maintain the IGF-I level in the range of 0.5–1.0 units/ml.

Other measures that may prevent undesired side effects when hGH treatment is used in experimental protocols can be suggested. First, subclinical carpal tunnel syndrome is common (15) and may be detected by Tinel's test and confirmed by EMG. Perhaps screening for subclinical carpal tunnel syndrome and excluding subjects with this condition would minimize the occurrence of hand and wrist symptoms during hGH treatment of elderly individuals. Second, elimination of those who screen positive for glucose intolerance, which occurs in 20%–50% of the elderly (16), would probably lower the frequency of hyperglycemic reactions to hGH. Third, plasma prolactin rises and the androgen/estrogen ratio falls in some, but not all, elderly men (17, 18). Future studies should examine the possible relationships between prolactin, androgen, and estrogen levels and hGH-induced gynecomastia.

Two methods are currently available to measure serum IGF-I. The Furlanetto assay is a direct RIA without preliminary fractionation of the plasma (2); in contrast, the Daughaday assay separates IGF-I from its binding proteins by an acid-ethanol extraction before the RIA (5). Although the values by the two assays were found to be proportional to each other, the correlation coefficient was only 0.55. The lack of a stronger correlation may reflect inter-individual differences in the system of IGF binding proteins (13). Consequently, the relationship between intra-treatment plasma IGF-I level, adverse hGH reactions, and body composition responses, which were established with the Furlanetto IGF-I values in the present study, should not be extrapolated to Daughaday values by using the conversion equation derived from Figure 10.6. A desirable range of serum IGF-I using the Daughaday method will have to be established in future studies of hGH in the elderly.

Three clinical trials with hGH in adults, which lasted 6 months or longer, have previously been reported (19–21). These were compared to the present study with regard to adverse hormone reactions (Table 10.6). The following points are noteworthy: (i) The doses used in the previous trials were 2–6 times higher than in the current study; (ii) in the two studies with younger adults, carpal tunnel syndrome, gynecomastia, and

TABLE 10.6. Comparison of previous and present trials of hGH lasting 4 months or longer in adult subjects with regard to adverse hormone reactions.

Author (ref.)	Description of subjects	No. of subjects	Dose (units/wk for man 5′10″, 166 lb)	Duration of treatment	Adverse hormone reactions
Christiansen et al. (19)	GH-deficient; avg age 24; age range 18–39; M/F, 14/8	16	27	12 months	Fluid retention (4)
Aloia et al. (20)	Osteoporotic; age range 53–76; M/F, 1/7	8	15	12 months	Carpal tunnel syndrome, 4/8; hyperglycemia, 2/8
Salomon et al. (21)	Avg age 39; age range 21–54; M/F, 8/4	12	37	6 months	Fluid retention, 6; carpal tunnel syndrome, 1; hypertension, 1
Present study	Hyposomatomedinemic; avg age 70; age range 55+; all male	21	6.8	12 months	Carpal tunnel syndrome, 10/19; gynecomastia, 4/19; hyperglycemia, 3/19

diabetes mellitus were infrequent or absent (0%–8% during 6 or 12 months of treatment); and (iii) in Aloia's experiment with elderly subjects (almost all women), at a dose level 2 times higher than in the present study, the 12-month incidence of carpal tunnel syndrome was 50% and that of hyperglycemia 25% (19). In the present trial, the occurrence of carpal tunnel syndrome during 12 months of treatment was 24% and that of gynecomastia 9%. It may be that elderly men are more prone to carpal tunnel syndrome and gynecomastia as these side effects of hGH than are younger adults, and the occurrence of these side effects may be related to the dose of hGH.

The finding of gynecomastia in the course of hGH treatment should prompt caution in regard to older women. Nabarro, in his series of 256 cases of acromegaly, reported increased malignancy in females, entirely attributable to cancer of the breast (22). No excess malignancies were observed in males. Recent studies have demonstrated IGF-I receptors in breast cancer cell lines, as well as a mitogenic effect of IGF-I on a number of epithelial breast cancer lines (23–25).

A final comment concerns the rate of loss of lean body mass in group II. In the general population of men over age 60, the annual rate of loss has been recorded by Flynn et al. to average −0.6% per year (95% confidence interval, 0%–1.4%) (6). In contrast, the untreated hyposomatomedinemic men in group II lost lean body mass at an average rate of −2.3% per year. This comparison is consistent with the hypothesis that diminished secretion of endogenous GH in late adulthood may be a contributory cause to the geriatric atrophy of lean body mass (1). To confirm this conclusion, it will be necessary to measure the annual rate of loss in eusomatomedinemic elderly men.

Acknowledgments. The authors are indebted to the following personnel of the North Chicago VA Medical Center: Uma Dilip Shah, M.D., Katherine K. Hahn, M.D., and Bhupendra K. Agrawal, M.D., who provided the nerve conduction studies; Ms. Stephanie A. Traynoff, Technical Supervisor of the Hematology Unit; Osvaldo Rubinstein, M.D., Chief of the Laboratory Service; Sharon Goeske, Program Coordinator and Secretary; Deborah Bourdo, Program Assistant for R&D Service; Dan Jackson, RN; and Mrs. Norma Caindec, R.D. Ms. Esther Carlton made the IGF-I assays available at Nichols Institute. Dr. Ruth Hartmann, of the Milwaukee VA Medical Center, provided valuable advice in the preparation of the manuscript.

References

1. Rudman D, Feller AG, Nagraj HS, et al. Effects of human growth hormone in men over sixty years old. N Engl J Med 1990;323:1–6.

2. Furlanetto RW, Underwood LE, Van Wyk JJ, D'Ercole AJ. Estimation of somatomedin C levels in normals and patients with pituitary disease by radioimmunoassay. J Clin Invest 1977;60:648–57.
3. Gellman H, Gelberman RM, Tan AM, Botte MJ. Carpal tunnel syndrome: an evaluation of the provocative diagnostic tests. J Bone Joint Surg 1986; 68A:735–7.
4. Stevens JC. AAEE minimonograph #26. The electrodiagnosis of carpal tunnel syndrome. Muscle Nerve 1987;10:99–113.
5. Daughady WH, Mariz IK, Blethen SL. Inhibition of access of bound somatomedin to membrane receptor and immunobinding sites: a comparison of radioreceptor and radioimmunoassay of somatomedin in native and acid-ethanol-extracted serum. J Clin Endocrinol Metab. 1980;51:781–8.
6. Flynn MA, Nolph GB, Baker AS, Martin WM, Krause G. Total body potassium in aging humans: a longitudinal study. Am J Clin Nutr 1989;50: 713–7.
7. Mazess RB, Barden HS, Bisek JP, Hanson J. Dual energy x-ray absorptiometry for total body and regional bone mineral and soft tissue composition. Am J Clin Nutr 1990;51:1106–12.
8. National Diabetes Data Group. Classification and diagnosis of diabetes mellitus and other categories of glucose intolerance. Diabetes 1979;28: 1039–57.
9. Clemmons DR, Van Wyk JJ, Ridgway EC, Kliman B, Kjellberg RN, Underwood LE. Evaluation of acromegaly by radioimmunoassay of somatomedin-C. N Engl J Med 1979;301:1138–42.
10. O'Duffy JD, Randall RV, MacCarty CS. Median neuropathy (carpal tunnel syndrome) in acromegaly. A sign of endocrine overactivity. Ann Intern Med 1973;78:379–83.
11. Jadresic A, Banks LM, Child DF, et al. The acromegaly syndrome. Q J Med, New Series, 1982;202:189–204.
12. Rudman D, Nagraj HS, Mattson DE, et al. Hyposomatomedinemia in the men of a Veterans Administration nursing home: prevalence and correlates. Gerontology 1987;33:307–14.
13. Donahue LR, Hunter SJ, Sherblom AP, Rosen C. Age-related changes in serum insulin-like growth factor-binding proteins in women. J Clin Endocrinol Metab 1990;71:575–9.
14. Unterman T, Dempsey S, Jentel JJ, Oehler DT, Rudman D. Alterations in insulin-like growth factor binding proteins in aging men. Clin Res 1992; 40:395A.
15. Spinner RJ, Bachman JW, Amadio PC. The many faces of carpal tunnel syndrome. Mayo Clin Proc 1989;64:829–36.
16. Davidson MB. The effect of aging on carbohydrate metabolism: a review of the English literature and a practical approach to the diagnosis of diabetes mellitus in the elderly. Metabolism 1979;28:688–705.
17. Blackman MR, Kowatch MA, Wehmann RE, Harman SM. Basal serum prolactin levels and prolactin responses to constant infusions of thyrotropin releasing hormone in healthy aging men. J Gerontol 1986;41:699–705.
18. Carlson HE. Gynecomastia: pathogenesis and therapy. Endocrinologist 1991;1:337–42.

19. Christiansen JS, Jorgensen JO, Pedersen SA, Miller J, Jorgensen J, Skakkeboek NE. Effects of growth hormone on body composition in adults. Horm Res 1990;4:61–4.
20. Aloia JF, Zanzi I, Ellis K, et al. Effects of growth hormone in osteoporosis. J Clin Endocrinol Metab 1976;43:992–9.
21. Salomon F, Cuneo RC, Hesp R, Sonksen PH. The effects of treatment with recombinant human growth hormone on body composition and metabolism in adults with growth hormone deficiency. N Engl J Med 1989;321:1797–803.
22. Nabarro JDN. Acromegaly. Clin Endocrinol (Oxf) 1987;26:481–512.
23. Cullen KJ, Yee D, Sly WS, et al. Insulin-like growth factor receptor expression and function in human breast cancer. Cancer Res 1990;50:48–53.
24. Daughaday WH. The possible autocrine/paracrine and endocrine roles of insulin-like growth factors of human tumors. Endocrinology 1990;127:1–4.
25. Essat S, Melmed S. Are patients with acromegaly at increased risk for neoplasia? J Clin Endocrinol Metab 1991;72:245–9.
26. Nagraj HS, Gergans GA, Mattson DE, Rudman IW, Rudman D. Osteopenia in the men of a Veterans Administration nursing home. Am J Clin Nutr 1990;51:100–6.

11

Effects of Growth Hormone in Older People

ROBERT MARCUS, LEAH HOLLOWAY, AND GAIL E. BUTTERFIELD

Pituitary *growth hormone* (GH) is a classical endocrine hormone with profound effects on somatic growth and body composition. Circulating concentrations of GH decline with advancing age, as do GH secretion rates and pituitary GH responsiveness to a variety of provocative stimuli. These declines are accompanied by reduced levels of *insulin-like growth factors* (IGFs), the putative mediators of many of the hormone's actions. Normal human aging is associated with important alterations in body composition that are also characteristic of GH-deficient children. These include increased adiposity and reduced muscle mass and strength and, of course, loss of bone mineral. Thus it seems reasonable to ask whether some age-related changes in body composition are brought about by a relative degree of GH deficiency and, further, whether GH therapy might have clinical utility in reversing these changes.

Recent evidence suggests an important role for IGF-I as an osteotrophic agent, and an argument can be made that IGF-I is a common mediator of skeletal response to multiple hormones. Like parathyroid hormone, GH activates cell surface receptors in cultured osteoblast-like cells to increase local IGF-I production (1–3). Estradiol promotes accumulation of IGF-I and IGF-II in cultured osteosarcoma cells. GH and estradiol both stimulate proliferation of osteoblast-like cells, an effect that is obliterated by the presence of anti-IGF-I immunoglobulin. In osteoblasts raised in serum-free medium, recombinant human GH stimulates proliferation and increases the relative amount of newly synthesized type-I collagen (4). In vivo, GH and IGF-I both increase bone turnover. These various results invite the conclusion that GH or IGF-I might specifically activate osteoblast proliferation and differentiated function to repair bone mineral deficits characteristic of osteoporosis. Until recently, GH treatment of other than GH-deficient children was limited by hormone supply. With the availability of *recombinant human GH* (rhGH), therapy of adults has become a practical, albeit expensive, strategy. In this chapter we review the current status of growth hormone therapy for older men and women. Emphasis is placed on recent studies using rhGH rather than earlier studies with pituitary hormone.

Short-Term Administration of rhGH to Healthy Elders

Our research group (5) reported the effects of 7 days of rhGH administration to 16 healthy older men and women (>60 years). Recombinant hGH produced a brisk rise in circulating IGF-I that was associated with striking changes in nitrogen retention, sodium excretion, and in the parathyroid-vitamin D axis. Twenty-four-hour urinary nitrogen and sodium excretion decreased by 38% and 50%, respectively, whereas urine calcium excretion markedly increased. Thus, rhGH uncoupled the usual tight relationship between sodium and calcium excretion. Significant increases were observed in circulating osteocalcin and in urinary hydroxyproline, suggesting that bone remodeling had been activated.

Sustained Treatment of Elderly Men and Women

The most widely publicized growth hormone trial to date was reported by Rudman and colleagues (6). These authors conducted a randomized, placebo-controlled intervention trial in 21 elderly men. GH (0.03 mg/kg 3 times per week) was given for 6 months and was found to produce significant increases in lean mass by ^{40}K analysis. Bone density was assessed at 9 different sites by dual photon absorptiometry, and a 1.6% increase in lumbar spine mineral density was reported. No significant change in skin-fold estimates of adiposity was observed. The authors state that "the effects of 6 months of hGH on lean body mass and adipose-tissue mass were equivalent in magnitude to the changes incurred during 10 to 20 years of aging." This statement has been transformed subsequently by news media to suggest that GH reverses the effects of aging.

Although the results of this experiment were provocative and interesting, concerns about their interpretation must be raised. The ^{40}K data provide convincing evidence of a true increase in lean mass. However, the stated changes in adiposity may have been confounded by fluid retention. The changes in bone mass were marginally significant at best and pose questions about the methods of analysis. The authors measured bone mass at 9 different sites. Even if these measurements were truly independent of one another, it would not be surprising for one site to achieve significance at the $P < 0.05$ level. Clearly, bone mass at one site is not independent from that at other sites, so the confounding effect of multiple comparisons may be substantial.

We have recently completed a randomized placebo-controlled intervention trial of rhGH (0.025 mg/kg/day) in 22 healthy elderly women.

Eight women received rhGH as a daily injection for 1 year. Compared to the placebo group, circulating levels of IGF-I were significantly elevated and maintained throughout the intervention. Recombinant hGH clearly stimulated a persistent increase in bone turnover. Sustained elevations were observed in circulating osteocalcin, type-I procollagen peptide, and bone alkaline phosphatase, as well as in urinary excretion of hydroxyproline in the treatment group. These elevations reverted to baseline values by 3 months after stopping treatment. On the other hand, no significant changes were observed in bone mineral density at either the lumbar spine or proximal femur. It is of interest that bone mineral density at the femoral trochanter and Ward's triangle decreased significantly in the placebo group, but did not change in the hormone-treated subjects. Thus, although rhGH did not increase bone mass, there appears to have been a maintenance effect at the hip.

No significant changes in lean mass or adiposity were observed by hydrostatic weighing, although an increase in lean mass was suggested by skin-fold thickness. Basal metabolic rate was significantly raised at 6 months in the treatment group, but reverted to baseline levels by 12 months. In a limited number of subjects (4 treatment, 6 placebo), no changes in nitrogen or calcium balance were observed at 6 or 12 months.

Similar to the results of our short-term trial, rhGH produced a sustained decrease in urinary sodium excretion that persisted throughout the 12-month trial. At the dose of hormone used, this effect caused only mild symptoms of fluid retention in a few subjects. However, related changes in fluid balance may have produced alterations in total and extracellular water compartments that may confound the body composition measurements. For example, we observed the same changes in skin thickness that were reported by Rudman et al. within 7 days of starting rhGH injections. At this time interval such a result must reflect water changes rather than a true loss of fat mass. This issue has not been adequately addressed and must be resolved in future studies.

With respect to safety of rhGH, results of this trial were fairly reassurring. A transient increase in insulin resistance was observed several weeks after starting therapy, but this resolved by 6 weeks and remained at baseline levels for the duration. No significant changes were observed at any time in blood pressure, lipoprotein constituents, thyroid function status, or fibrinogen levels. Thus, with respect to a panel of cardiovascular risk factors, rhGH administration appears to have been a safe intervention. One woman from the hormone treatment group developed classical manifestations of temporal arteritis during the course of the study. It is important to note that the margin of safety may not be very large for rhGH. An attempt to treat a different group of older women with rhGH at 0.05 mg/kg/day led to intolerable fluid retention in most subjects and symptoms of carpal tunnel compression in two cases.

Role of GH in Therapy of Osteoporosis

The osteotrophic actions of GH in growing skeletons were described above and suggest a rationale for considering this hormone to increase bone mass in patients with osteopenia. Aloia et al. (7) reported an experience with GH in several osteoporotic patients, but the results did not support the clinical utility of this agent. In a subsequent report from this same group (8), combined sequential therapy with GH and calcitonin produced a small, but persistent, increase in whole body calcium content of osteoporotic patients. Unfortunately, neither the results of Rudman et al. (6) nor our own experience, outlined above, permit enthusiasm that rhGH *given as single daily monotherapy* will offer a means to achieve meaningful increases in bone mass. Although a maintenance effect may have occurred at the hip, a variety of antiresorptive agents currently offer similar protection, and it would be hard to justify the use of an expensive, injectable protein hormone to achieve this result. Since the doses employed are close to maximally tolerated levels, it is unlikely that a simple adjustment of dose will rectify this situation.

The results do clearly establish that rhGH promptly induces a sustained increase in bone remodeling. Since bone remodeling is an inherently inefficient process, anything that activates remodeling should aggravate bone loss. If there is a role for GH in osteoporosis, therefore, it will likely involve a complex therapeutic strategy. A number of questions require clarification before such a role can be established. Should treatment be reserved for patients with low levels of IGF-I? What are the effects of rhGH on bone remodeling as defined by dynamic histomorphometry? Should GH be used cyclically in combination with an antireorptive agent? Recombinant IGF-I is itself nearing availability for clinical studies. Some of the adverse metabolic effects of GH may be obviated with IGF-I, so the relative actions of this agent on skeletal turnover will be of great interest.

Finally, it should be remembered that GH, at least transiently, promotes nitrogen retention in adults. Since decreased muscle mass and strength contribute to the risk for falls, and therefore to the risk for hip fracture, a therapeutic role for rhGH might be defined that is independent of bone mass per se. Although our own results did not confirm a persistent effect of rhGH on nitrogen balance, it is conceivable that it may synergize with an exercise program to improve muscle mass and strength.

Summary

Many of the changes in body composition resemble those associated with GH deficiency, and GH secretion decreases with normal human aging. With the availability of rhGH, it is now possible to test the metabolic and structural effects of GH administration to adults. Administration of rhGH to healthy elders acutely promotes nitrogen retention and activates bone remodeling. Chronic administration may increase lean mass in men, but appears to give less impressive results in women. Changes in bone remodeling are not accompanied by dramatic increases in bone mass. Thus, it is not likely that daily rhGH as a single agent will prove to be an effective means for regaining deficits in bone, but it may be possible to incorporate rhGH into a multi-agent strategy. In healthy elders GH can be given without obvious adverse effects on cardiovascular risk factors.

References

1. Stracke H, Schultz A, Moeller D, Rossol S, Schatz H. Effect of growth hormone on osteoblasts and demonstration of somatomedin C/IGF-1 in bone organ culture. Acta Endocrinol (Copenh) 1984;107:16–24.
2. Chenu C, Valentin-Opran A, Chavassieux P, et al. Insulin like growth factor I hormonal regulation by growth hormone and by 1,25$(OH)_2D_3$ and activity on human osteoblast-like cells in short-term cultures. Bone 1990;11:81–6.
3. Barnard R, Ng KW, Martin TJ, Waters MJ. Growth hormone (GH) receptors in clonal osteoblast-like cells mediate a mitogenic response to GH. Endocrinology 1991;128:1459–64.
4. Ernst M, Froesch ER. Growth hormone dependent stimulation of osteoblast-like cells in serum-free cultures via local synthesis of insulin-like growth factor I. Biochem Biophys Res Commun 1988;151:142–7.
5. Marcus R, Butterfield G, Holloway L, et al. Effects of short term administration of recombinant human growth hormone to elderly people. J Clin Endocrinol Metab 1990;70:519–27.
6. Rudman D, Feller AG, Nagraj HS, et al. Effects of human growth hormone in men over 60 years old. N Engl J Med 1990;323:1–6.
7. Aloia JF, Zanzi I, Ellis K, et al. Effects of growth hormone in osteoporosis. J Clin Endocrinol Metab 1976;43:992–9.
8. Aloia JF, Vaswani A, Kapoor A, Yeh JK, Cohn SH. Treatment of osteoporosis with calcitonin, with and without growth hormone. Metabolism 1985;34:124–9.

12

Alteration of Growth Hormone Secretion in Aging: Peripheral Effects

THIERRY ABRIBAT, NATHALIE DESLAURIERS, ISABELLE REEVES, DOMINIQUE R. GARREL, AND PAUL BRAZEAU

In humans and animals, the aging process is associated with major endocrine changes. The view that these changes are responsible for the aging of an organism is undoubtedly oversimplistic (1). However, strong evidence indicates that they participate in the progressive loss of important physiologic functions, such as reproduction, immune function, maintenance of somatic anabolism, thyroid function, and so forth (2). In this process, synthesis and secretion of pituitary hormones are differently affected (3), being either increased (prolactin) or decreased (gonadotropins and growth hormone). Among endocrine alterations that develop early in aging, the blunting of basal and stimulated *growth hormone* (GH) secretion has been extensively documented in human and animal models (4–9). Numerous experimental observations of the aging somatotropic axis by our laboratory and others support the view that these alterations originate at the level of GH regulatory factors. Some important mechanisms that lead to this impairment are reviewed in the first part of this chapter.

Modifications of body composition occur in parallel to these endocrine changes (10). Recent data (11) showing that a GH-replacement therapy in aged patients can reverse these changes confirm that the diminution of GH secretion is one of the main factors that affect metabolism and body composition in aging. However, from a dynamic point of view, aging is also characterized by a diminished resistance of the organism to an external aggression, either in the form of injury or disease. Both the immune reaction (13) and the capacity of tissue reconstruction (14) are altered in aging. GH is known to act on the immune system (15, 16), and both GH and *insulin-like growth factor I* (IGF-I) have been shown to stimulate protein synthesis (17) and tissue regeneration in various models of wound and muscle healing (18–22). Therefore, it could be hypothesized that the GH impairment that occurs in aging is, in part, responsible for the altered resistance of aged patients. As a first attempt to test this hypothesis, we have evaluated the effect of aging on IGF-I levels during the recovery period in patients with acute myocardial infarction and in patients with severe burns. The objective of these preliminary studies was to determine whether aging affects not only basal IGF-I levels, but also their variations following injury or disease.

Mechanisms of GH Impairment in Aging

GH secretion is known to be primarily regulated by *growth hormone releasing factor* (GRF) (23) and *somatostatin* (SRIF) (24), whereas GH synthesis is exclusively under the control of GRF (25). Data from humans and experimental animal models clearly indicate that both the GRF-GH axis and the SRIF-GH axis are altered in aging.

Pituitary Growth Hormone in Aging

Although both GH synthesis and pituitary GH content are progressively reduced in aging rat (7, 9, 26), a body of evidence indicates that the aging pituitary somatotroph cells are still able to release as much GH as cells from young organisms if adequately stimulated. In 20-month-old male rats, we reported (9) lower pituitary rGH content when compared to 8-month-old rats, but higher rGH content compared to 2-month-old rats. Since GRF-stimulated GH secretion was higher in 2-month-old rats than in 20-month-old rats, we concluded that the diminution in pituitary rGH content cannot solely account for the in vivo diminished GH secretion. In aged humans, a coadministration of arginine and GRF has been shown to reverse impaired GRF-induced GH secretion (27). Theophylline, which acts directly at the *cyclic adenosine monophosphate* (cAMP) level, has been reported to restore GH responsivness to GRF in aged men (28). Finally, in aged rats, a co-injection of *GH releasing peptide* (GHRP) and GRF released significantly more rGH than in young rats (29). Therefore, based on these data, it becomes apparent that regulatory factors of GH secretion, rather than GH production itself, are responsible for decreased GH secretion in aging.

The GRF-GH Axis in Aging

Dramatic alterations of GRF synthesis, release, and bioactivity have been reported in aging humans and animals. In rats, both basal GRF gene expression and GRF immunoreactive material are reduced up to 50% in old vs. young rats (30, 31). Hypothalamic GRF release by L-Dopa has also been shown to be blunted in aged humans (32). Moreover, it has recently been suggested that the negative feedback of GH on GRF gene expression that occurs in young rats is impaired in old animals (33). It seems, therefore, that all mechanisms that regulate GRF gene expression are blunted in aging.

In parallel, a gradual resistance to GRF biological action on the pituitary occurs. The in vitro GRF-stimulated GH secretion is progressively impaired (9), which has been shown to be a receptorial alteration, because (i) GRF, but not *vasointestinal polypeptide* (VIP)-induced adenylate cyclase activity is impaired (34), and (ii) we have established that the number of high-affinity GRF binding sites, which are responsible for the biological effects of GRF (35, 36), decreases from 2 to 18 months of age in rats, whereas the number of low-affinity binding sites increases (36). This receptorial alteration is considered to be an initiating factor that leads to blunted GRF in sensitivity because it precedes it in time (36). In addition, this phenomenon is specific for aging because the GRF-stimulated GH blunting that occurs in obesity is not associated with changes in pituitary GRF binding parameters (37).

The biochemical features responsible for GRF receptor alterations in aging are currently under investigation. In preliminary studies, we have shown that saturation experiments performed in young rat pituitaries in the presence of guanyl nucleotides resulted in dramatic changes in binding parameters, mimicking those found in old pituitaries, whereas the same nucleotides had no effect on GRF binding in old pituitaries (38). These results suggest that a progressive GRF receptor/G-protein uncoupling is responsible for the decreased GRF responsiveness observed in aged rats. Similar conclusions have been drawn by others studying the effect of GTP analogs on GRF-stimulated adenylate cyclase activity in young vs. old rat pituitary membranes (39). Namely, these authors reported higher basal adenylate cyclase activity and a diminished sensitivity to GRF in old compared to young rat pituitary membranes.

Whether alterations of hypothalamic GRF synthesis and pituitary GRF receptors are related is so far unknown. In their initial experiments on adenylate cyclase activity, Parenti et al. (40) reported that an in vivo pretreatment with native rGRF in rats resulted in a decrease of basal adenylate cyclase activity in old but not young rats. Moreover, in middle-aged, 12-month-old rats that had been chronically treated for 4 months with a potent GRF analog ([desamino Tyr^1; $DAla^2$; Ala^{15}]hGRF[1–29]-NH_2; 7.5 mg per 3 weeks in a subcutaneous slow-releasing preparation), we observed a higher in vitro rGRF (6.25 pM)-stimulated GH response by isolated, perifused pituitary cells from GRF-treated vs. control rats (net rGH increase: +1.48 μg/ml vs. + 0.96 ± 0.06 μg/ml; $n = 4$ experiments; $P < 0.01$) without any modification in pituitary rGH content (1.01 ± 0.09 mg/pituitary in GRF-treated rats vs. 0.96 ± 0.14 mg/pituitary in control rats; $n = 4$ measurements per group; NS) (unpublished results).

This increased in vitro GRF responsiveness without any change in pituitary rGH content suggests that GRF could regulate its own pituitary binding sites in aged animals.

The SRIF-GH Axis in Aging

SRIF is known to inhibit GH secretion from somatotroph cells. Hypothalamic SRIF synthesis has been shown to be decreased in aging (41), whereas hypothalamic SRIF content has been reported to be unchanged (31) or slightly diminished (9). However, experimental data from many laboratories have documented a general increased somatostatinergic tone in aging. Basal and K^+-stimulated SRIF release from rat hypothalamus fragments have been shown to increase in aging up to 14 months of age (42), and in vitro SRIF secretion from hypothalamic neurons of aged rats has been shown to be increased when compared to young rats (43). Since pituitary SRIF binding sites seem to be unchanged in old rats (44), we reported as a consequence an increased pituitary SRIF accumulation (9). This increased SRIF secretion appears to be important in aging, as demonstrated by pharmacological manipulations that suppress SRIF tone. Namely, the administration of arginine and passive immunization against SRIF have been shown to dramatically improve the GRF-induced GH response in aged humans and rats, respectively (27, 45).

Integrated Model

Most available experimental data indicate that the decreased GH secretion that develops in aging mainly results from increased SRIF secretion and diminished GRF sensitivity (Fig. 12.1). Whether this last phenomenon results from or induces the documented diminution of GRF synthesis and secretion is so far uncertain. Whether relationships exist between alterations of the SRIF-GH axis and those of the GRF-GH axis has also to be investigated. Nevertheless, a growing body of evidence suggests that a progressive uncoupling between GRF receptor and stimulatory G-proteins is one of the main events that leads to blunted GH secretion in aging.

Peripheral Consequences

IGF-I is a major mediator of the peripheral biological actions of GH. Its blood levels do not exhibit significant diurnal variations and are considered to reflect the integrated daily GH secretion (46). Its measurement is therefore a useful tool to assess both GH secretion and bioactivity in a given pathophysiological condition. In the following studies, we have used a formic acid-acetone procedure to separate IGFs from their binding proteins before measurement by radioimmunoassay, as previously de-

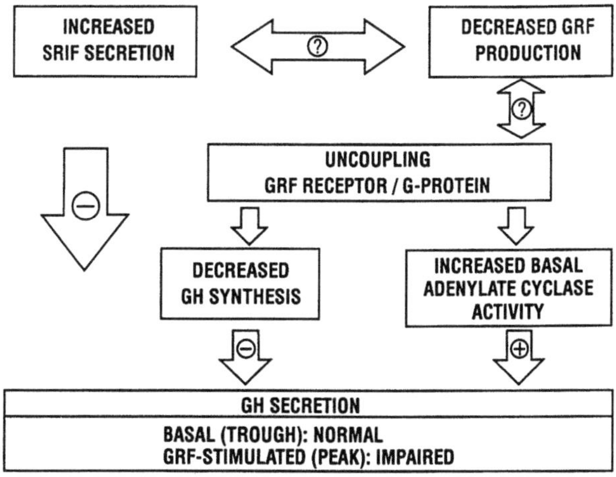

FIGURE 12.1. Proposed integrated model for blunted GH secretion in aging. The increased basal adenylate cyclase activity, diminished GH synthesis, and increased SRIF tone lead to unchanged basal GH secretion, which is even elevated in vitro, when SRIF influence is suppressed (9). Peaks of GH secretion, which are GRF-dependent, are blunted, mainly due to decreased GRF sensitivity (alterations at the GRF receptor level).

scribed and validated by others (47). Detailed methodology and results will be reported elsewhere (manuscripts submitted or in preparation).

Basal IGF-I Levels and Body Composition

The diminution of GH secretion in aging is associated with a diminished half-life of GH in circulation (48) that might be due to the gradual lowering of the blood levels of its specific binding proteins (49). This leads to a marked diminution of 24-h-integrated GH bioactivity and progressive diminution of circulating IGF-I concentrations (Table 12.1). In both men and women, this diminution occurs as early as 30 years of age (Table 12.1), and as we did, many authors have reported a negative correlation between age and IGF-I concentrations in normal humans (reviewed in 50). Moreover, in the course of aging, IGF-I, but not IGF-II levels decrease (data not shown), indicating that the decrease in IGF-I levels is directly related to diminished GH secretion. In parallel, alterations in nitrogen, lipid, and phosphocalcic metabolism result in gradual changes of body composition in the course of aging (10). These changes include increased fat deposition, muscle loss, skin thinning, and bone demineralization. Chronic treatment with *recombinant human growth hormone* (rhGH) has recently been shown to reverse these changes in healthy elderly subjects (11).

TABLE 12.1. Effects of age and sex on IGF-I serum concentration in normal human subjects.

Age (years)	Sex	n	IGF-I (ng/ml) (mean ± SEM)
20–29	M	34	317 ± 13
	F	39	324 ± 12
30–39	M	14	240 ± 15[c]
	F	18	252 ± 20[b]
40–49	M	20	243 ± 12[c]
	F	20	257 ± 17[c]
50–59	M	17	191 ± 12[c]
	F	25	220 ± 12[c]
60–75	M	4	185 ± 21[c]
	F	4	198 ± 44[a]

[a] $P < 0.02$.
[b] $P < 0.01$.
[c] $P < 0.001$, when compared to the 20–29 years group.
Note: Values are mean ± SEM and n indicates the number of subjects per group. The overall correlation between IGF-I levels and age of subject was $r = -0.60$ ($P < 0.01$). There was no difference in IGF-I concentrations between males (M) and females (F) at any age range.

IGF-I Variations in Patients During Recovery from Injury or Disease

Both GH and IGF-I are known to facilitate healing through their actions on nitrogen metabolism and tissue regeneration (17–22). As an example of tissue regeneration, they both have been shown to increase granulation tissue formation in wound healing (18–20). This effect is due to their actions on cell proliferation and extracellular matrix synthesis. It can therefore be suspected that the somatotropic axis plays a major role in tissue reconstruction following an injury or a disease. We have evaluated the effect of aging on IGF-I variations following two categories of severe injury for which tissue regeneration is needed for proper recovery: myocardial infarction and burns.

Follow-Up of Patients with Myocardial Infarction

We have monitored daily IGF-I concentrations in 19 patients aged 19–71 years who were admitted to the critical care unit of Notre Dame Hospital for myocardial infarction. Infarction was diagnosed based on clinical symptoms and blood elevation of *creatinine kinase MB* (CKMB). IGF-I was measured within 24 h post-admission, then daily for 3 days. Results are summarized in Table 12.2. We observed a significant negative cor-

TABLE 12.2. Effect of aging on IGF-I concentrations and their variations following myocardial infarction.

	IGF-I (ng/ml) Day 1	IGF-I (ng/ml) Day 4	Δ IGF-I (ng/ml) Day 1–4
Patients <50 years (*n* = 8) (CKMB: 2287 ± 482 U/l)	200 ± 16	222 ± 35	+22 ± 27 NS
Patients >50 years (*n* = 11) (CKMB: 2010 ± 446 U/l)	151 ± 15	108 ± 12	−43 ± 8 $P < 0.001$
Correlation (*r*) IGF-I/age	−0.48 $P < 0.05$	−0.68 $P < 0.01$	−0.62 $P < 0.01$

Note: Values are expressed as mean ± SEM, and Δ IGF-I is the variation in IGF-I concentration from day 1 to day 4, also expressed as mean ± SEM. Note that both groups (<50 and >50 years of age) exhibited infarctions of similar intensity, as shown by CKMB levels within 24 h.

relation at each sampling day between IGF-I levels and age of patients. In addition, IGF-I variations for each individual patient during the study period were also negatively correlated with age of patients (Δ IGF-I, Table 12.2). In the first 4 days following infarction, younger patients (<50 years of age) showed no significant variation in IGF-I levels, whereas older patients (>50 years of age) had a significant decrease in their IGF-I levels ($P < 0.001$, Table 12.2).

Follow-Up of Burned Patients

Eleven severely burned patients aged 22–55 years were sampled every 3 days for IGF-I measurement for 25 days following admission in the burn intensive care unit of Hotel Dieu Hospital, Montreal. When considering all patients, there was a significant effect of time on IGF-I levels, with IGF-I levels being low during the first week and gradually rising from day 4 to day 25 post-injury (data not shown). Despite aggressive enteral and parenteral nutrition, all patients remained in negative nitrogen balance during the study period (cumulative nitrogen loss: −107 ± 25 g). As shown in Table 12.3, all patients exhibited low IGF-I concentrations 4 days following burn when compared to healthy subjects (Table 12.1), but only younger patients (<40 years of age) had a significant increase in their IGF-I concentrations between day 4 and day 25 post-injury. As a result of these variations, there was a significant negative correlation between age of patients and IGF-I levels on day 25, but not on day 4 (Table 12.3). Of great interest is the fact that younger patients were able to markedly increase their IGF-I levels despite a severe catabolic state. Since IGF-I levels are regulated by both metabolism and GH secretion, it

TABLE 12.3. Effect of aging on IGF-I concentrations and their variations following severe burns.

	IGF-I (ng/ml) Day 4	IGF-I (ng/ml) Day 25	Δ IGF-I (ng/ml) Day 4–25
Patients <40 years ($n = 5$)	117 ± 20	303 ± 13	+186 ± 32 $P < 0.01$
Patients >40 years ($n = 6$)	97 ± 24	140 ± 37	+43 ± 38 NS
Correlation (r) IGF-I/age	−0.29 NS	−0.77 $P < 0.01$	−0.63 $P < 0.05$

Note: Values are expressed as mean ± SEM, and Δ IGF-I is the variation in IGF-I level from day 4 to day 25, also expressed as mean ± SEM.

could be hypothesized that this IGF-I increase in the second phase of recovery was GH dependent.

Discussion

In myocardial infarction, IGF-I levels remained stable in younger patients, whereas they decreased significantly in older patients. In the first month following major burns, IGF-I levels increased markedly in younger patients, whereas they remained low and stable in older patients. These data suggest that in both situations, older individuals are less able to produce IGF-I or to maintain its blood levels in the recovery period than younger individuals. Given the biological actions of IGF-I on nitrogen metabolism and tissue regeneration, it can be expected that patients exhibiting the highest IGF-I levels and variations—that is, younger patients—have a better chance to recover and heal properly and more quickly than older patients. Based on these results, we hypothesize that the diminution of GH secretion is, at least in part, responsible for the delayed healing and recovery that develops in aging.

Conclusion

GH secretion is progressively blunted in aging. Early and major defects that lead to this impairment are increased SRIF secretion and alterations of pituitary GRF binding sites, which might result from a progressive uncoupling between GRF receptors and stimulatory G-proteins. In healthy subjects, this decreased GH secretion induces an early diminution in basal IGF-I blood levels that is responsible for the observed modifications in body composition. Following injury (myocardial infarction or burn), the aging process affects not only basal IGF-I levels, but also their variations in the recovery period. Given the actions of both GH and IGF-I on metabolism and tissue regeneration, we suggest that the altered GH secretion participates in the impairment of healing in older patients. In these conditions, GH therapy could be proposed as a promoter of tissue regeneration and healing in elderly patients.

Acknowledgment. The authors wish to thank Maryse Richer for the preparation of the manuscript.

References

1. Hayflick C. Theories of biological aging. Exp Gerontol 1985;20:145–59.
2. Meites J, Goya R, Takahashi S. Why the neuroendocrine system is important in aging processes. Exp Gerontol 1987;22:1–15.
3. Zbuzek V, Arnetz BB, Eneroth P, Zbuzek VK. Age-related differences in concomitant hormone release from the superfused rat adenopituitary. Gerontology 1991;37:253–61.
4. Finkelstein JW, Roffwang P, Boyar RM, Kream J, Hellman L. Age-related change in the twenty-four-hour spontaneous secretion of growth hormone. J Clin Endocrinol Metab 1972;35:665–70.
5. Ho KJ, Evans WS, Blizzard RM, et al. Effects of sex and age on the 24-hour profile of growth hormone secretion in man: importance of endogenous estradiol concentrations. J Clin Endocrinol Metab 1987;64:51–8.
6. Shibasaki T, Shizume K, Nakahara M, et al. Age-related changes in plasma growth hormone response to growth hormone-releasing factor in man. J Clin Endocrinol Metab 1984;58:212–4.
7. Sonntag WE, Steger RW, Forman LJ, Meites J. Decreased pulsatile release of growth hormone in old male rats. Endocrinology 1980;107:1875–9.
8. Ceda G, Valenti G, Butterini U, Hoffman AR. Diminished pituitary response to growth hormone-releasing factor in aging male rats. Endocrinology 1986;118:2109–14.
9. Deslauriers N, Gaudreau P, Abribat T, Renier G, Petitclerc D, Brazeau P. Dynamics of growth hormone responsiveness to growth hormone-releasing factor in aging rats: peripheral and central influences. Neuroendocrinology 1991;53:439–46.
10. Rudmann D. Growth hormone, body composition, and aging. J Am Geriatr Soc 1985;33:800–7.
11. Rudmann D, Feller AG, Nagraj HS, et al. Effects of human growth hormone in men over 60 years old. N Engl J Med 1990;323:1–6.
12. Marcus R, Butterfield G, Holloway L, et al. Effects of short term administration of recombinant human growth hormone to elderly people. J Clin Endocrinol Metab 1990:519–27.
13. Weksler ME. Immune senescence in man. In: Fabris N, ed. Immunology and ageing. The Hague: Martinus Nijhoff, 1982:165–73.
14. Goodson WH, Hunt TK. Wound healing and aging. J Invest Dermatol 1979;73:88–91.
15. Khansari DN, Gustad T. Effects of long-term, low dose growth hormone therapy on immune function and life expectancy of mice. Mech Ageing Dev 1991;57:87–100.

16. Wiedermann CJ, Niedermühlbichler M, Beimpold H. Braunsteiner H. In vitro activation of neutrophils of the aged by recombinant human growth hormone. J Infect Dis 1991;164:1017–20.
17. Sonntag WE, Hylka VW, Meites J. Growth hormone restores protein synthesis in skeletal muscle of old male rats. J Gerontol 1985;40:689–94.
18. Jorgensen PH, Andreassen TT. A dose-response study of the effects of biosynthetic human growth hormone on formation and strengh of granulation tissue. Endocrinology 1987;121:1637–41.
19. Garrel DR, Gaudreau P, Zhang L, Reeves I, Brazeau P. Chronic administration of growth hormone-releasing factor increases wound strength and collagen maturation in granulation tissue. J Surg Res 1991;51:297–302.
20. Mueller RV, Spencer EM, Sommer A, et al. The role of IGF-I and IGFBP3 in wound healing. In: Spencer EM, ed. Modern concepts of insulin-like growth factors. Elsevier, 1991:185–92.
21. Ullman M, Oldfors A. Effects of growth hormone on skeletal muscle, I. Studies on normal adult rats. Acta Physiol Scand 1989;135:531–6.
22. Ullman U, Alameddine H, Skottner A, Oldfors A. Effects of growth hormone on skeletal muscle, II. Studies on regeneration and denervation in adult rats. Acta Physiol Scand 1989;135:537–43.
23. Guillemin R, Brazeau P, Bohlen P, et al. Growth hormone-releasing factor from a human pancreatic tumor that caused acromegaly. Science 1982; 218:585–7.
24. Brazeau P, Vale W, Burgus R, et al. A hypothalamic polypeptide that inhibits the secretion of pituitary growth hormone. Science 1973;179:77–9.
25. Gick GG, Zeytin FN, Brazeau P, Ling NC, Esch FS, Bancroft FC. Growth hormone-releasing factor regulates growth hormone mRNA in primary cultures of rat pituitary cells. Proc Natl Acad Sci USA 1984;81:1553–5.
26. Martinoli MG, Ouellet J, Rhéaume E, Pelletier G. Growth hormone and somatostatin gene expression in adult and aging rats as measured by quantitative in situ hybridization. Neuroendocrinology 1991;54:607–15.
27. Ghigo E, Goffi S, Nicolosi M, et al. Growth hormone (GH) responsiveness to combined administration of arginine and GH-releasing hormone does not vary with age in man. J Clin Endocrinol Metab 1990;71:1481–5.
28. Coiro V, Volpi R, Cavazzini U, et al. Restoration of normal growth hormone responsiveness to GHRH in normal aged men by infusion of low amounts of theophylline. J Gerontol 1991:M155–8.
29. Walker RF, Yank SW, Bercu BB. Robust growth hormone (GH) secretion in aged female rats co-administered GH-releasing hexapeptide (GHRP-6) and GH-releasing hormone (GHRH). Life Sci 1991;49:1499–504.
30. De Gennaro Colonna V, Zoli M, Cocchi D, et al. Reduced growth hormone releasing factor (GHRF)-like immunoreactivity and GHRF gene expression in the hypothalamus of aged rats. Peptides 1989;10:705–8.
31. Morimoto N, Kawakami F, Makino S, Chihara K, Hasegawa M, Ibata Y. Age-related changes in growth hormone-releasing factor and somatostatin in the rat hypothalamus. Neuroendocrinology 1988;47:459–64.
32. Bando H, Zhang C, Takada Y, Yamasaki R, Saito S. Impaired secretion of growth hormone-releasing hormone, growth hormone and IGF-I in elderly men. Acta Endocrinol (Copenh) 1991;124:31–6.

33. De Gennaro Colonna V, Fidone F, Cocchi D, Müller EE. The negative feedback effect of growth hormone-releasing hormone is not evident in aged rats [Abstract]. 9th Int Cong Endocrinol, Nice, France, Aug. 30–Sept. 5, 1992:20.04.011.
34. Robberecht P, Gillard M, Waelbroeck M, Camus JC, De Neef P, Christophe J. Decreased stimulation of adenylate cyclase by growth hormone-releasing factor in the anterior pituitary of old rats. Neuroendocrinology 1986;44: 429–32.
35. Abribat T, Boulanger L, Gaudreau P. Characterization of [^{125}I-Tyr^{10}] human growth hormone-releasing factor (1–44) amide binding to rat pituitary: evidence for high and low affinity classes of sites. Brain Res 1990;528:291–9.
36. Abribat T, Deslauriers N, Brazeau P, Gaudreau P. Alterations of pituitary growth hormone-releasing factor binding sites in aging rats. Endocrinology 1991;128:633–5.
37. Abribat T, Finkelstein JA, Gaudreau P. Alteration of somatostatin but not growth hormone-releasing factor pituitary binding sites in obese Zucker rats. Regul Pept 1991;36:263–70.
38. Lefrançois L, Abribat T, Boulanger L, Gaudreau P. Effect of 5′-guanylylimidodiphosphate and sulfhydryl-modifying reagents on [^{125}I-Tyr^{10}]h growth hormone-releasing factor (1–44)NH_2 binding in rat pituitary. Society for Neuroscience, New Orleans, Louisiana, Nov. 1991.
39. Parenti M, Cocchi D, Ceresoli G, Marcozzi C, Müller EE. Age-related changes of growth hormone secretory mechanisms in the rat pituitary gland. J Endocrinol 1991;131:251–7.
40. Parenti M, Dall'Ara A, Rusconi L, Cocchi D, Müller EE. Different regulation of growth hormone-releasing factor-sensitive adenylate cyclase in the anterior pituitary of young and aged rats. Endocrinology 1987;121:1649–53.
41. Sonntag WE, Boyd RL, Booze RM. Somatostatin gene expression in hypothalamus and cortex of aging male rats. Neurobiol Aging 1990;11:409–16.
42. Ge F, Tsagarakis S, Rees LH, Besser GM, Grossman A. Relationship between growth hormone-releasing hormone and somatostatin in the rat: effects of age and sex on content and in vitro release from hypothalamic explants. J Endocrinol 1989;123:53–8.
43. Sonntag WE, Gottschall PE, Meites J. Increased secretion of somatostatin-28 from hypothalamic neurons of aged rats in vitro. Brain Res 1986;380: 229–34.
44. Spik K, Sonntag WE. Increased pituitary response to somatostatin in aging male rats: relationship to somatostatin receptor number and affinity. Neuroendocrinology 1989;50:489–94.
45. Sonntag WE, Forman LJ, Miki N, et al. Effects of CNS active drugs and somatostatin antiserum on growth hormone release in young and old male rats. Neuroendocrinology 1981;33:73–8.
46. Clemmons DR, Van Wyk JJ. Factors controlling blood concentration of somatomedin C. J Clin Endocrinol Metab 1984;13:113–43.
47. Bowsher RR, Lee W, Apathy JM, O'Brien PJ, Ferguson AL, Henry DP. Measurement of insulin-like growth factor II in physiological fluids and tissues, I. An improved extraction procedure and radioimmunoassay for human and rat fluids. Endocrinology 1991;128:805–14.

48. Iranmanesh A, Lizarralde G, Veldhuis JD. Age and relative adiposity are specific negative determinants of the frequency and amplitude of growth hormone (GH) secretory bursts and the half life of endogenous GH in healthy men. J Clin Endocrinol Metab 1991;73:1081–8.
49. Hattori N, Kurahachi H, Ikekubo K, et al. Effect of sex and age on serum GH binding protein levels in normal adults. Clin Endocrinol (Oxf) 1991; 35:295–7.
50. Hammerman MR. Insulin-like growth factors and aging. Endocrinol Metab Clin 1987;16:995–1011.

13

Effects of Growth Hormone Releasing Peptides on Stimulated Growth Hormone Secretion in Old Rats

RICHARD F. WALKER, SEI-WON YANG, RYUJI MASUDA, CHENG-SHIH HU, AND BARRY B. BERCU

Relationship of Reduced GH Secretion to Senescence

Degenerative changes in body composition characterized by reduced muscle and bone mass decrease physical performance in the elderly. These decremental somatic changes correlate with reduced serum *growth hormone* (GH) concentrations. Limited availability of cadaver-derived hormone before the advent of recombinant gene technology precluded testing the functional relationship between GH deficiency and senescence. However, when bioengineered GH became available for experimentation, its administration to old men significantly increased serum IGF-I concentrations, urinary nitrogen retention, body weight gain, lean body mass, bone density, renal function, and improved quality of life (1–3). Thus, aberrant patterns of spontaneous GH secretion and low serum GH concentrations during aging (4–7) did not necessarily represent loss of GH efficacy. Furthermore, the experimental observations suggested a causal relationship between anatomical senescence and GH insufficiency (3, 4, 8, 9).

Involution of central elements of the endocrine, immune, and excretory systems, such as the pituitary, thymus, and kidney, respectively, increases the risk for terminal illness and also correlates with reduced GH secretion during aging. The thymus effect has significant biomedical relevance because of its relationship to reduced T cell-dependent immune function and immunodeficiency diseases in the elderly (10). Adult-onset involution of the thymus and subsequent decrements in T cell function may be GH related because thymic size and GH secretion are maximal at puberty and decline in parallel with advancing age (11–13). Furthermore, GH administration to immunocompromised mice increased their thymic weights (14–16), increased *interleukin-2* (IL-2) secretion (17), augmented splenocyte proliferation and *natural killer* (NK) activity, and reversed age-related defects in macrophage priming responses to interferon γ (18, 19). These data suggest that immune function in the old animals was not inherently defective, but deficient in supporting factors, one of which may be GH. Thus, insufficient GH secretion during aging may contribute to thymic involution and decreased support for T cell development/functional maturation (20–22).

Although serum GH concentrations decline during aging (4, 5), somatic responsiveness to GH seems not to be affected, since *insulin-like growth factor I* (IGF-I; somatomedin C) decrements are readily reversed by administration of exogenous GH (23, 24) or by stimulation of endogenous GH (9). This response suggests that reduced pituitary GH secretion may be responsible for age-related decrements in GH-dependent functions.

Except for two reports (24, 25), there is general consensus that GH secretion in response to *growth hormone releasing hormone* (GHRH) administration in vivo declines with advancing age (26–33). Deficits in pituitary GH concentrations are not likely the cause of age-related deficits in GH secretion, since they did not vary significantly with age when measured in humans at autopsy (34). Although pituitary GH concentrations were initially low in old rats, robust secretion of endogenous GH was sustained in response to coadministration of GH secretagogues for 45 consecutive days, and subsequently, pituitary GH reserves were restored to youthful values (35). During aging, pituitary responsiveness to GHRH is reduced (26, 29, 31, 36, 37). The progressive hyposensitivity to GHRH is accompanied by reduced hypothalamic GHRH-like immunoreactivity and reduced hypothalamic GHRH mRNA concentrations (36, 38). GHRH secretion from perifused hypothalamic tissues from old rats was also lower than from young rats (39). Thus, reduced pituitary GH secretion in old animals could result from inadequate stimuli. However, pituitary hyposensitivity to GHRH in old individuals is somewhat surprising in light of a putative decrement in the GH secretagogue. Generally, deficiencies of trophic factors lead to up-regulation or hypersensitivity of their target tissues. The answer to this paradox may lie in our recent findings that coadministration of GHRH and GH releasing hexapeptide (GHRP-6) to old rats caused GH hypersecretion, while individual administration of the peptides caused GH hyposecretion (9). The results of that study suggest that GH secretory mechanisms do not decay during aging. Rather, a reduction seems to occur in the amount of available and/or appropriate secretagogues. Since GHRH responsiveness in aging rats was restored by coadministration with GHRP-6, the data suggest that the reduced efficacy of GHRH is due to the absence or reduced concentrations of yet undefined, endogenous cosecretagogue(s). This hypothesis is supported by a recent report in which Goth et al. (40) suggested that "at least three factors operating via distinct receptors are able to increase GH secretion" and that the data "ascribe a potential physiological role for hitherto putative hypophysiotropic factors."

The decline in GHRH efficacy during aging is accompanied by deterioration of its pituitary binding sites. GHRH binding sites with high and low affinities have been identified on rat pituitary membranes (41). The high- and low-affinity sites were of low and high capacity, respectively. These binding characteristics progressively changed during aging, culminating in loss of the high-affinity site in 18-month-old rats (42). Since decreased capacity and loss of the high-affinity site correlated with reduced GH secretion during aging, it was suggested that the high-affinity binding site mediates GHRH action in somatotrophs. Prior to this GHRH binding study, we reported that GHRP-6 has two binding sites on pituitary and hypothalamic membranes (43). GHRH did not compete for these GHRP-6 binding sites, suggesting that the two peptides bind different entities. However, the GHRH and GHRP-6 binding sites may be functionally related, since somatotrophs lacking GHRH receptors are unresponsive to GHRP-6 administered alone (44).

Functional linkage has been demonstrated many times in the ability of GHRP-6 to enhance GHRH efficacy (45, 46), and we recently showed that passive immunization against GHRH reduced GHRP-6 activity approximately 90% (47). The ability of GHRP-6 to increase GHRH activity in aging rats (9) may result from facilitation of GHRH binding through cooperativity/allosteric interactions. Changes in intracellular transduction of GHRH has also been associated with aging. GHRH-activated cAMP was lower in pituitary cells from old rats than from young ones (48). However, this apparent deficit was overcome by coadministration of GHRH and GHRP-6 in vivo (9), again suggesting cooperativity between the peptides and deficiency or loss of a cosecretagogue(s) for GHRH during aging.

Another potential cause for age-related loss of GHRH efficacy in the pituitary is increased exposure to somatostatin (31–33), a peptide that inhibits GH secretion. Sonntag et al. (37) reported that in anesthetized male rats, GH release in response to administration of GHRH was significantly lower in old rats than in young rats. Furthermore, passive immunization with *somatostatin* (SRIF) antiserum increased GH release more in old rats than in young rats (32, 49), and reduced pulsatile GH secretion in aging female rats correlated with SRIF hypersecretion (50). In a clinical study, administration of the acetylcholinesterase inhibitor pyridostigmine to block endogenous SRIF release partially restored GH responsiveness to GHRH in elderly subjects (27). Direct measurement of hormones released from perifused hypothalamic explants confirmed increased SRIF secretion from tissues of old rats (51). It has been proposed that the activity of GHRP-6 is greater in vivo than in vitro because the hexapeptide is a functional antagonist of SRIF, perhaps acting at the pituitary to prevent SRIF hyperpolarization of the somatotroph (40, 52).

No matter the cause, progressive GH deficiency in aging has the potential to be significantly involved in the process of senescence. This deficiency seems to result from changes in pituitary sensitivity to GH secretagogues characterized by reduced individual activity and enhanced combined activity. Thus, the apparent biomedical relevance of GH deficiency in the elderly and the evidence suggesting that it results from secretory decrements make age-related changes in pituitary sensitivity to GH secretagogues an important biomarker for evaluating the progression of senescence at different chronological ages. Characterization of these changes in animal models and humans is simple and nonlethal, requiring only a single blood sample following a simple, provocative test for GH secretion. This biomarker also has significant potential for therapeutic intervention since the data would allow prediction of when exogenous GH supplementation or stimulated endogenous GH secretion might oppose some of the catabolic and dysfunctional changes in the elderly. Thus, in addition to its theoretical implications, sensitivity to GH secretagogues has significant practical application in geriatric medicine. Accordingly, the purpose of the present study was to determine the effects of chronic coadministration of GHRH and GHRP-6 on structural and functional senescence in rats.

Methods and Materials

Animal Husbandry

Fischer 344 female rats (2 and 18 months of age) were purchased from the NIA colony at Harlan Industries (Indianapolis, IN) and acclimated to local conditions for 6 months before being used in the present study. The older rats (24 months old) weighed between 251 g and 298 g, while the younger rats (8 months old) weighed between 201 g and 231 g when the experiment began. Female rats were chosen for study because they respond more consistently and reproducibly to GHRP-6 than do conscious males (53). Housing conditions included alternating 12-h periods of light and darkness (0600 h–1800 h), 60% ± 5% relative humidity, 72 ± 2°F ambient temperature, constantly available food (Purina Rat Chow; Ralston Purina Company, Kalamazoo, MI), and tap water. The animal facility at All Children's Hospital, in which this study was performed, is a fully accredited institutional member of the American Association of Laboratory Animal Sciences and complies with federal statutes and regulations for animal use in research.

Blood Sampling

Samples for sequential analysis of plasma GH concentrations were collected from anesthetized rats between 0900 h and 1100 h, except when surgical procedures and sampling consumed a full day (approximately 0800 h to 1800 h). When collection intervals were extended, samples were drawn alternately from the different groups so as to control for potential variation due to time of day. Sequential blood samples for analysis of temporal GH secretory profiles in anesthetized female rats were collected via cannulae that were surgically placed within the femoral artery after administration of ketamine (60 mg/kg, IM) and pentobarbital (40 mg/kg, intraperitoneal [IP]). All rats in a given experiment were surgically prepared en masse, and treatments were alternated between individual animals so that anesthesia times for all groups were approximately the same. Solutions of peptides (1 ml/kg) were administered as 30-sec intra-arterial infusions, and blood samples for GH analysis were collected at designated times into heparinized tubes.

Single-point determination of GH secretagogue activity was accomplished by administering the peptides via IP injection. Fifteen minutes later, the rats were decapitated and trunk blood was collected in heparinized centrifuge tubes for subsequent analysis of plasma GH concentrations. Rats awaiting decapitation were isolated in an adjacent room to avoid potential stress resulting from sounds and smells associated with the procedure. When administered IP, GHRP-6 is rapidly absorbed and stimulates GH release in a manner that is indistinguishable from that following intravenous (IV) administration (53), precluding the need to restrain the rats for IV administration via the tail.

Experimental Background for the Present Study

At the core of the present study are data from a pilot study in which we investigated GHRP-6 and GHRH-mediated GH release in aged rodents (9). This study was based on GHRP-6's ability to potentiate GHRH-efficacy in vivo and in vitro (45, 46, 54). Since GHRH-efficacy is reduced 50%–75% in old rats (37), we tested the effect of GHRP-6 alone and in combination with GHRH on GH release in old rats. After acclimation to local conditions, the rats were pre-anesthetized with ketamine (60 mg/kg IM), then administered pentobarbital (35 mg/kg IP) to provide stable profiles of plasma GH by eliminating the influences of endogenous SRIF (55). The right femoral vein and artery were cannulated in the anesthetized animals for peptide administration and blood collection, respectively. Blood samples (200 μl) were drawn at 0, 5, 10, 15, 20, 30, 45, and 60 min after injection of GHRP-6 (30 μg/kg) and/or GHRH (2 μg/kg), and plasma GH in the samples was estimated by radioim-

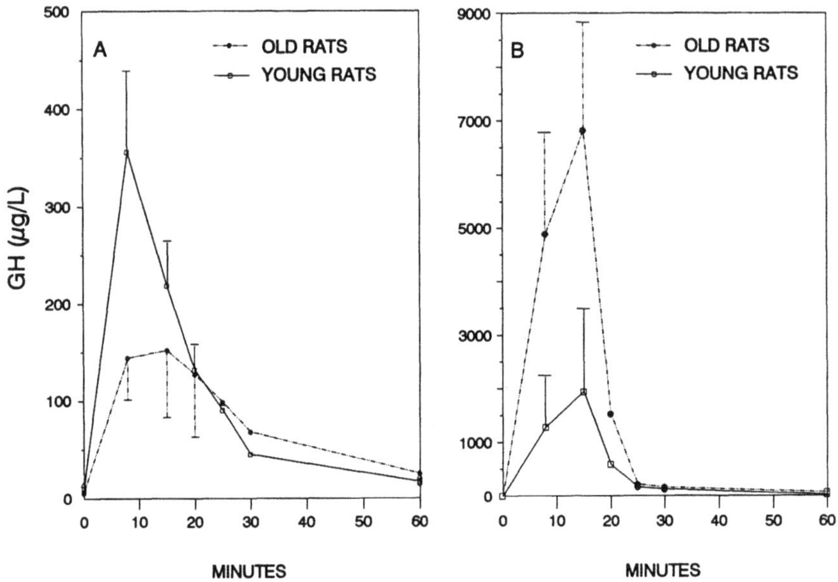

FIGURE 13.1. Comparison of GH secretion in 3- and 19-month-old anesthetized rats administered GHRP-6 (30 μg/kg) alone (*A*) or in combination (*B*) with GHRH (2 μg/kg). Note the scale difference on the ordinate of each graph. Reprinted from *Life Sciences* 49:1499–1504, 1991, with kind permission from Pergamon Press Ltd, Headington Hill Hall, Oxford OX3 OBW, UK (9).

munoassay. As seen in Figure 13.1, peak plasma GH concentrations resulting from GHRP-6 administration in old rats were approximately 60% less than in young rats. In contrast, peak plasma GH concentrations were greater in old rats than in young rats administered GHRP-6 + GHRH, suggesting that GH secretory deficits in the aged animals were due to insufficient or inappropriately transduced GH releasing stimuli.

Since stimulated endogenous GH secretion was low in rats administered GHRP-6 alone and high in rats coadministered GHRP-6 and GHRH, a functional interaction between the two peptides was suggested. This hypothesis was supported by our recent report in which young rats were passively immunized against endogenous GHRH or administered an antagonist to the pituitary GHRH receptor [N-Ac-Tyr^1, D-Arg^2]-GRF 1-29. In the first instance, GHRP-6 activity was significantly attenuated in young female rats administered GHRH antiserum or α-methyl-ρ-tyrosine to reduce endogenous GHRH concentrations (Fig. 13.2).

Preadministration of 5–50 μg/kg of [N-Ac-Tyr^1, D-Arg^2]-GRF 1-29 amide, which blocks pituitary GHRH receptors, also attenuated GHRP-6 activity. However, 150 μg/kg of [N-Ac-Tyr^1, D-Arg^2]-GRF 1-29 amide

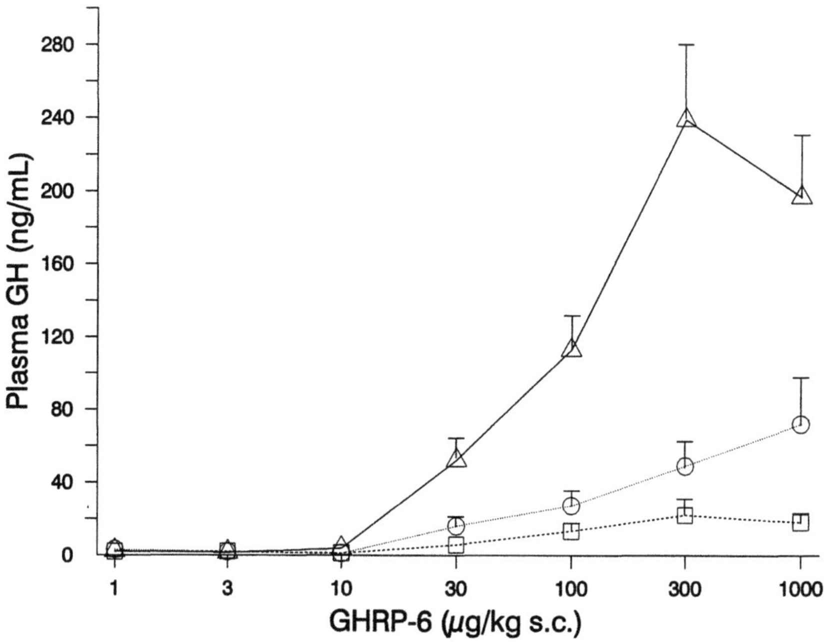

FIGURE 13.2. Dose response curve for GHRP-6 in rats pretreated wth vehicle (triangles), α-methyl-ρ-tyrosine (circles), or GHRH antiserum (squares). Values represent means ± SEM (8 rats per dose and treatment) in plasma collected 15 min after administration of GHRP-6. Reprinted with permission from Bercu, Yang, Masuda, and Walker (47), © The Endocrine Society, 1992.

potentiated GHRP-6 activity, presumably due to partial agonist activity of the GHRH receptor agonist at the higher dose (Fig. 13.3).

These data suggest that activation of pituitary GHRH receptors contributes to full expression of GHRP-6 activity in vivo. During aging, maladaptive changes in GHRH binding sites have been reported (42) and may thus contribute to reduced sensitivity of old rats to individually administered GHRP-6 (9).

Design of the Present Study

Old (24 months) and younger (8 months) rats were randomly assigned to 3 groups (12 rats per group) designated to receive daily, subcutaneous injections of GHRH and GHRP-6 (1 group of old rats) or saline (1 group of old and 1 group of younger rats). Eight-month-old rats were used rather than younger (2- or 3-month-old) adults, since animals for the present study were purchased together at 2 and 18 months of age and thus had the same history of acclimation to local conditions. More importantly, the 8-month-old rats were still young enough not to be adversely affected

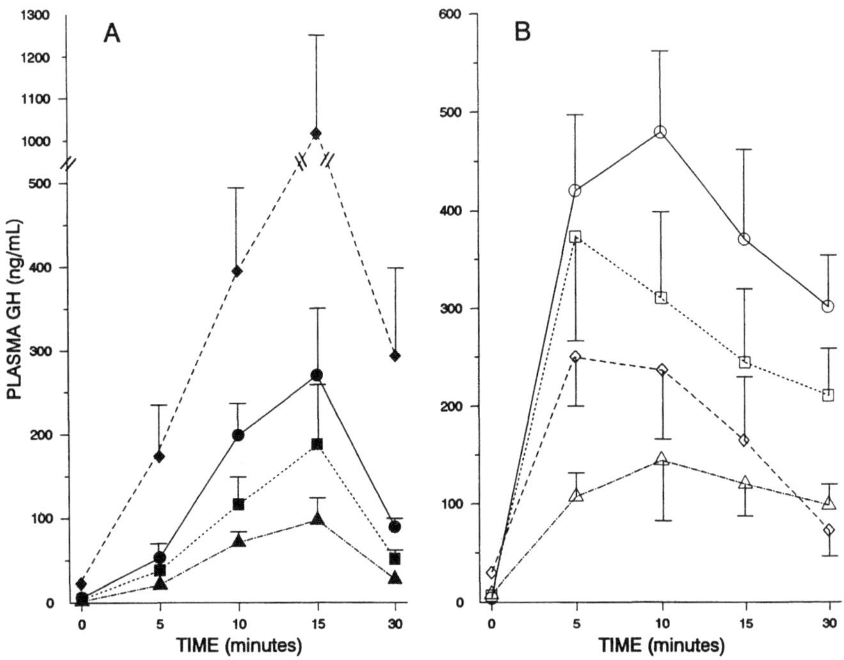

FIGURE 13.3. Effect of [N-Ac-Tyr[1], D-Arg[2]]-GRF 1-29 amide, a GH receptor antagonist, on GH release in response to (*A*) GHRP-6 (30 μg/kg)or (*B*) morphine (1.5 mg/kg). The GHRH receptor antagonist was administered in doses of 0, (circle), 5 (square), 50 (triangle), or 150 (diamond) μg/kg. Values represent mean ± SEM for 8 rats per group. GHRP-6-stimulated GH secretion as a function of time was significantly reduced ($P < 0.05$) by 50 mg/kg GHRH receptor antagonist and significantly increased ($P < 0.01$) by 150 mg/kg GHRH receptor antagonist. Morphine-stimulated GH secretion was significantly reduced ($P < 0.05$) by 50 and 150 mg/kg GHRH receptor antagonist. Reprinted with permission from Bercu, Yang, Masuda, and Walker (47), © The Endocrine Society, 1992.

by aging, and they displayed no overt signs of structural and functional senescence. Solutions (1 ml/kg) of GHRP-6 and rat GHRH (300 μg/kg and 3 μg/kg, respectively) or saline were administered subcutaneously (sc) at approximately 0900 h for 60 consecutive days. In a preliminary dose-range study, 300 μg/kg GHRP-6 was the ED_{100} in conscious female rats (47). The dose of GHRH (3 μg/kg) was selected from published data showing that the peptide was efficacious when administered at a dose of 10 μg/kg, sc (56). The lower dose was used in the present study because of the known synergistic activity of GHRH and GHRP-6 (46, 57).

An interim analysis of coadministered GHRH and GHRP-6 efficacy in old rats was performed after 4 consecutive days of treatment. A saline-treated group of old rats was also used to corroborate prior findings of hypersensitivity of naive old rats to the coadministered peptides. On the fifth day, rats (6 rats/group) were anesthetized and the right femoral artery was cannulated. Peptides (GHRP-6, 30 μg/kg; GHRH, 1 μg/kg) were administered, and blood samples (approximately 250 μl) were collected in heparinized tubes before and at selected intervals after administration of the GH secretagogues. The lower doses of GHRP-6 and GHRH were used because pentobarbital anesthesia suppresses SRIF secretion and thereby enhances the response to pituitary stimulation (55). The rats were exsanguinated during the last blood sample and their pituitary glands were rapidly removed, weighed, and frozen on dry ice for subsequent estimation of GH concentrations.

Little mice (lit/lit), which lack receptors for GHRH (44), and *dwarf rats* (dw/dw), which have defective GHRH-activated *cyclic adenosylmonophosphate* (cAMP) systems (58), were purchased from Jackson Laboratories (Bar Harbor, ME) and Harlan Industries (Indianapolis, IN), respectively, to compare the acute effects of coadministered GHRH and GHRP-6 on GH secretion. These particular mutant strains were selected because GHRH receptor and second messenger defects have been reported in old rats (42, 48). It was of interest to determine if the mutant rodents, like old rats, were also hyperresponsive to the coadministered GH secretagogues. Because of their small size, the little mice could not be fitted with cannulae for sequential blood sampling. Thus, they were decapitated 15 min after IP administration of the peptides, and trunk blood was collected in heparinized tubes. Peak GH plasma concentrations were subsequently estimated by radioimmunoassay. Responses of the naive dwarf rats were determined in repeated blood samples taken from the anesthetized animals via cannulae, as described above for use in old rats.

The remaining old rats and their younger controls received daily subcutaneous injections of saline or GHRP-6 and GHRH for a total period of 60 consecutive days. The rats were weighed at regular intervals, and lean body mass was determined approximately one week before the end of treatment. Lean body mass was estimated using the ratio of water displacement volume to dry body weight. The rats were immersed in a water-filled tank containing a series of concentric rings that acted as baffles to reduce surface turbulence resulting from swimming. Thus, at the tank perimeter, the water rose without turbulence and spilled through a channel into a collecting vessel. This apparatus allowed accurate and reproducible measurement of water displaced by each rat, which was then weighed (1 cc H_2O = 1 g) and divided by the animals dry body weight (vol H_2O displaced/body weight) to provide a *body composition index* (BCI) or estimate of lean body mass. Since fat is less dense than muscle, decreasing BCI indicated increasing lean body mass. A glucose tolerance test was performed after 50 days of treatment to determine the effects of daily, episodic release of endogenous GH on carbohydrate metabolism in old rats. A 30% aqueous solution of glucose (1 g/kg; 3 ml/kg) was administered as an intraperitoneal injection to unanesthetized rats from which food was withheld for approximately 8 h. Blood samples were taken from a tail vein before the glucose injection and at 15-min intervals thereafter. Blood glucose concentrations were determined using an ExacTech Blood Glucose Sensor (MediSense, Inc., Cambridge, MA). On the final day of treatment, all rats were anesthetized with ketamine/pentobarbital and administered an IP injection of GHRH (1 μg/kg) and GHRP-6 (30 μg/kg). The rats were decapitated 15 min after coadministration of the peptides (time to peak stimulated GH release), and trunk blood was collected in heparinized centrifuge tubes for subsequent analysis of plasma hormone concentrations. Each rat was then necropsied, and selected organs were removed and weighed. All procedures in this experiment were described in a protocol that was approved by the University of South Florida Institutional Animal Care and Use Committee.

Quantitation of Hormone Concentrations

Blood samples were centrifuged and plasma was collected for *radioimmunoassay* (RIA) analysis of GH, IGF-I, prolactin (PRL), insulin, and corticosterone concentrations. Pituitaries were homogenized in 0.1 N NaOH/saline, diluted, and aliquots were analyzed by RIA. GH and PRL concentrations were estimated in duplicate by RIA and expressed in terms of the reference preparations rGH-RP-2 and rPRL-RP-3, respectively, from the NIDDK. Radiolabeled rat GH and PRL were purchased from Chemicon International, Inc. (Temecula, CA), and precipitating antibody (goat anti-monkey γγ [P-4] and goat anti-rabbit γγ for GH and PRL, respectively) were purchased from Antibodies, Inc. (Davis, CA) and used at a concentration of 1:10. Intra-assay and interassay coefficients of variation were <5% and <8%, respectively. Corticosterone concentrations were estimated using a commercially available rat corticosterone ^{125}I kit (ICN Biomedical, Inc, Diagnostics Division, Costa Mesa CA). IGF-I concentrations were estimated after acid ethanol extraction from plasma using methods provided in a commercially available kit (Nichols Institute, Diagnostics, San Juan Capistrano, CA). Plasma insulin concentrations were determined according to the method of Yalow and Berson (59), using Lilly Pork Insulin PJ 5682 as a reference standard. Serum *free fatty acids* (FFA) and plasma cholesterol concentrations were measured using reagents purchased in kit form (Sigma Chemical Corporations, St. Louis, MO, cholesterol kit #352; Waco Laboratories, Waco, TX, FFA kit #990-75401).

Chemicals

GHRP-6 (His-D-Trp-Ala-Trp-D-Phe-Lys-NH_2) and rat GHRH were purchased from Peninsula Laboratories (Belmont, CA). Controls received saline administered in equal volume and by the same route as the peptides. GHRP-6 is a synthetic GH-releasing peptide with less efficacy than GHRH in vitro, but with comparable efficacy to GHRH in vivo (57). Its mechanism of action differs from GHRH. GHRP-6 was synthesized as a potential therapeutic agent following the discovery in 1977 that [D-Trp] methionine enkephalin amide released GH, but not PRL or other hormones from rat anterior pituitary glands in vitro (57). Trial-and-error substitutions aided by conformational energy calculations culminated in the discovery of GHRP-6, a potent and specific GH secretagogue in all species tested that bears no structural resemblence to enkephalins or GHRH. GHRP-6 specifically binds two sites in rat pituitary and hypothalamus (43). The high-affinity site (kd approximately 10 nM) represents the bioactive site as determined by in vitro studies in which GHRP-6 released GH from dispersed rat pituitary cells at an ED_{50} concentration of 3 nM. Different GH releasing mechanisms for GHRP-6 and GHRH are suggested by the facts that the peptides lack structural homology, have different binding sites, utilize different intracellular second messengers (60), and are synergistic. However, passive immunization against GHRH blocks GHRP-6-mediated GH release, and GHRP-6 has poor activity in dwarf mice lacking GHRH receptors, suggesting that GHRP-6 activity in vivo requires GHRH at some level (47). Thus, cooperativity may exist between the two molecules.

Data Analysis

Pituitary and plasma hormone concentrations, as well as tissue weights and body composition indices, were compared among the groups of young and old rats using analysis of variance with Duncan's multiple range test. Changes in peptide-stimulated plasma GH concentrations, plasma glucose, and body weight change over time were compared using repeated measures analysis. Results were expressed as means ± standard errors of the means. A P value of <0.05 was used to define statistically significant differences.

Results

Old rats were coadministered GHRH and GHRP-6 or saline, and young rats were administered saline by subcutaneous injection for 4 consecutive days. The data presented in Figure 13.4 compare plasma GH and IGF-I concentrations immediately after intravenous coadministration of a challenge dose of GHRH and GHRP-6 to rats in each of the 3 groups on the fifth day of treatment. As seen in the main portion of the figure, plasma GH concentrations following the provocative test were significantly greater ($P < 0.05$) in old rats pretreated with saline for 4 consecutive days than in age-matched rats pretreated with GHRH and GHRP-6 or young rats

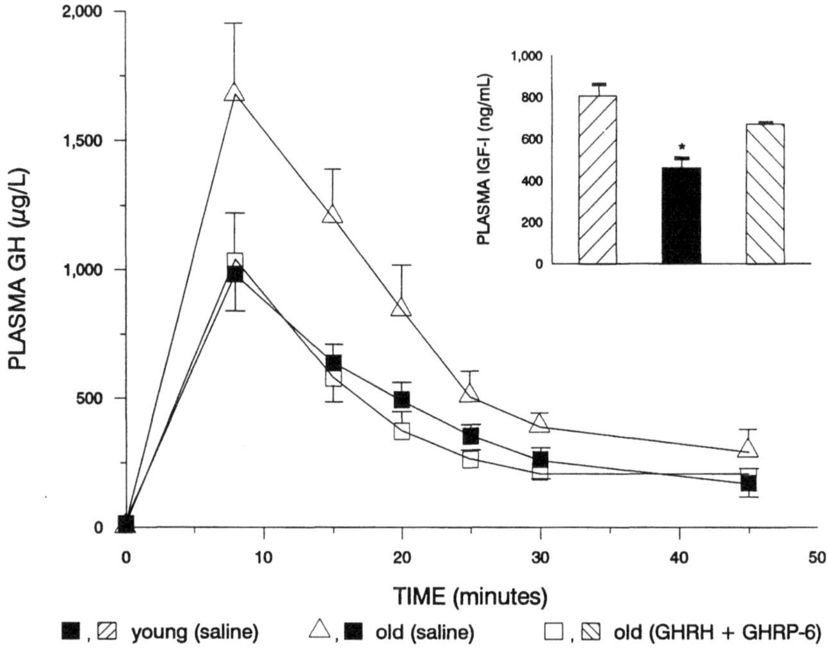

FIGURE 13.4. Plasma GH secretory profiles in old rats coadministered GHRH and GHRP-6 and in old rats and young rats administered saline for 5 consecutive days. Inset compares plasma IGF-I concentrations for each group. (N = 6 rats/group; * = significantly different from values for other groups.)

pretreated with saline. Plasma GH concentrations after the provocative test were not significantly different in the latter 2 groups. These data confirm our previous observation of GH hypersecretion in old rats naive to the synergistic effects of coadministered GHRH and GHRP-6 (9) and show that daily administration of the GH secretagogue combination rapidly down-regulated subsequent responses to youthful levels. The inset in Figure 13.4 compares plasma IGF-I concentrations in the 3 groups on the fifth day of treatment. The mean IGF-I concentration in old rats pretreated with GHRH and GHRP-6 was between those for the other 2 groups and significantly greater than that for old rats pretreated with saline, indicating that peptide-stimulated endogenous GH was bioactive. The data presented in Figure 13.5 show that pituitary GH concentrations were slightly increased after 4 consecutive days of GHRH and GHRP-6 coadministration, although the differences were not statistically significant.

Since maladaptive changes in GHRH receptors and GHRH-mediated cAMP have been reported in aging rats (42, 48), it was of interest to determine whether mutant animals with similar signal transduction defects were hyperresponsive to coadministered GHRH and GHRP-6 like naive

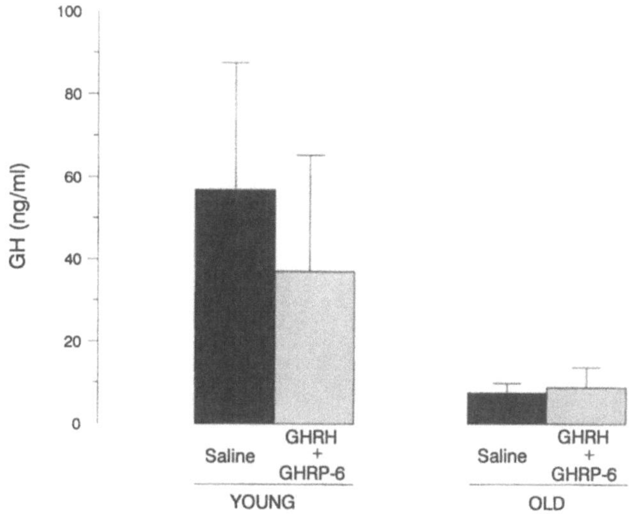

FIGURE 13.5. Pituitary GH concentrations in young and old rats after coadministration of GHRH and GHRP-6 or saline for 5 consecutive days. (N = 6 rats/group.) Differences between, but not within, age groups are significant ($P <$ 0.05).

old rats. Only 2 of 6 homozygous lit/lit mice lacking GHRH receptors (44) released GH above the detection limits of our assay, while GH secretion was potentiated in the heterozygous mice with normal growth phenotypes (mean ± SEM peak plasma GH; 1.65 ± 0.7 ng/ml vs. 316 ± 89.2 ng/ml, respectively). As seen in Figure 13.6, homozygous dw/dw rats with defective cAMP second messenger systems (58) were also hyporesponsive to coadministered GHRH and GHRP-6 compared to heterozygous (dw/?) rats of the same strain with normal phenotypes. These data suggest that previously reported, age-related, decremental changes in pituitary GHRH receptors (42) and GHRH-activated cAMP (48) are not functionally relevant to GH hyposecretion in old rats since coadministered GHRH and GHRP-6 caused GH hypersecretion. In contrast, receptor and second messenger defects in mutants were associated with GH hyposecretion rather than hypersecretion in response to the coadministered GH secretagogues.

Mean body weight change and BCI were monitored in the experimental subjects after approximately 40 days' coadministration of GHRH and GHRP-6 or saline. The data in Figure 13.7 show that mean body weight change from that recorded on reference day 1 was increased in old rats coadministered the GH secretagogues compared to age-matched, saline-treated rats whose mean body weight change was decreased. The positive body weight change in GH secretagogue-treated old rats was accompanied

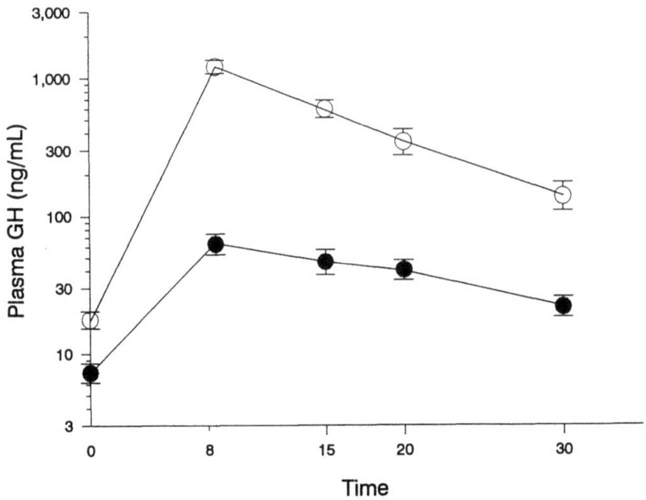

FIGURE 13.6. Comparison of temporal patterns of GH secretion in homozygous (dw/dw, closed circles) and heterozygous (dw/?, open circles) female Lewis dwarf rats coadministered GHRH and GHRP-6. Plasma GH concentrations in dwarf rats were significantly lower ($P < 0.001$) than those in normal-statured rats. (N = 6 rats/group.)

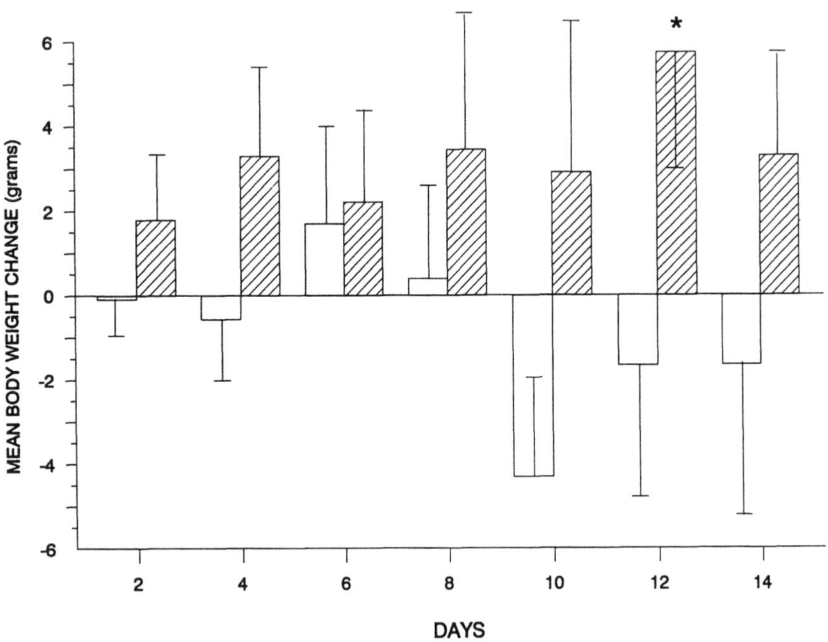

FIGURE 13.7. Changes in mean body weight in old rats chronically administered saline (open bars) or GHRH and GHRP-6 (hatched bars). Each value is compared to a reference taken on day 40 when the body weight comparisons were initiated. (* = $P < 0.05$.)

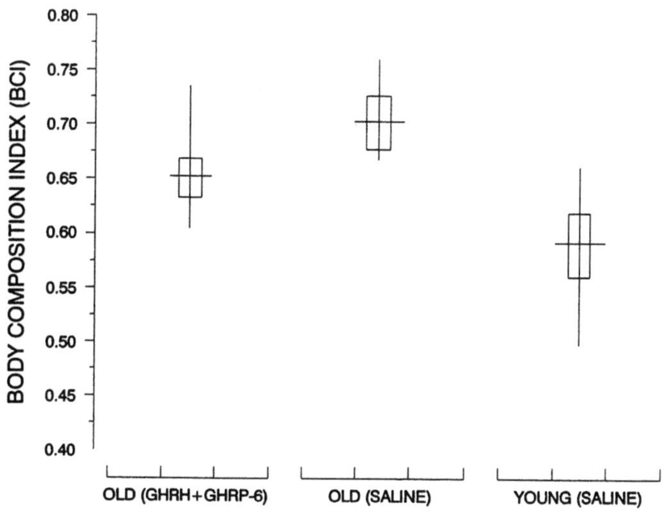

FIGURE 13.8. Comparison of body composition index in old rats administered saline or coadministered GHRH and GHRP-6 and young rats administered saline. BCI value of young rats is significantly less ($P < 0.05$) than that of old rats. (Horizontal line = mean, vertical line = range, and box = SEM determined for 6 rats/group.)

by a slight reduction in BCI, suggesting an increase in somatic density of the population. However, the difference between the groups was not statistically significant, whereas BCI of young rats was significantly lower ($P < 0.01$) than that of the old rats (Fig. 13.8).

Several differences in organ weights of GH secretagogue-treated and saline-treated old rats were detected at necropsy. The most significant involved the pituitary gland, which was enlarged and adenomatous in the majority (5 of 6) of saline-treated rats. In contrast, macroscopic adenomas were infrequent (1 in 6) in pituitaries from old rats coadministered GHRH and GHRP-6 for 60 consecutive days, and mean pituitary weight in this group was comparable to that of young rats (Table 13.1). In addition, GH secretagogue-treated old rats had higher thymus weights and lower adrenal and renal weights than age-matched, saline-treated controls. Kidneys of old saline-treated rats were enlarged and had a brownish tint that was frequently mottled with greyish white spots. The kidney surfaces were rough and granular compared to the smooth kidney surfaces observed in young rats or old rats coadministered GHRH and GHRP-6.

The data presented in Table 13.2 express the effects of chronic coadministration of GHRH and GHRP-6 on selected humoral factors in old female rats. Peak plasma GH values 15 min after IP administration of GHRH and GHRP-6 on day 60 of treatment were comparable in old

TABLE 13.1. Effects of chronic coadministration of GHRH and GHRP-6 for 60 days on selected structural parameters in old Fischer 344 female rats.

Structural parameter	Young rats (saline)	Old rats (saline)	Old rats (GHRH/GHRP-6)
Pituitary weight (mg)	11.3 ± 3.1	58.0 ± 11.1*	15.6 ± 6.9
Pituitary tumor incidence (%)	0	83*	17
Thymus : body weight ratio	0.87 ± 0.07	0.50 ± 0.05*	0.73 ± 0.03
Thymus weight (mg)	191 ± 24	140 ± 18.5*	208 ± 10.5
Adrenal weight (mg)	27.0 ± 0.5	33.2 ± 1.2*	29.0 ± 1.4
Kidney : body weight ratio	3.5 ± 0.07	4.8 ± 0.4*	3.7 ± 0.08
Kidney weight (mg)	755.4 ± 17.0	1289 ± 106*	1043 ± 22
Body weight (g)	215 ± 9*	273 ± 12	276 ± 10

Note: Values represent means ± standard errors of the means for duplicate determinations from 6 rats per group.
* Significantly different from at least one other group ($P < 0.05$).

TABLE 13.2. Effects of chronic coadministration of GHRH and GHRP-6 for 60 days on selected endocrine and metabolic parameters in old Fischer 344 female rats.

Endocrine parameter	Young (saline)	Old (saline)	Old (GHRH/GHRP-6)
Plasma GH (ng/ml)	683 ± 213	956 ± 225*	402 ± 127
Pituitary GH (μg/mg wet wt)	248 ± 35	39 ± 15*	275 ± 38
Plasma IGF-I (ng/ml)	839 ± 31	542 ± 32*	780 ± 24
Plasma prolactin (ng/ml)	22 ± 9	375 ± 121*	54 ± 8
Pituitary prolactin (ng/mg wet wt)	334 ± 65	169.5 ± 68*	422 ± 69
Plasma corticosterone (ng/ml)	806.2 ± 32.7*	592.2 ± 67.3	653.3 ± 21.8
Plasma cholesterol (mg/dl)	74.0 ± 1.1	199.8 ± 33.7*	108.7 ± 11.1
Plasma insulin (mIU/ml)	9.5 ± 0.8	15.6 ± 1.9*	13.7 ± 1.1
Blood glucose (mg/dl)	69.4 ± 5	70.0 ± 12	67.2 ± 8

Note: Values represent means ± standard errors of the means for duplicate determinations from 6 rats per group.
* Significantly different from at least one other group ($P < 0.05$).

peptide-pretreated rats and young saline-pretreated rats, but significantly lower ($P < 0.05$) than those in old saline-pretreated rats. These findings are consistent with the interim analysis and our original report (9) that naive old rats are hypersensitive to coadministered GHRH and GHRP-6. The data also show that aging did not adversely affect the bioavailability of these peptides when they were administered subcutaneously. The data in Table 13.2 also show that plasma IGF-I concentrations were significantly higher in old rats administered the GH secretagogues than in saline-treated, age-matched controls, and comparable to young rats. Thus, the mechanisms for chronic IGF-I production and secretion seemed not to deteriorate during aging, and daily coadministration of GHRH and GHRP-6 caused IGF-I concentrations to be reestablished and to sustain youthful levels, presumably in response to daily episodes of stimulated endogenous GH secretion.

Coincident with reduced pituitary weight and tumor incidence were increased pituitary GH concentrations to youthful values. The values were approximately 7 times higher than those in pituitaries from saline-treated old rats, and since daily GH secretion probably occurred, the data suggest that chronic administration of the GH secretagogues stimulated pituitary GH synthesis as well as release. Pituitary PRL concentrations were significantly higher in old rats administered the combination of GH secretagogues compared to those administered saline, and elevated concentrations of plasma PRL were lowered to values of young rats. These changes corresponded with reduced pituitary weight and adenoma incidences, suggesting that chronic coadministration of GHRH and GHRP-6 reduced lactotroph secretory activity.

Fasting blood glucose levels were comparable in old and young rats. However, plasma insulin was significantly higher in both groups of old rats compared to the group of young rats. Furthermore, relative glucose intolerance was observed in old rats coadministered GHRH and GHRP-6 compared to saline-treated old and young rats, suggesting that daily episodes of stimulated GH secretion were diabetogenic and that age-related deficits in pancreatic function/insulin resistance were exacerbated by treatment (Fig. 13.9). Plasma cholesterol was significantly reduced in old rats coadministered GHRH and GHRP-6, while FFA were slightly increased. Plasma corticosterone concentrations in the group of young rats were significantly greater than those in either group of aged rats (Table 13.2).

Discussion

The present study was designed to investigate the relationship between endogenous GH and degenerative changes in form and function associated with aging. The available literature attests to the biomedical relevance of

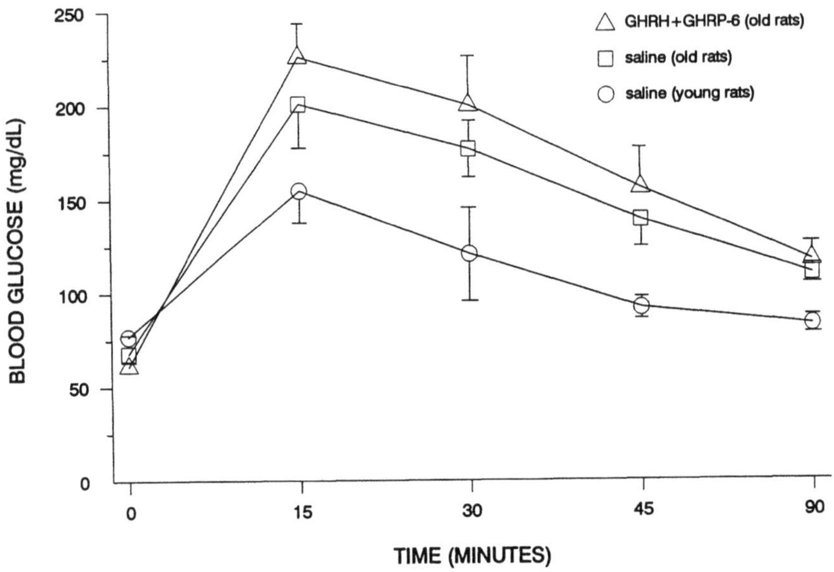

FIGURE 13.9. Glucose tolerance tests in old rats coadministered GHRH and GHRP-6 (triangles) or old (squares) and young (circles) rats administered saline for 50 consecutive days. Blood glucose clearance after intraperitoneal administration of glucose (30% solution, 1 g/kg body weight) was significantly slower ($P < 0.05$) in old rats coadministered the GH secretagogues than in young rats.

such an effort. Although patterns of spontaneous GH secretion deteriorate and serum GH concentrations decline during aging (4, 5, 6), somatic responsiveness to GH seems to remain intact, since IGF-I (somatomedin C) decrements and certain physical/physiological deficits are readily reversed by administration of exogenous GH (24). Our study was framed within this background literature and designed to determine the effects of GH secretagogues on endogenous GH secretion in aging rats. Hypothetically, reversal or prevention of certain aspects of senescence with endogenous GH would be more desirable than with exogenous GH, since a more physiological pattern of hormone secretion could be simulated by stimulating the pituitary, and more of the GH neuroendocrine axis would be activated. Administration of exogenous GH would suppress GH axial function, further promoting atrophy of the pituitary through negative feedback influence, whereas stimulation of the pituitary to increase endogenous GH output might improve pituitary function. Thus, a pharmacological agent(s) with the potential to improve endogenous GH secretion in the elderly, opposing some of the catabolic and dysfunctional effects of aging, would have significant biomedical relevance and practical application in geriatric medicine (61).

The significant finding of this study was that age-related deficits in GH secretion and corresponding degeneration of certain structures and functions were, at least in part, reversed by pharmacological stimulation of the pituitary with the synergistic GH secretagogues GHRH and GHRP-6. Despite low GH concentrations in pituitaries of the old rats, sufficient reserve existed to sustain daily responses to the potent pharmacological stimuli. Although actual measurements of GH secretion were not made each day, interim and terminal measurements in association with beneficial structural and functional changes in the old treated rats provided circumstantial, but compelling, evidence to support the hypothesis that GH deficiency contributes to many maladaptive, age-related changes.

After the initial hypersecretory responses to coadministered GHRH and GHRP-6, subsequent stimulated plasma GH concentrations in old rats were normalized to youthful levels and correlated with significantly increased plasma concentrations of IGF-I. This correlation may be functionally relevant because IGF-I is a negative feedback modulator of GH release (62). Furthermore, the initial, transient hypersensitivity to GHRH and GHRP-6 suggested that deficits in GH regulatory hormones other than GHRH contribute to the age-related decline in GH secretion. Deficits in pituitary GH concentrations are not likely the cause of low GH secretion in the elderly because robust secretion of endogenous GH was sustained in response to coadministration of GH secretagogues for 60 consecutive days despite low values at start of the study. Subsequently, pituitary GH reserves were restored to youthful values. The previously reported progressive and age-related hyposensitivity to GHRH is accompanied by reduced hypothalamic GHRH-like immunoreactivity, secretion, and hypothalamic GHRH mRNA concentrations (36, 38, 39). Thus, reduced pituitary GH secretion in old animals could result from inadequate stimuli. However, pituitary hyposensitivity to GHRH in old individuals is somewhat surprising in light of a putative decrement in the GH secretagogue. Generally, deficiencies of trophic factors lead to upregulation or hypersensitivity of their target tissues. The answer to this paradox may be that there is more than one physiologically relevant GH secretagogue. Since GHRH responsiveness in aging rats was restored by coadministration with GHRP-6, the data suggest that the reduced efficacy of GHRH was due to the absence or reduced concentrations of one or more as yet undefined endogenous cosecretagogue(s). This hypothesis is supported by a recent report in which Goth et al. (40) suggested that "at least three factors operating via distinct receptors are able to increase GH secretion" and that (their data) "ascribe a potential physiological role for hitherto putative hypophysiotropic factors."

Age-related, maladaptive changes in GHRH receptors (42) and GHRH-activated cAMP (48) have been reported. Since coadministration of GHRH and GHRP-6 caused robust GH secretion presumably in the presence of one or both of these defects (9), they are either irrelevant to GH secretion or reflect inadequate/inappropriate pituitary stimulation in old rats. The first alternative seems unlikely since the coadministered GH secretagogues were relatively ineffective in mutants with permanent GHRH receptor or second messenger defects. Because of the immediate efficacy of the secretagogue combination in old rats, it is more likely that receptor (42) and/or second messenger decrements (26, 48) in tissues from these animals were more apparent than real, reflecting a natural history of inadequate or inappropriate stimulation prior to measurement in vitro.

Support for the hypothesis of age-related, inadequate stimulation also derives from changes in pituitary GH concentrations during the course of the present study. Low pituitary GH concentrations were most likely not due to defects in somatotroph gene transcription or translation since mean pituitary GH concentrations were slightly higher than controls after only 4 consecutive days of GHRH and GHRP-6 coadministration and significantly higher, reaching youthful values, after 60 consecutive days of treatment. Remarkably, this increase in GH concentrations was associated with a statistically significant reduction in pituitary weight and tumor incidence. The data are insufficient to identify a mechanism for these pituitary effects; however, they may reflect the potential for phenotype interconversion of mammotrophs and somatotrophs, as proposed by Frawley and Boockfor (63). During aging reduced hypothalamic catecholamines (64, 65) may diminish the release of GH secretagogues while lowering the pituitary concentration of dopamine, which inhibits PRL secretion. This change in catecholamine tone could increase the activity of mammotrophs, shifting the pool of somatomammotrophs and/or somatotrophs to the mammotroph phenotype, promoting hyperprolactinemia and possibly mammotroph hypertrophy, hyperplasia, and/or adenoma formation. The significant increase in pituitary PRL, as well as reductions in pituitary weights, adenomas, and plasma PRL, in the present study suggests that daily somatotroph stimulation by coadministration of GHRH and GHRP-6 opposed this trend, shifting the equilibrium away from the mammotroph phenotype and reducing their activity, number, and/or size. Because of the animals' advanced age when treatment was begun and the fact that approximately 85% of controls had pituitary tumors, the peptides were probably therapeutic, causing regression, rather than prevention, of the lesions. This finding may have novel clinical relevance since the present treatments for lactotroph adenomas are surgical or pharmacological, involving dopamine administration (66).

Finally, the results of this study demonstrate that pharmacological stimulation of endogenous GH secretion has beneficial effects on structure and function in old rats. The data show that, like exogenous bioengineered GH (2), endogenous GH concentrations elevated for as little as 90 min daily significantly affected several aspects of structure and function. Body weight and lean body mass were maintained and slightly increased, while they decreased in age-matched, saline-treated controls. Although lean body mass was not significantly increased, more notable results may have occurred with a longer duration of treatment. On the other hand, thymus weight increased significantly, reflecting the trophic effect of GH on this tissue, consistent with the observations of others (17). Since GH stimulates T cell proliferation, which is significantly reduced during aging (10, 18, 20), and shifts the ratio of helper and killer T cells (67), coadministered GHRH and GHRP-6 may cause thymic recrudescence and immunoreconstitution in old animals and thus have a potential role in treatment of certain age-related diseases.

Adrenal weights were significantly reduced in old rats administered GHRH and GHRP-6 by a mechanism that may be related to pituitary changes because mammotrophic tumors are sometimes also adrenocorticotrophic (15). The changes are probably beneficial since adrenal hypertrophy and dysfunction are maladaptive. However, plasma corticosterone concentrations were not restored to youthful values in the old animals, suggesting that dynamic parameters of adrenocortical hormone secretion were not improved by treatment. Although renal atrophy commonly occurs during aging, kidney enlargement has been reported in old rats (68). The large kidneys in saline-treated old rats are attributed to renal infection and inflammation that occur in most Fischer 344 rats by 2 years of age (69). This hypothesis is supported by the rough, granular surface of kidneys from saline-treated old rats, suggesting a polycystic cortical nephropathy. Although unconfirmed histologically, reduced kidney weight in the treated rats possibly reflected improved kidney function and health. Reduced plasma cholesterol in rats coadministered GHRH and GHRP-6 suggests a beneficial effect of stimulated endogenous GH secretion on lipid metabolism, although it may also result from secondary factors related to improved renal function. The only negative effect of coadministered GHRH and GHRP-6 in old rats was a relatively mild glucose intolerance that accompanied elevated insulin levels. Glucose intolerance in old rats coadministered GHRH and GHRP-6 is attributed to SRIF secretion, which occurs in response to GHRH and GHRP-6 coadministration (53), and the increased pancreatic sensitivity to SRIF in old rats that reduces stimulated insulin secretion (70).

In conclusion, this study has shown that old rats derived certain benefits from daily coadministration of GHRH and GHRP-6. Hypothetically, reversal or prevention of certain aspects of senescence with endogenous GH would be more desirable than with exogenous GH, since a more physiological pattern of hormone secretion could be simulated by stimulating the pituitary, and more of the GH neuroendocrine axis would be reactivated. Thus, pharmacological agent(s) with the potential to improve endogenous GH secretion in the elderly, opposing some of the catabolic and dysfunctional effects of aging, would have significant biomedical relevance and practical application in geriatric medicine.

Summary

The purpose of this study was to determine the effects of chronically coadministered GHRH and GHRP-6, an efficacious combination of GH secretagogues, on structural and functional senescence in rats. Interim analysis showed that GH hypersecretion in 24-month-old rats coadministered GHRH and GHRP-6 was attenuated within the first 5 days of treatment, correlating with a significant increase in plasma IGF-I, a negative feedback modulator of GH secretion. Coadministration of GHRH and GHRP-6 to the old rats for 60 consecutive days was associated with significantly lower pituitary, adrenal, and kidney weights, a reduced incidence of pituitary adenomas, and higher thymus weights compared to saline-treated, age-matched controls. Structural changes were accompanied by beneficial endocrine and metabolic changes, including significantly increased concentrations of pituitary GH, plasma IGF-I, and reduced concentrations of plasma PRL and cholesterol. However, plasma corticosterone was lower and insulin was higher in old rats than in young rats, suggesting that age-related adrenal and pancreatic dysfunctions were not reversed by daily administration of GHRH and GHRP-6. In conclusion, the results of this study suggest that low pituitary and plasma GH concentrations in old age can be attributed to inappropriate or inadequate trophic signals rather than to inherent defects in pituitary function. The data also suggest that pharmacological stimulation of endogenous GH secretion may correct certain maladaptive changes of aging.

References

1. Kaiser FE, Silver AJ, Morley JE. The effect of recombinant human growth hormone on malnourished older individuals. J Am Geriatr Soc 1991;39:235–40.
2. Marcus R, Butterfield G, Holloway L, et al. Effects of short term administration of recombinant human growth hormone to elderly people. J Clin Endocrinol Metab 1990;70:519–27.
3. Rudman D, Feller AG, Hoskote SN, et al. Effects of human growth hormone in men over 60 years old. N Engl J Med 1990;323:1–6.
4. Rudman D, Kutner MH, Rogers CM, Lubin MF, Fleming GA, Bain RP. Impaired growth hormone secretion in the adult population. J Clin Invest 1981;67:1361–9.
5. Finkelstein JW, Roffwarg HP, Boyar RM, Kream J, Hellman L. Age-related change in the twenty-four-hour spontaneous secretion of growth hormone. J Clin Endocrinol Metab 1972;35:665–70.

6. Ho KY, Evans WS, Blizzard RM, et al. Effects of sex and age on the 24-hour profile of growth hormone secretion in man: importance of endogenous estradiol concentrations. J Clin Endocrinol Metab 1987;64:51–7.
7. Sonntag WE, Meites J. Decline in growth hormone secretion in aging animals and man. In: Everett AV, Walton JR, eds. Regulation of neuroendocrine aging. Basel: Karger AG, 1988:111–24.
8. Root AW, Oski FA. Effects of human growth hormone in elderly males. J Gerontol 1969;24:97–104.
9. Walker RF, Yang S-W, Bercu BB. Robust growth hormone (GH) secretion in aged female rats co-administered GH-releasing hexapeptide (GHRP-6) and GH-releasing hormone (GHRH). Life Sci 1991;49:1499–504.
10. Kay MMB. Effect of age on T-cell differentiation. Federation Proc 1978; 37:1241–4.
11. Denckla WD. Systems analysis of possible mechanisms of mammalian aging. Mech Ageing Dev 1977;6:143–53.
12. Finkelstein JW, Roffwarg HP, Boyar RM, Kream J, Hellman L. Age-related change in the twenty-four-hour spontaneous secretion of growth hormone. J Clin Endocrinol Metab 1972;35:665–70.
13. Weksler ME. The immune system and the aging process in man. Proc Soc Exp Biol Med 1980;165:200–35.
14. Shrewsbury MM, Reinhardt WO. Effect of pituitary growth hormone on lymphatic tissues, thoracic duct lymph flow, lymph protein and lymphocyte output in the rat. Endocrinology 1959;65:858–60.
15. Takemoto H, Yokoro K, Furth J, Cohen AI. Adrenotropic activity of mammo-somatotropic tumors in rats and mice. Cancer Res 1962;22:917–24.
16. Pierpaoli W, Baroni C, Fabris N, Sorkin E. Immunology 1969;16:217–30.
17. Kelley KW, Brief S, Westly JH, et al. GH_3 pituitary adenoma cells can reverse thymic aging in rats. Proc Natl Acad Sci 1986;83:5663–7.
18. Davila DR, Brief S, Simon J, Hammer RE, Brinster RL, Kelly KW. Role of growth hormone in regulating T-dependent immune events in aged, nude, and transgenic rodents. J Neurosci Res 1987;18:108–16.
19. Davila DR, Edwards CK, Arkins S, Simon J, Kelly K. Interferon-gamma-induced priming for secretion of superoxide anion and tumor necrosis factor/alpha declines in macrophages from aged rats. FASEB J 1990;4:2906–11.
20. Kay M. Immunological aspects of aging: early changes in thymic activity. Mech Ageing Dev 1984;28:193–218.
21. Fabris N, Pierpaoli W, Sorkin E. Lymphocytes, hormones and ageing. Nature 1972;240:557–9.
22. Price GB, Makinodan T. Immunologic deficiencies in senescence, II. Characterization of extrinsic deficiencies. J Immunol 1972;198:413–7.
23. Johanson AJ, Blizzard RM. Low somatomedin-C levels in older men rise in response to growth hormone administration. Johns Hopkins Med J 1981; 149:115–9.
24. Pavlov EP, Harman SM, Merriam GR, Gelato MC, Blackman MR. Responses of growth hormone (GH) and somatomedin-C to GH releasing hormone in healthy aging men. J Clin Endocrinol Metab 1986;62:595–600.
25. Wehrenberg WB, Ling N. The absence of an age-related change in the pituitary response to growth hormone-releasing factor in rats. Neuroendocrinology 1983;37:463–6.

26. Ceda GP, Valenti G, Butturini U, Hoffman AR. Diminished pituitary responsiveness to growth hormone-releasing factor in aging male rats. Endocrinology 1986;118:2109–14.
27. Ghigo E, Goffi S, Arvat E, et al. Pyridostigmine partially restores the GH responsiveness to GHRH in normal aging. Acta Endocrinol (Copenh) 1990;123:169–74.
28. Iovino M, Monteleone P, Steardo L. Repetitive growth hormone-releasing hormone administration restores the attenuated growth hormone (GH) response to GH-releasing hormone testing in normal aging. J Clin Endocrinol Metab 1989;69:910–3.
29. Lang I, Kruz R, Geyer G, Tragl KH. The influence of age on human pancreatic growth hormone releasing hormone stimulated growth hormone secretion. Horm Metab Res 1988;20:574–8.
30. Lang I, Schernthaner G, Pietschmann P, Kurz R, Stephenson JM, Templ H. Effects of sex and age on growth hormone response to growth hormone-releasing hormone in healthy individuals. J Clin Endocrinol Metab 1987; 65:535–40.
31. Shibasaki T, Shizume K, Nakahara M, et al. Age-related changes in plasma growth hormone response to growth hormone-releasing factor in man. J Clin Endocrinol Metab 1984;58:212–4.
32. Sonntag WE, Gough MA. Growth hormone releasing hormone induced release of growth hormone in aging male rats: dependence on pharmacological manipulation of endogenous somatostatin release. Neuroendocrinology 1988; 47:482–8.
33. Sonntag WE, Steger RW, Forman LJ, Meites J. Decreased pulsatile release of growth hormone in old male rats. Endocrinology 1980;107:1875–9.
34. Gershberg H. Growth hormone content and metabolic actions of human pituitary glands. Endocrinology 1957;61:160–5.
35. Bercu BB, Yang S-W, Masuda R, Hu C-S, Walker RF. Structural and functional responses of aged female rats to chronic co-administration of growth hormone releasing hormone (GHRH) and GH releasing hexapeptide (GHRP-6) [Abstract]. Endocr Soc annu meet, San Antonio, 1992.
36. De Gennaro Colonna V, Zoli M, Cocchi D, et al. Reduced growth hormone releasing factor (GHRF)-like immunoreactivity and GHRF gene expression in the hypothalamus of aged rats. Peptides 1989;10:705–8.
37. Sonntag WE, Hylka VW, Meites J. Impaired ability of old male rats to secrete growth hormone in vivo but not in vitro in response to hpGRF(1–44). Endocrinology 1983;113:2305–7.
38. Morimoto N, Kawakami F, Makino S, Chihara K, Hasegawa M, Ibata Y. Age-related changes in growth hormone releasing factor and somatostatin in the rat hypothalamus. Neuroendocrinology 1988;47:459–64.
39. Ono M, Miki N, Shizume D. Release of immunoreactive growth hormone-releasing factor (GRF) and somatostatin from incubated hypothalamus in young and old male rats. Neuroendocrinology 1986;43:111.
40. Goth MI, Lyons CE, Canny BJ, Thorner MO. Pituitary adenylate cyclase activating polypeptide, growth hormone (GH)-releasing peptide and GH-releasing hormone stimulate GH release through distinct pituitary receptors. Endocrinology 1992;130:939–44.

41. Abribat T, Boulanger L, Gaudreau P. Characterization of [^{125}I-Tyr^{10}] human growth hormone-releasing factor (1–44) amide binding to rat pituitary: evidence for high and low affinity classes of sites. Brain Res 1990;528:291–9.
42. Abribat T, Desiauriers N, Brazeau P, Gaudreau P. Alterations of pituitary growth hormone-releasing factor binding sites in aging rats. Endocrinology 1991;128:633–5.
43. Codd EE, Shu AYL, Walker RF. Binding of a growth hormone releasing hexapeptide to specific hypothalamic and pituitary binding sites. Neuropharmacology 1989;28:1139–44.
44. Jansson J-O, Downs TR, Beamer WG, Frohman LA. The dwarf "little" (LIT/LIT) mouse is resistant to growth hormone releasing peptide (GHRP-6) as well as to GH-releasing hormone (GRH) [Abstract]. Endocr Soc, 1986.
45. Cheng K, Chan WW-S, Barreta A, Convey DM, Smith RG. The synergistic effects of HIS-D-TRP-ALA-TRP-D-PHE-LYS-NH_2 on growth hormone (GH)-releasing factor-stimulated GH release and intracellular adenosine 3′,5′-monophosphate accumulation in rat primary pituitary cell culture. Endocrinology 1989;124:2791–8.
46. Bowers CY, Reynolds GA, Durham D, Barrera CM, Pezzoli SS, Thorner MO. Growth hormone (GH)-releasing peptide stimulates GH release in normal men and acts synergistically with GH-releasing hormone. J Clin Endocrinol Metab 1990;70:975–82.
47. Bercu BB, Yang S-W, Masuda R, Walker RF. Role of selected endogenous peptides in growth hormone-releasing hexapeptide activity: analysis of growth hormone releasing hormone, thyroid hormone-releasing and gonadotropin-releasing hormone. Endocrinology 1992:2579–96.
48. Parenti M, Dall'ara A, Rusconi L, Coochi D, Muller EE. Different regulation of growth hormone-releasing factor-sensitive adenylate cyclase in the anterior pituitary of young and aged rats. Endocrinology 1987;121:1649–53.
49. Sonntag WE, Forman LJ, Miki N, et al. Effects of CNS active drugs and somatostatin antiserum on growth hormone release in young and old male rats. Neuroendocrinology 1981;33:73–8.
50. Takahashi S, Gottshall PE, Quigley KL, Goya RG, Meites J. Growth hormone secretory patterns in young, middle-aged and old female rats. Neuroendocrinology 1987;46:137–42.
51. Ge F, Tsagarakis S, Rees LH, Besser GM, Grossman A. Relationship between growth hormone-releasing hormone and somatostatin in the rat: effects of age and sex on content and in-vitro release from hypothalamic explants. J Endocrinol 1989;123:53–8.
52. Jaffe CA, Ho PJ, DeMott Friberg R, Bowers CY, Barkan AL. Effect of prolonged growth hormone releasing peptide infusion on GH secretion in normal men. Endocr Soc annu meet, San Antonio, 1992.
53. Bercu BB, Weideman CA, Walker RF. Sex differences in growth hormone secretion by rats administered growth hormone releasing hexapeptide. Endocrinology 1991;129:2592–8.
54. Blake AD, Smith RG. Desensitization studies using perifused pituitary cells show that growth hormone-releasing hormone and His-D-Trp-Ala-Trp-D-Phe-Lys-NH_2 stimulate growth hormone release through distinct receptor sites. J Endocrinol 1991;129:11–9.

55. Chihara K, Arimura A, Schally V. Immunoreactive somatostatin in rat hypophyseal portal blood: effects of anesthetics. Endocrinology 1979;104:1434–41.
56. Shulman DI, Sweetland M, Duckett G, Root AW. Age-related differences in the growth hormone secretory response to hGHRH 1–44 in male rats from infancy through puberty. In vivo and in vitro studies. Acta Endocrinol (Copenh) 1987;116:138–44.
57. Bowers CY, Sartor AO, Reynolds GA, Badger TM. On the actions of growth hormone-releasing hexapeptide, GHRP. Endocrinology 1991;128:2027–35.
58. Downs TR, Frohman LA. Evidence for a defect in growth hormone-releasing factor signal transduction in the dwarf (dw/dw) rat pituitary. Endocrinology 1991;129:58–67.
59. Yalow RS, Berson SA. Immunoassay of plasma insulin. Methods Biochem Anal 1964;12:69–76.
60. Cheng K, Chan WW-S, Butler B, Barreto A Jr, Smith RG. Evidence for a role of protein kinase-C in His-D-Trp-Ala-Trp-D-Phe-Lys-NH_2-induced growth hormone release from rat primary pituitary cells. Endocrinology 1991;129:3337–42.
61. Thorner MO, Vance ML, Rogol AD, et al. Growth hormone releasing hormone and growth hormone releasing peptide as potential therapeutic modalities. Acta Paediatr Scand Suppl 1990;327:29–30.
62. Berelowitz M, Szabo M, Frohman LH, Fireston S, Chu L. Somatomedin-C mediates growth hormone negative feedback by effects on both the hypothalamus and pituitary. Science 1981;212:1279–81.
63. Frawley LS, Boockfor FR. Mammosomatotropes: presence and functions in normal and neoplastic pituitary tissue. Endocr Rev 1992;12:337–55.
64. Meites J. Remembrance: neuroendocrinology and aging. A perspective. Endocrinology 1992;130:3107–8.
65. Walker RF. Impact of age-related changes in serotonin and norepinephrine metabolism on reproductive function in female rats. An analytical review. Neurobiol Aging 1984;5:121–39.
66. Bevan JS, Webster J, Burke CW, Scanlon MF. Dopamine agonists and pituitary tumor shrinkage. Endocr Rev 1992;13:220–40.
67. Li YM, Brunke DL, Dantzer R, Kelley KW. Pituitary epithelial cell implants reverse the accumulation of $CD4^-CD8^-$ lymphocytes in thymus glands of aged rats. Endocrinology 1992;130:2703–9.
68. Goya RG, Castelletto L, Sosa YE. Plasma levels of growth hormone correlate with the severity of pathologic changes in the renal structure of aging rats. Lab Invest 1991;64:29–34.
69. Coleman GL, Barthold SW, Osbaldiston GW, Foster SJ, Jonas AM. Pathological changes during aging in barrier-reared Fischer 344 male rats. J Gerontol 1977;32:258–78.
70. Casad RC Jr, Adelman C. Aging enhances inhibitory action of somatostatin in rat pancreas. Endocrinology 1992;130:2420–2.

14

Changes in Growth Hormone Secretory Dynamics in Chronic Renal Failure and in Adaptation to Moderately High Altitude Living

GERMAN RAMIREZ

Several studies have shown abnormalities in the hypothalamic-hypophyseal axes in patients on chronic hemodialysis: blunted *thyroid stimulating hormone* (TSH) response to exogenous *thyrotropin releasing hormone* (TRH) (1–4), hyperprolactinemia (5, 6), elevated basal levels of *luteinizing hormone* (LH) (7–9), attenuated gonadotropin secretory response to exogenous *gonadotropin releasing hormone* (GnRH) (9), lack of *adrenocorticotropin* (ACTH) release in response to insulin-induced hypoglycemia (10), and blunted ACTH response following exogenous *corticotropin releasing hormone* (CRH) administration (11). The secretory pattern of *growth hormone* (GH) is also abnormal. Hypoglycemia fails to release GH (12) and hyperglycemia produces a paradoxical increase in GH (12). *Growth hormone releasing hormone* (GHRH) produces an exaggerated release of pituitary GH in uremic children (13) and adults (14). These observations suggest that there is a differential hypothalamo-hypophyseal regulation, possibly due to alterations in hypothalamic neurotransmitters, which can affect secretion of the hypothalamic hormones.

Anemia is invariably present in patients with renal failure. Correction of the anemia with the utilization of *recombinant human erythropoietin* (rHuEPO) in patients with end-stage renal disease on chronic dialysis has been associated with improvement in appetite, exercise tolerance, phagocytosis index, libido, and potency (15). Because the endocrine function modulates erythropoiesis under physiological conditions, a study was conducted to evaluate the hypothalamic-hypophyseal axes in men on chronic hemodialysis, both before and after the correction of anemia (16). After the correction of anemia, there was normalization of the TSH response to TRH. Basal growth hormone (GH) and prolactin levels that were abnormally elevated became normal after the correction of the anemia (16). In addition, the blunted *follicle stimulating hormone* (FSH) response to GnRH in anemic dialysis patients was normalized. However, prolactin response to exogenous TRH remained absent (16) and the low basal testosterone levels increased slightly after the correction of the anemia, but the testes did not respond to the administration of the hypothalamic hormone GnRH (16). Furthermore, patients on chronic dialysis that had an exaggerated response to the administration of exogenous GHRH manifested by a significant increment in the normal growth hormone response (14) continued with an exaggerated response despite the correction of the anemia (Ramirez et al., unpublished observations). These persistently elevated levels may be due to an increased half-life of GH, as has been demonstrated in rats (17).

We, as well as others (14, 18), have previously reported elevated plasma concentration of *insulin-like growth factor* (IGF-I) in the dialysis patients. However, low values have also been reported (19). In uremic children (20), the IGF-I levels by radioimmunoassay have been found to be normal, but its bioactivity in sera decreased. This decrement in bioactivity was found to be due to accumulation of a low molecular weight IGF binding protein in the uremic sera; when removed by affinity chromatography, the bioactivity of the IGF-I increased in the sera. Therefore, the increased IGF-I level found in uremic patients may be a response to less "available" free IGF-I. These previous studies demonstrate that chronic anemia of the dialysis patients may play a role in the abnormal responses of selective hormones after stimulation with exogenous hypothalamic hormones.

In renal failure there are other hormonal abnormalities in addition to hypothalamic hormone dysregulation. Basal cortisol levels in patients on chronic hemodialysis have been reported to be similar to the controls (21–24) or elevated (25–30). Aldosterone levels are elevated in renal failure patients, but this abnormality has been explained by the lack of specificity of the radioimmunoassay for aldosterone because after elimination of the inactive metabolites, the levels are similar to controls (24, 31). However, the response to endogenous ACTH for cortisol after administration of CRH has been found to be blunted (Ramirez et al., unpublished observations).

Our observation that a number of abnormal hypothalamic-pituitary secretory responses in uremia could be reversed by correction of anemia led us to explore the underlying mechanisms. A hypoxic effect on hormonal regulation was one of the possibilities entertained, and there have been a number of studies exploring the effects of hypoxia on hypothalamic-pituitary dynamics and other hormonal axes. Acute hypoxemia produces a dissociation of the renin and aldosterone secretion. Acute hypoxemia has been found to reduce aldosterone levels in seated adults (32) and ameliorate the expected rise in plasma aldosterone concentration during exercise (33) and following the administration of physiologic doses of ACTH (34).

Originally it was thought that hypoxemia induced the release of *atrial natriuretic factor* (ANF) (33), a known inhibitor of aldosterone release (35) and, therefore, it was expected that diuresis should be present during hypoxemic conditions. This hypothesis was tested by the administration of a standardized salt load to normal subjects during normoxemia at sea level, acute hypoxemia at sea level, and hypoxemia at moderately high altitude (3,000 m) (36). A natriuresis and diuresis was only observed during acute hypoxemia at sea level. These changes were accompanied by a decrease in plasma aldosterone levels, but did not correspond to changes in plasma aldosterone levels, ANF, or other hormones (36).

These findings are very perplexing because acute hypoxemia is a known stimulus for the secretion of ANF in vivo (37) and in vitro (38). In addition, chronic hypoxemia and hypobaria in rats, leading to hypoxia-induced pulmonary hypertension, have been associated with increased production and plasma levels of ANF (39). Similar findings have been observed in patients with *chronic obstructive pulmonary disease* (COPD), where levels of ANF have been proportionally correlated with right ventricular afterload volume (40).

Curiously, during this study it was found that with mobilization to moderately high altitude, the *anti-diuretic hormone* (ADH) did not increase in concentration with increased serum osmolality (36). The same finding was present in subjects born and raised at moderately high altitude (3,000 m) (41). At high altitude, humans experience a loss of body water associated with decreased voluntary fluid intake and usually increased urinary output (42). This phenomenon has been suggested to be of importance to the health of high-altitude sojourners (43). The absence of this physiological event has been associated with severe acute mountain sickness (44). Furthermore, a lack of response of vasopressin to increased osmolality has been demonstrated in mountaineers after long-time exposure to *high altitude* (HA) (43). Interestingly, baseline ADH levels were significantly higher at lower osmolality in HA natives than in unadapted patients mobilized to 3,000 m (41). This abnormal finding in the neuroendocrine secretion of ADH led us to suspect that the hypothalamic-hypophyseal axis for the anterior pituitary hormones was altered.

Low-altitude natives adapted to HA conditions and HA-adapted natives had higher GH levels than individuals not acclimatized to HA living (45). In animal studies, mice made hypoxemic either by bleeding or by exposure to 0.5 atmospheric pressure showed increased GH secretion (46). Rats adapted to HA have higher GH basal levels, and with exercise the rate of GH increment is faster and the response is higher than in low-altitude animals after mobilization to HA or exposure to induced hypoxia (46). However, rats raised at simulated HA (3,800 m) had reduced pituitary

GH content (47). High-altitude natives appear to have a resistance to the action of GH because of their short stature (48), despite an exaggerated response to GHRH with a hypersecretion of GH in comparison to sea-level controls, and elevated levels of IGF-I, and normal concentration of IGF binding protein-3 (Ramirez et al., unpublished observations).

Interestingly, IGF-I has been implicated as a cofactor for the proliferation of various hematopoietic cells (49), and it is intriguing to suggest that increments in IGF-I levels in HA dwellers may, in part, add to the persistent erythrocytosis in the absence of persistently high levels of erythropoietin (50). Therefore, in another study, we have demonstrated abnormal hypothalamic-hypophyseal GH axis in dialysis patients and an interesting adaptation to HA living with hypersecretion of GH after exogenous administration of GHRH, but in the presence of an apparent GH resistance, as judged by the short stature in HA-born subjects. Therefore, GH and IGF-I secretory patterns may change as an adaptive mechanism to different environmental conditions. Alteration of its secretory pattern may be a physiological response rather than a pathological process.

References

1. Ramirez G, O'Neill WM, Jubiz W, Bloomer HA. Thyroid dysfunction in uremia: evidence for thyroid hypophyseal abnormalities. Ann Intern Med 1976;84:672–6.
2. Dandona P, Newton D, Champion A, Platts MM. Thyroid function in chronic renal failure. Proc Eur Dialysis Trans Assoc 1976;12:268–71.
3. Gonzalez-Barcena D, Kastin AJ, Schalch DS, et al. Response to thyrotropin releasing hormone in patients with renal failure and after infusion in normal man. J Clin Endocrinol Metab 1973;36:117–20.
4. Lim VS, Fang VS, Katz AI, Refetoff S. Thyroid dysfunction in chronic renal failure: a study of the pituitary thyroid axis and peripheral turnover kinetics of thyroxine and triiodothyronine. J Clin Invest 1977;60:522–34.
5. Peces R, Horcajada C, Lopez Novoa JM, Frutos MA, Casado S, Hedrnando L. Hyperprolactinemia in chronic renal failure: impaired responsiveness to stimulation and suppression. Nephron 1981;28:11–6.
6. Ramirez G, O'Neill WM, Bloomer HA, Jubiz W. Abnormalities in the regulation of prolactin in patients with chronic renal failure. J Clin Endocrinol Metab 1977;45:658–61.
7. Holdsworth S, Atkin RC, DeKretser DM. The pituitary testicular axis in men with chronic renal failure. N Engl J Med 1977;296:1245–9.
8. Phadke AG, MacKinnon KJ, Dossetor JB. Male fertility in uremia: restoration by human allografts. Can Med Assoc J 1970;102:607–9.
9. Lim VS, Fang VS. Gonadal dysfunction in uremic man: a study of the hypothalamic-pituitary testicular axis before and after renal transplantation. Am J Med 1975;58:655–62.
10. Ramirez G, Gomez-Sanchez C, Meikle WA, Jubiz W. Evaluation of the hypothalamic hypophyseal adrenal axis in patients recieving longterm hemodialysis. Arch Intern Med 1982;142:1448–52.

11. Luger A, Lang I, Kovarik J , Stummvoll HK, Templ H. Abnormalities in the hypothalamic-pituitary-adrenocortical axis in patients with chronic renal failure. J Kidney Dis 1987;9:51–4.
12. Ramirez G, O'Neill WM, Bloomer HA, Jubiz W. Abnormalities in the regulation of growth hormone in chronic renal failure. Arch Intern Med 1978;138:267–72.
13. Bessarione D, Perfumo F, Giusti M, et al. Growth hormone response to growth hormone-releasing hormone in normal and uremic children. Comparison with hypoglycemia following insulin administration. Acta Endocrinol 1987;114:5–11.
14. Ramirez G, Bercu BB, Bittle PA, Ayers CW, Ganguly A. Response to growth hormone-releasing hormone in adult renal failure patients on hemodialysis. Metabolism 1990;39:764–8.
15. Eschbach JW, Egrie JC, Downing MR, Browne JK, Adamson JW. Correction of the anemia of end-stage renal disease with recombinant human erythropoietin. N Engl J Med 1987;316:73–8.
16. Ramirez G, Bittle PA, Sanders H, Bercu BB. Hypothalamo-hypophyseal thyroïd and gonadal function before and after erythropoietin therapy in dialysis patients. J Clin Endocrinol Metab 1992;74:517–24.
17. Johnson V, Maak T. Renal excretion, filtration, absorption and catabolism of growth hormone. Am J Physiol 1977;233:F185–96.
18. Enberg G, Hall K. Immunoreactive IGF-2 in serum of healthy subjects and patients with growth hormone disturbances and uremia. Acta Endocrinol 1984;107:164–70.
19. Goldberg AC, Trivedi B, Delmez JA, Harter HR, Daughaday WH. Uremia reduces insulin growth factor I, increases insulin growth factor II and modifies their serum protein binding. J Clin Endocrinol Metab 1982;55:1040–5.
20. Blum WF, Ranke MB, Kietzmann K, Tonshoff B, Mehls O. Growth hormone resistance and inhibition of somatomedin activity by excess of insulin-like growth factor binding protein in uremia. Pediatr Nephrol 1991;5:539–44.
21. Luger A, Lang I, Kovarik J, Stummvoll HK, Templ H. Abnormalities in the hypothalamic-pituitary-adrenocortical axis in patients with chronic renal failure. J Kidney Dis 1987;9:51–4.
22. Ramirez G, Brueggemeyer C, Ganguly A. Counterregulatory hormonal responses to insulin-induced hypoglycemia in patients on chronic hemodialysis. Nephron 1988;49:231–6.
23. Watschinger B, Watzinger U, Templ H, Spona J, Graf H, Luger A. Effect of recombinant human erythropoietin on anterior pituitary function in patients on hemodialysis. Horm Res 1991;36:22–6.
24. Ramirez G, Ganguly A, Brueggemeyer C, Gomez-Sanchez C. Aldosterone response to insulin-induced hypoglycemia in hemodialysis patients. Am J Nephrol 1989;9:215–21.
25. Kokot F, Wijecek A, Grzeszczak W, Kiln M. Influence of erythropoietin treatment on function of the pituitary-adrenal axis and somatotropin secretion in hemodialyzed patients. Clin Nephrol 1990;33:241–6.
26. Siamopoulos KC, Eleftheriades EG, Pappas M, Sferopoulos G, Tsolas O. Ovine corticotropin-releasing hormone stimulation test in patients with chronic renal failure: pharmacokinetics properties, and plasma adrenocorticotropic hormone and serum cortisol responses. Horm Res 1986; 30:17–21.

27. Siamopoulos KC, Dardamanis M, Kyriaki D, Pappas M, Sferopoulos G, Alevisou V. Pituitary adrenal responsiveness to corticotropin-releasing hormone in chronic uremic patients. Peritoneal Dialysis Int 1990;10:153–6.
28. Carlstrom K, Posette A, Stege R, Lindholm A. Serum hormone levels in men with end stage renal disease. Scand J Urol Nephrol 1990;24:75–8.
29. Ferraris JR, Ramirez JA, Goldberg V, Rivarola MA. Glucocorticoids and adrenal androgens in children with end stage renal disease. Acta Endocrinol (Copenh) 1991;124:245–50.
30. Ivic MA, Stefanovic V. Does high cortisol in uremic patients influence their glucagon levels? Exp Clin Endocrinol 1988;91:362–4.
31. Koshida H, Miyamori I, Miyazaki R, Tofuku Y, Takeda R. Falsely elevated plasma aldosterone concentration by direct radioimmunoassay in chronic renal failure. J Lab Clin Med 1989;114:294–300.
32. Colice GL, Ramirez G. Hypoxemia's effect on the renin-angiotensin-aldosterone system in humans. J Appl Physiol 1985;58:724–30.
33. Shigeaoka JW, Colice GL, Ramirez G. Effects of hypoxemic and normoxemic exercise on renin and aldosterone. J Appl Physiol 1985;59:142–8.
34. Ramirez G, Bittle PA, Hammond M, Ayers CW, Dietz JR, Colice GL. Regulation of aldosterone secretion during hypoxemia at sea level and at moderately high altitude. J Clin Endocrinol Metab 1988;67;1162–5.
35. Chartier L, Schiffrin E, Thibault G, Garcia R. Atrial natriuretic factor inhibits the effects of angiotensin II, ACTH and potassium on aldosterone secretion in vitro and angiotensin II-induced steroidogenesis in vivo. Endocrinology 1984;115:2026–8.
36. Ramirez G, Hammond M, Agosti SJ, Bittle PA, Dietz JR, Colice GL. Effects of hypoxemia at sea level and high altitude on sodium excretion and hormonal levels. Aviat Space Environ Med 1992;63:891–8.
37. Baertschi AJ, Adams JM, Sullivan MP. Acute hypoxemia stimulates atrial natriuretic factor secretion in vivo. Am J Physiol 1988;255:H295–300.
38. Baertschi AJ, Hausmaninger C, Walsh RS, Mentzer RM, Wyatt DA, Pence RA. Hypoxia-induced release of atrial natriuretic factor (ANF) from the isolated rat and rabbit heart. Biochem Biophys Res Comm 1986;140:427–33.
39. McKenzie JC, Tanaka I, Inagami T, Misono KS, Klein RM. Alterations in atrial and plasma atrial natriuretic factor (ANF) content during development of hypoxia-induced pulmonary hypertension in the rat. Proc Soc Exp Biol Med 1986;181:459–63.
40. Kurihara N, Fujimoto S, Kohno M, Ohta K, Hirata K, Takeda T. Exercise induced hypoxemia and exercise tolerance in patients with COPD and the benefits of oxygen supplementation. Jpn J Thoracic Dis 1989;27:155–62.
41. Ramirez G, Bittle PA, Agosti SJ, Dietz J, Colice GL. Salt loading test in a population adapted to moderately high altitude living (3000 m). Aviat Space Environ Med (in press).
42. Blume FD, Boyer SJ, Braverman LE, Cohen A, Dirkse J, Mordes JP. Impaired osmoregulation at high altitude. JAMA 1984;252:524–6.
43. Harber MJ, Williams JD, Morton JJ. Antidiuretic hormone excretion at high altitude. Aviat Space Environ Med 1981;52:38–40.
44. Richalet JP, Rathat C, Keromes A, et al. Plasma volume, body weight and acute mountain sickness. Lancet 1983;1:525.

45. Bouissou P, Fiet J, Guezennec CY, Pesquies PC. Plasma adrenocorticotrophin and cortisol responses to acute hypoxia at rest and during exercise. Eur J Appl Physiol 1988;57:110–3.
46. Nessi AC, Bozzini CE. Ultrastructural changes in somatotropic cells induced by anemic or hypoxic hypoxia. Acta Physiol Latinoam 1982;32:175–83.
47. Nelson ML, Cons JM. Pituitary hormones and growth retardation in rats raised at simulated high altitude (3800 m). Environ Physiol Biochem 1975; 5:273–8.
48. Yip R, Binkin NJ, Trowbridge FL. Altitude and childhood growth. J Pediatr 1988;113:486–9.
49. Aron DC. Insulin-like growth factor I and erythropoiesis. Biofactors 1992; 3:211–6.
50. Milledge JS, Cotes PM. Serum erythropoietin in humans at high altitude and its relation to plasma renin. J Appl Physiol 1985;59:360–4.

Part IV

Novel Aspects of Growth Hormone Research II

15

Atypical Growth Hormone Releasing Peptides

C.Y. Bowers, K. Veeraragavan, and K. Sethumadhavan

For several years we considered the new class of small synthetic peptides, *GH releasing peptides* (GHRPs) that consist of 5 to 7 amino acids, as an oddity and a technical achievement. Their artificiality was underscored by the fact that they were invented rather than isolated (1, 2). This conclusion was reinforced when it became increasingly apparent that the GHRPs did not release GH via the GHRH receptor and acted via different endocrine and molecular mechanisms than GHRH to release GH (3–18). Subsequently, however, this more technical viewpoint of the GHRPs has gradually changed.

Some reasons for a less technical, more physiological viewpoint of the GHRPs are their high potency, specific action, activity in multiple animal species, including humans, regardless of age or gender, and the fact that they act in a very complementary way with GHRH to release GH (8, 12, 13, 19–22). Select examples of the latter are recorded in Figures 15.1–15.4. GHRP-1 in low dosages very effectively releases GH in normal younger and older men and women (Figs. 15.1–15.2), and this effect is significantly greater than the GH release induced by GHRH (Fig. 15.2). Studies are in progress to understand why GH secretion is less in older than in younger subjects and to determine whether chronic administration of the GHRPs to older subjects could up-regulate the GH response to the GHRPs. Since the GHRPs and GHRH have a complementary action, the GHRPs may also up-regulate the GH response to GHRH. Noteworthy is the finding that the GHRPs release considerably more GH than the maximal dosage of GHRH as well as the finding of the synergistic release of GH when GHRP(s) and GHRH are administered together, as shown in Figures 15.2 and 15.3, because both strongly support the uniqueness of the GHRPs and the independent but complementary effects of the GHRPs and GHRH. The very high potency of the GHRPs is especially evident from the results in Figure 15.3 since 0.1 μg/kg GHRP-6 in com-

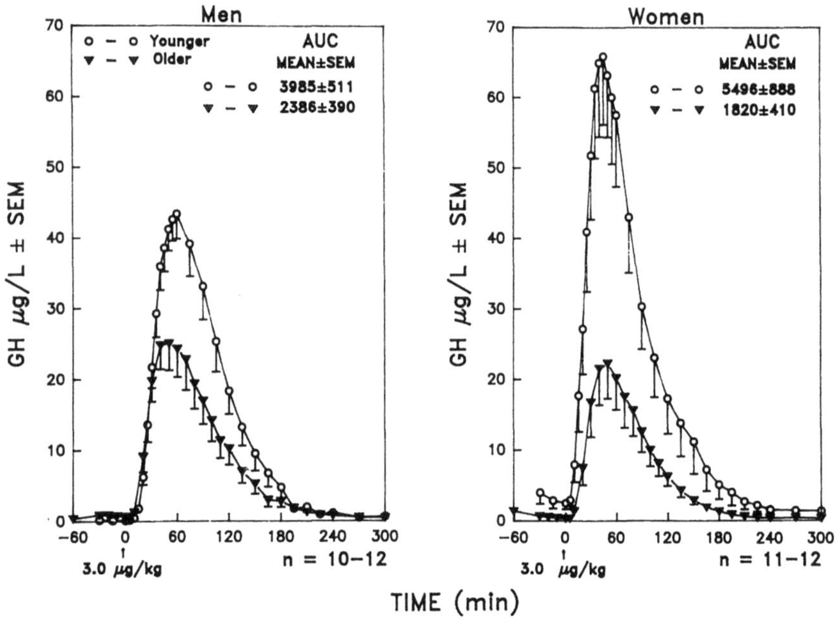

FIGURE 15.1. Mean serum GH levels in normal younger and older men and women after sc administration of GHRP-1. The mean ages (years) of the younger and older men and women were 27 ± 0.9, 63 ± 2.2, 25 ± 1.2, and 64 ± 1.8, respectively; BMIs were 25 ± 0.4, 25 ± 0.9, 21 ± 0.8, and 28 ± 1.5, respectively; serum IGF-I levels (μg/l) were 338 ± 31, 144 ± 13, 257 ± 24, and 125 ± 18, respectively.

bination with 1.0 μg/kg GHRH(1-44)NH_2 synergistically released GH. Already two new GHRPs, GHRP-1 and GHRP-2, have been developed that are 3–6 times more potent than GHRP-6 (23, 24), respectively. Our evidence indicates that except for differences in potency, the results of the different GHRPs produce the same effects and, thus, the results of the various GHRPs can be directly equated with each other in terms of their qualitative effects. Special insight into the GH releasing action(s) of the GHRPs is revealed from the results recorded in Figure 15.3 because it is evident that GHRP-6 + GHRH can synergistically release GH even when GHRP-6 itself releases little, if any, GH. As discussed below, GHRPs act on both the pituitary and hypothalamus, and the results in Figure 15.3 may reflect mainly GHRP's action on the hypothalamus. The lack of GHRP + GHRH synergism plus GH release in vitro (see above and below) is another salient reason for concluding that GHRP's hypothalamic action plays a key role in synergism.

Results in Figure 15.4 reveal that GHRPs also release GH in children as well as adults (21, 25) and that this release can be induced by oral administration of the peptide (21, 26). Other studies demonstrate that

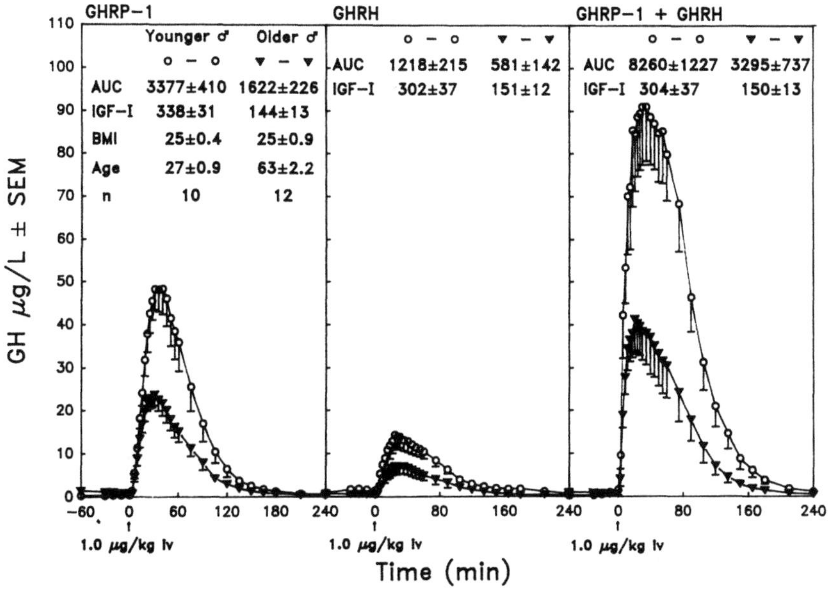

FIGURE 15.2. Comparative mean GH responses to GHRP-1, GHRH*, and GHRP-1 + GHRH in normal younger and older men. (Somatobiss®, GHRH(1-44)NH_2, Bissendorf Peptide, Germany.)

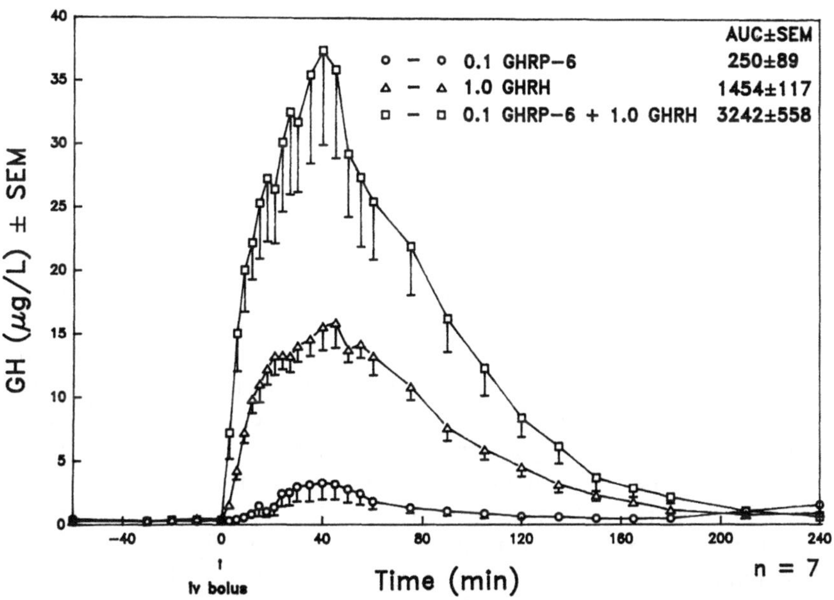

FIGURE 15.3. Comparative mean GH responses to GHRP-6, GHRH, and GHRP-6 + GHRH in normal younger men. Reprinted with permission from Bowers, Reynolds, Durham, Barrera, Pezzoli, and Thorner (19), © The Endocrine Society, 1990.

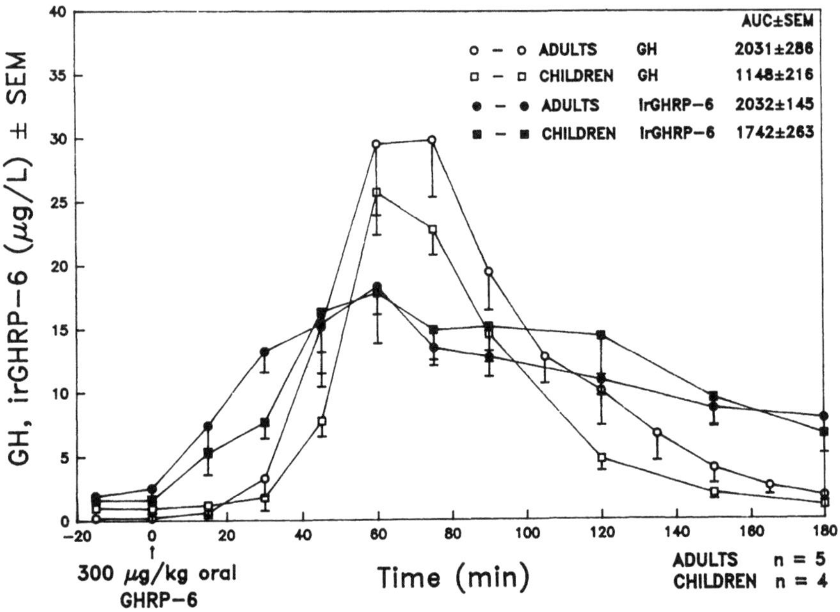

FIGURE 15.4. Comparative mean serum GH and irGHRP-6 levels in normal younger men and GH-deficient children. GHRP-6 was administered at 8:00 A.M. in the fasting state. Reprinted with permission from Bowers, Alster, and Frentz (21), © The Endocrine Society, 1992.

GHRPs release GH in humans after nasal administration (27) and enhance the normal pulsatile secretion of GH during continuous administration (28, 29).

From these and other results, the conclusion that a putative GHRP-like ligand and receptor may naturally exist has become increasingly more compelling to us, as well as to other investigators. Current work in our laboratory involves the isolation and identification of the putative GHRP-like ligand from porcine hypothalami, as well as characterization of the pituitary and hypothalamic GHRP receptor. Relatively high affinity binding sites for GHRPs have been demonstrated in the hypothalamus and the pituitary, and they appear to have the same characteristics (6, 7).

Because of the strong indirect data that has been and is still accumulating in support of the existence of a putative endogenous GHRP-like ligand, it seems reasonable to describe the putative ligand as a peptide that should exist, rather than a peptide that could exist. Furthermore, if one becomes convinced to this degree that a natural ligand for the GHRPs must exist, proposing conceptual models on how a putative GHRP-like ligand might act to regulate GH secretion will need increasing focus. It is well established that the major physiological regulators of GH secretion are GHRH and SRIF and, thus, it is only within this context that conceptual models that include a putative GHRP-like ligand should be proposed.

An immediate difficulty in reaching a more definitive stage is the still incomplete understanding of how the GHRPs act to release GH. Nevertheless, certain selective results already established should be of considerable help in proposing various models, and as relevant new data is obtained the models can be modified.

At this time, the best approach in constructing conceptual models seems to be the proposal of the model and then the determination of how the facts already established about the action of the GHRPs support the model. Although the three conceptual models proposed have been categorized as "pituitary" "hypothalamic," and "hypothalamic-pituitary" in type, it must be remembered that the hypothalamus and pituitary function as a unit and that both anatomic parts are required regardless of the model type. The models proposed emphasize the primary anatomic site of action of the putative GHRP-like ligand.

Pituitary Model

The seemingly simplest, most straightforward conceptual physiological model is a pituitary model, in which the GHRP-like ligand along with GHRH is secreted from the median eminence of the hypothalamus directly into the hypothalamic-hypophyseal portal blood in order to stimulate the pituitary somatotrophs. When GHRP(s) + GHRH is administered together to humans and rats, the synergistic release of GH is dramatic (8, 19). To explain this synergism, the question arises of whether this effect is mediated via the pituitary alone or due to a dual effect on the pituitary plus the hypothalamus. It is envisioned that the two peptides would act complementarily rather than independently of each other to produce a special effect on GH release. Important potential effects of the two peptides via a direct pituitary action could include not only the synergistic release of GH by a direct pituitary action of the peptides, but also the more ready reversal or prevention of SRIF's action on the pituitary to inhibit GH release. Since each of the three peptides, GHRH, GHRPs, and SRIF, act via different pituitary receptors and intracellular molecular mechanisms, both of these possibilities seem not only reasonable, but also desirable in that new dimensions would be added to the GH regulatory system via either of these effects.

In vitro results in Figure 15.5 indicate that the GHRP-6 + GHRH direct action on the pituitary has essentially an additive rather than a synergistic effect on the GH release in monolayer culture of dispersed rat pituitary cells. Also, the same additive effects have been obtained using

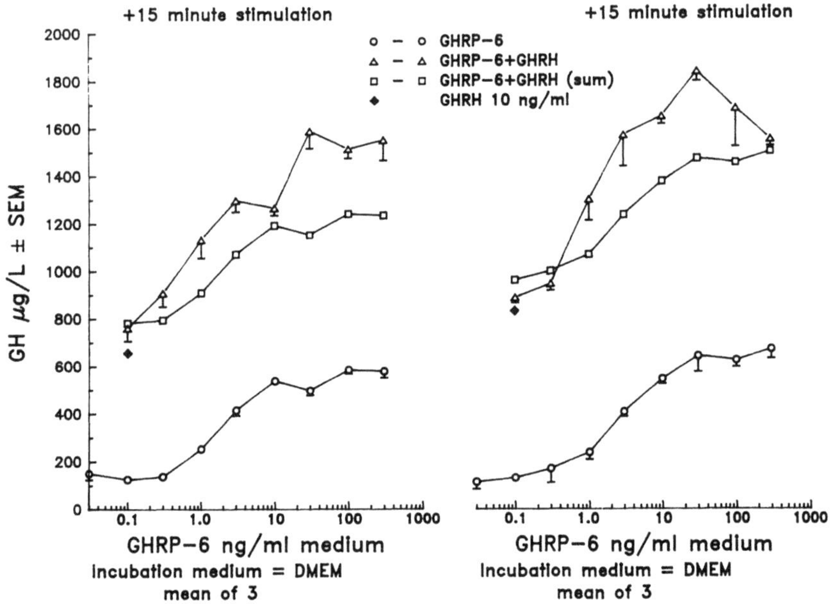

FIGURE 15.5. The interaction of GHRP-6 and GHRH on GH release in dispersed pituitary cell culture. Essentially additive effects of the two peptides occurred, since the greatest differences were 38% (left panel) and 28% (right panel) at the 30 µg dosage of GHRP-6. Reprinted with permission from Bowers, Sartor, Reynolds, and Badger (8), © The Endocrine Society, 1991.

pituitary incubates and pituitary perifusion (8). Only one report has indicated that in vitro GHRP-6 + GHRH releases GH synergistically (5). In vitro results in Figure 15.6 also show that GHRP-1 increased GH without increasing pituitary cAMP levels, while GHRH increased both (16). The combined responses were greater than the arithmetic sum responses of the individual peptides for cAMP and GH by 25% and 35%, respectively. As recorded below, the GH response of the combined peptides is much greater in vivo than in vitro. Because of the latter results, this synergistic GH response has been concluded to be primarily due to GHRP's action on the hypothalamus rather than the pituitary.

So far, some of the variables that we have studied on the possible interaction of the GHRP(s), GHRH, and SRIF at the pituitary level are the following. The relative effectiveness of SRIF-14, and occasionally SRIF-28, on the inhibition of the GH response to GHRPs vs. GHRH or GHRP(s) + GHRH has been determined both in vitro and in vivo. Whether a difference occurs in vivo compared to in vitro was considered important because of GHRP's hypothalamic action, which may play a direct role in the SRIF, GHRP-like ligand interaction. Also, whether a difference occurs because SRIF's actions are being interfered with,

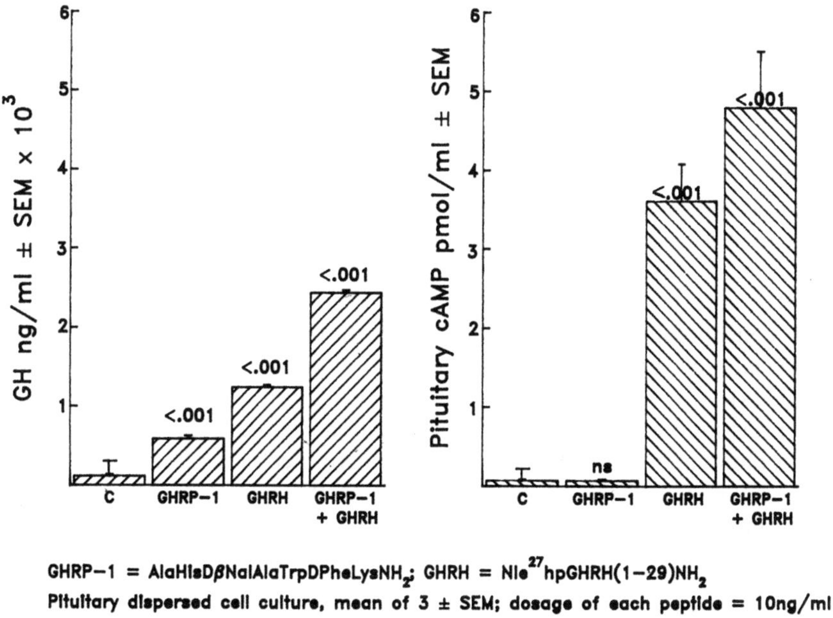

FIGURE 15.6. Concomitant effects of GHRP-1 on release of GH and pituitary cAMP levels in vitro.

reversed, or prevented was considered to be of potential importance. The actual number of permutations of these relationships are many and, thus, only a select portion of them have been studied.

Recorded in Table 15.1 are results of the relative SRIF IC_{50} levels for the GH responses to GHRP-1, $Nle^{27}GHRH(1\text{-}29)NH_2$ and a combination of the two peptides using a monolayer culture of rat pituitary cells in vitro. Noteworthy is that the IC_{50} SRIF values for the individual peptides GHRP-1 and $Nle^{27}GHRH(1\text{-}29)NH_2$ were the same, indicating that GHRPs do not interfere with the SRIF pituitary inhibitory action on GH

TABLE 15.1. In vitro inhibition of the GH response to GHRP-1, $Nle^{27}GHRH(1\text{-}29)NH_2$, and GHRP-1 + $Nle^{27}GHRH(1\text{-}29)NH_2$ by SRIF-14 and SRIF-28.

	Dose (pmol/ml)	SRIF (IC_{50} pmol)
GHRP-1	10	0.17
GHRP-1	10	0.12*
$Nle^{27}GHRH(1\text{-}29)NH_2$	3	0.10
$Nle^{27}GHRH(1\text{-}29)NH_2$	3	0.10*
GHRP-1 + $Nle^{27}GHRH(1\text{-}29)NH_2$	10 ± 3	0.48
GHRP-1 + $Nle^{27}GHRH(1\text{-}29)NH_2$	30 ± 3	0.33

Note: GHRP-1 = $AlaHisD\beta NalAlaTrpDPheLysNH_2$; pituitary dispersed cell culture.
* SRIF-28.

release in a way that is more effective than GHRH. Also apparent is that SRIF-14 and SRIF-28 have the same IC_{50} values. However, when the IC_{50} for GHRP-1 + GHRH was calculated, it was found to be 3–4 times greater, raising the potentially important possibility that if the GHRP-like ligand and GHRH are secreted together into portal blood, the two peptides could stimulate GH release not only by directly stimulating the somatotrophs to release GH, but also by directly interfering with the inhibitory action of SRIF on the somatotrophs. This effect on SRIF's action would only seem to be relevant if the GHRP-like ligand and GHRH were secreted concomitantly from the hypothalamus.

A conceptual pituitary model in which the GHRP-like ligand has a special and unique physiological role does seem possible; that is, GHRP(s) + GHRH interfere with SRIF's pituitary action. However, other possibilities that will need consideration are the GHRP-like ligand's effect alone and with GHRH on the negative feedback effects of GHRH and/or IGF-I on GH synthesis, as well as the effects on the pituitary paracrine and/or autocrine factors that may influence GH secretion.

Hypothalamic Model

A seemingly more obtuse conceptual physiological model is a hypothalamic model because this maximizes GHRP's hypothalamic action, which is still poorly defined, and minimizes GHRP's pituitary action, which is well established and definitely more completely characterized. As shown in Figure 15.7, in vitro evidence indicates that GHRP-6 acts on both the hypothalamus and pituitary. Additionally, relatively high affinity binding sites for GHRPs have been demonstrated (6, 7) in the hypothalamus and in the pituitary (Table 15.2). In vivo evidence also supports that GHRPs act on the hypothalamus. Whether GHRP's action is primarily on the hypothalamus or pituitary is a very basic question still unresolved, and this certainly adds to the difficulty of proposing conceptual physiological models. To be considered is that the physiological action of the putative GHRP-like ligand could involve primarily the pituitary, while the pharmacological action of exogenously administered GHRPs could involve both the hypothalamus and pituitary.

A potential role for GHRP-like ligand to regulate SRIF and/or GHRH release from the hypothalamus is certainly a possibility because, if found to occur, it would be an important component of the hypothalamic physiological conceptual model. Our in vitro results obtained from hypothalamic-pituitary incubate studies utilizing tissue of rats, however, were unsatisfactory and did not help to resolve this point more directly. The hypothalamic in vitro studies of Hoa et al. (30) indicate that GHRP-6 increases SRIF release from the hypothalamus, which would be antagonistic rather than facilitory to increase the release of GH when GHRP(s)

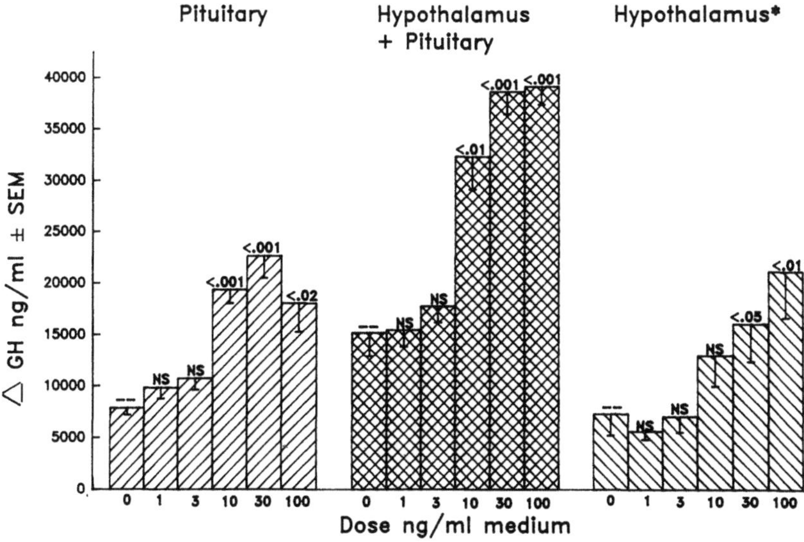

FIGURE 15.7. In vitro effect of GHRP-6 on release of GH in pituitary vs. hypothalamus + pituitary incubates. Reprinted with permission from Bowers, Sartor, Reynolds, and Badger (8), © The Endocrine Society, 1991.

TABLE 15.2. Dissociation constants and maximal binding capacities of Tyr-Ala-GHRP binding sites from Scatchard plot analysis.

	Pituitary		Hypothalamus	
	pH 5.0	pH 7.4	pH 5.0	pH 7.4
Kd1(nM)	125 ± 50	152 ± 20	162 ± 40	80 ± 5
Kd2(μm)	43.0 ± 13.5	38.0 ± 19.0	17.0 ± 2.4	15.0 ± 1.8
B*max1*(pmol/mg)	13.3 ± 6.2	8.6 ± 2.7	42.8 ± 12.0	3.8 ± 0.15*
B*max2*(pmol/mg)	5000 ± 500	560 ± 300**	3700 ± 350	640 ± 100**

Note: The affinity and binding sites at pH 5.0 and 7.4 were compared using unpaired Student's t-test.
* $P < 0.01$.
** $P < 0.001$.
Source: Reprinted with permission from Sethumadhavan, Veeraragavan, and Bowers (7).

is administered in vivo and, thus, this particular in vitro effect is unlikely to occur in vivo. Furthermore, results of in vivo studies are strongly against the possibility that GHRP(s) inhibits SRIF release from the hypothalamus (8). For example, when agents that presumably inhibit the release of SRIF (pentobarbital, opiates, and pyridostigmine) or immunoneutralize circulating SRIF (SRIF antiserum, see below) are given with GHRP(s), the GH response to GHRP(s) is increased. It was reasoned that if GHRP(s) inhibits SRIF's release or action the agents would not have further increased the GHRP-stimulated GH response, but the fact that they did indicates that GHRP(s) alone does not appear to inhibit the release or action of SRIF. In contrast, GHRPs in combination with GHRH do appear to be able to inhibit SRIF's pituitary inhibitory action on GH release in vitro and especially in vivo. Thus, the physiological and pharmacological actions of the GHRP-like ligand or GHRPs may in part involve interference with the pituitary action of SRIF by GHRP + GHRH.

GHRP's release of GHRH via its hypothalamic action is seemingly less straightforward. A combination of in vitro and in vivo findings indicates that the GHRP(s) and GHRH relationship can be both dependent and independent, as well as complementary. GHRH immunoneutralization in vivo studies in rats (8) indicate a dependency on the presence of GHRH, but this does not necessarily mean GHRP(s) releases GHRH (Fig. 15.8). The finding that GHRPs release GH in vitro without the presence of

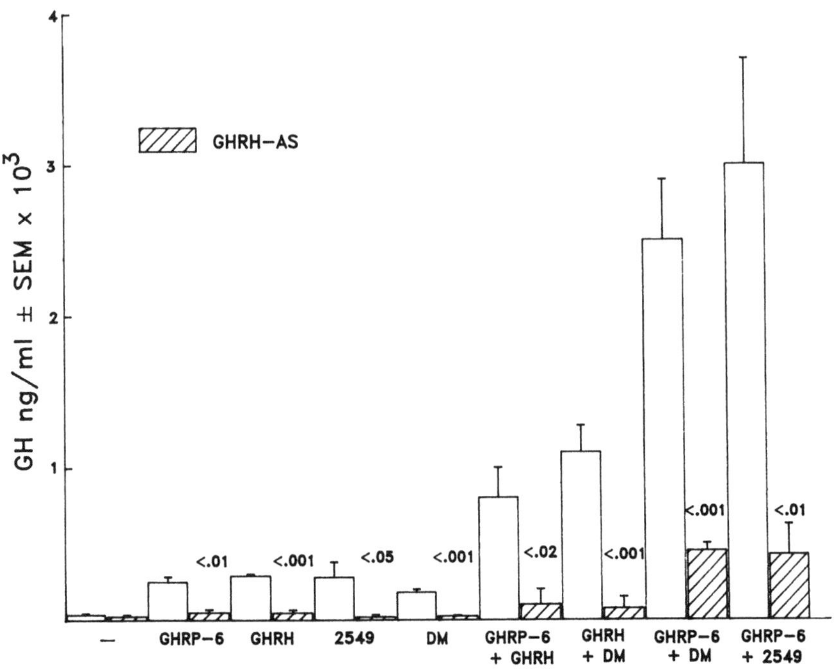

FIGURE 15.8. In vivo effect of rat GHRH antiserum pretreatment on the GH response in immature rats. (GHRH = rat GHRH[1-43]OH [10 μg]; DM = TyrDAlaPheGlyTyrProSerNH$_2$ [100μg]; 2549 = benzomorphan opiate compound [10 μg]; GHRP-6 = HisDTrpAlaTrpDPheLysNH$_2$ [10 μg]; mean of 6 ± SEM.) Reprinted with permission from Bowers, Sartor, Reynolds, and Badger (8), © The Endocrine Society, 1991.

GHRH demonstrates an independent action (2, 8). A complementary action of the two peptides is demonstrated by the marked in vivo synergistic release of GH induced by administration of GHRP + GHRH.

Our in vivo studies of the concomitant effects on GH release and pituitary cAMP levels after administration of GHRP(s) or GHRH alone and in combination (16) have been particularly instructive in revealing the independency, but also the complementary actions of the peptides (Table 15.3). In vivo results shown in Table 15.3 indicate that Ala^1GHRP released essentially the same amount of GH as GHRH, but GHRP, in contrast to GHRH, did not increase or only slightly increased pituitary cAMP levels, thus indicating the independent action of the two peptides. Also, these results are one of the major reasons for concluding that endogenous GHRH release plays a permissive role in GHRP's release of GH and that GH released by GHRP(s) does not occur because it releases GHRH from the hypothalamus. If the latter did occur, GHRP(s) would have been expected to raise pituitary cAMP levels concomitantly with the increase of GH release. That GH can be markedly released in vivo via nondependent pituitary intracellular cAMP pathways is also demonstrated by the results of the tetrapeptide *dermorphin* (DM) analog and the DM analog + GHRP. Most of the pituitary cAMP rise induced by GHRH + DM analog or GHRP + GHRH + DM analog can be accounted for by GHRH. The above results again indicate that GHRP and GHRH act via different receptors, as well as different molecular and endocrine mechanisms.

Another possibility for the hypothalamic model besides release of GHRH or inhibition of SRIF release from the hypothalamus is the hypothesis of *U-factor* (unknown factor) that may be released by GHRP's hypothalamic action (8, 16). The role(s) and action(s) of U-factor are discussed below under "Hypothalamic-Pituitary Model."

In the purely hypothalamic model, the action of the GHRPs would be relegated to function as a hypothalamic paracrine, autocrine factor involved solely in the release of U-factor. Despite this possible unique hypothalamic action, GHRP's pituitary action has too many well-established properties, including selective pituitary binding sites, selective intracellular biochemical pathways, independent action of GHRH, and so on, to disregard its action on the pituitary as being superfluous in the physiological action of the GHRP-like ligand. Because of the latter point, a hypothalamic model seems very unlikely.

Hypothalamic-Pituitary Model

Just as outlined for the hypothalamic model, the most important component in the conceptual hypothalamic-pituitary model may involve the release of the putative U-factor via GHRP's action on the hypothalamus.

Table 15.3. Concomitant effects of Ala[1]-GHRP on release of GH and pituitary cAMP levels in rats.

Peptide	Dosage (μg, IV)	Serum GH (ng/ml ± SEM) +10 min	P value	Pituitary cAMP (pmol/mg protein ± SEM) +10 min	P value
Control	—	13 ± 1	—	6 ± 1	—
Ala[1]-GHRP	1	42 ± 7	<0.01	10 ± 2	NS
Ala[1]-GHRP	10	144 ± 37	<0.01	14 ± 3	NS
GHRH	1	79 ± 12	<0.001	41 ± 8	<0.001
GHRH	10	108 ± 13	<0.001	67 ± 9	<0.001
DM Analog	30	125 ± 42	<0.02	4 ± 1	NS
DM Analog	100	308 ± 53	<0.001	8 ± 2	NS
Ala[1]-GHRP + GHRH	1 + 1	815 ± 108	<0.001	78 ± 11	<0.001
Ala[1]-GHRP + GHRH	10 + 10	2109 ± 369	<0.001	124 ± 20	<0.001
Ala[1]-GHRP + DM Analog	1 + 100	1131 ± 132	<0.001	6 ± 1	NS
Ala[1]-GHRP + DM Analog	10 + 30	1918 ± 132	<0.001	28 ± 7	<0.01
GHRH + DM Analog	1 + 100	3932 ± 420	<0.001	57 ± 6	<0.001
GHRH + DM Analog	10 + 30	2513 ± 167	<0.001	89 ± 10	<0.001
Ala[1]-GHRP + GHRH + DM Analog	1 + 1 + 100	4750 ± 322	<0.001	62 ± 10	<0.001
Ala[1]-GHRP + GHRH + DM Analog	10 + 10 + 30	3344 ± 357	<0.001	120 ± 21	<0.001

Note: Ala[1]GHRP = AlaHisDTrpAlaTrpDPheLysNH_2; GHRH = Nle^{27}GHRH(1-29)NH_2; DM analog = TyrDArgPheGlyNG_2; values are mean of 8 ± SEM of 26-day-old female rats.

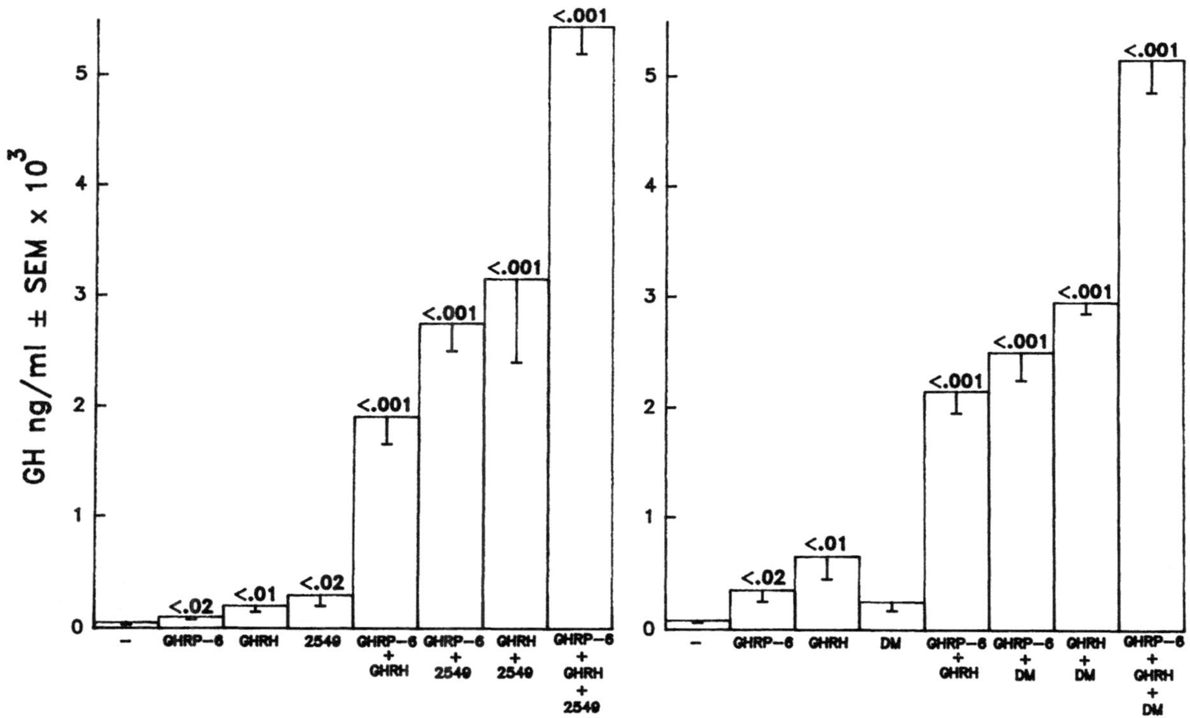

FIGURE 15.9. In vivo studies on the synergistic release of GH in immature rats. (Mean of 6 ± SEM; peptides and 2549 dosages are the same as described in Figure 15.8.) Reprinted with permission from Bowers, Sartor, Reynolds, and Badger (8), © The Endocrine Society, 1991.

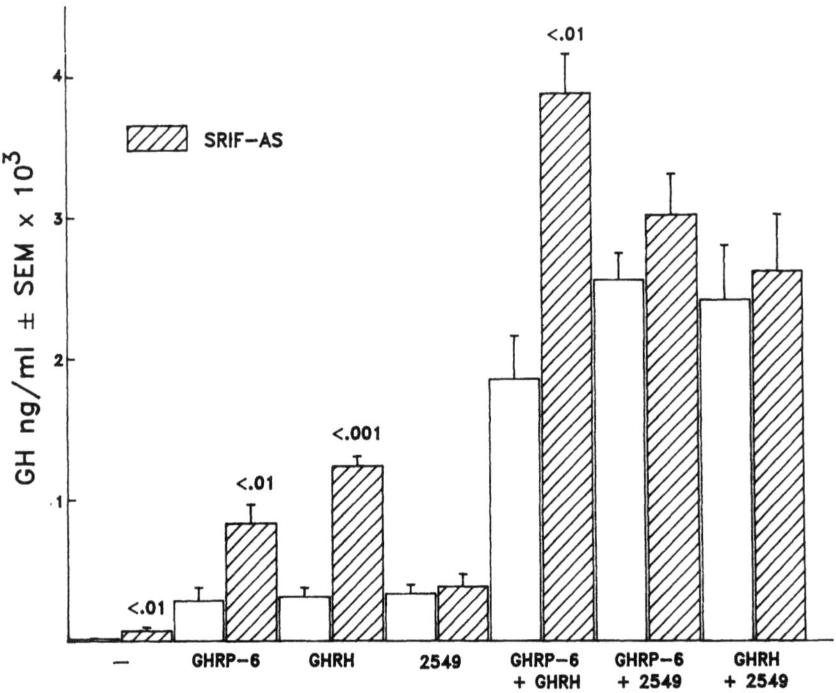

FIGURE 15.10. In vivo effect of SRIF antiserum pretreatment on the GH response in immature rats. (Mean of 6 ± SEM; peptide dosages are the same as described in Figure 15.8.) Reprinted with permission from Bowers, Sartor, Reynolds, and Badger (8), © The Endocrine Society, 1991.

The combined effects of U-factor and GHRH on the pituitary are envisioned to release GH synergistically from the pituitary somatotrophs. The U-factor hypothesis was proposed in order to explain the synergistic GH response that is induced when GHRP(s) + GHRH is administered in vivo (8). The in vivo synergistic GH response just does not appear to be explained by an action of GHRP(s) + GHRH on the pituitary or by an action of GHRP(s) on the hypothalamus either to increase the release of GHRH or to decrease the release of SRIF from the hypothalamus (8). Thus, U-factor has been proposed and should be carefully considered as a possible component of the conceptual hypothalamic-pituitary model, as well as the hypothalamic model (8, 16).

As recorded in Figures 15.9 and 15.10, when GHRP-6 + GHRH was administered in vivo, the synergistic release of GH was marked. The results in Figure 15.10 indicate that the synergistic GH response to GHRP-6 + GHRH occurs without SRIF being present (SRIF antiserum results). Thus, even though GHRP-6 + GHRH interferes with SRIF's action on the pituitary, this effect cannot fully explain the GH synergistic action of these two peptides, and therefore U-factor has been proposed.

In vivo studies in rats have been performed to assess further the possible interrelationship between the action of SRIF and the peptide combination, GHRP-1 + GHRH, on GH release. As shown in Table 15.4, results indicate that the SRIF IC_{50} for inhibition of the GH response to GHRP-1 or GHRH in pentobarbital-pretreated immature female rats was essentially the same—that is, 9.5 and 13.5 nmol, respectively. In contrast, when GHRP-1 + GHRH was administered to conscious immature female rats, the SRIF IC_{50} for the combined peptides was definitely higher than for the individual peptides since the combination was 169 nmol, while the SRIF IC_{50} for GHRP-1 alone was only 9.8 nmol. Thus, the in vitro and in vivo results are in agreement with each other. The combined, but not the individual, peptides (GHRP-1 or GHRH) much more effectively interfere with SRIF's pituitary inhibitory action on GH release. This effect on SRIF could reasonably and logically form the basis of how the GHRP-like ligand would function as a physiological regulator of GH secretion in addition to the regulation by GHRH and SRIF.

In contrast to the pituitary model, the hypothalamic-pituitary model seems inordinately complex. The physiological hypothalamic-pituitary model would require that the GHRP-like ligand have a paracrine, autocrine hypothalamic, as well as pituitary, action and that both actions be related and coordinated with each other. Nevertheless, both the hypothalamic and pituitary anatomical sites of action do appear to be involved in the release of GH induced by exogenous GHRPs because of the strong evidence that supports the existence of a putative hypothalamic U-factor. At present, this latter action of exogenous GHRPs is considered

TABLE 15.4. In vivo inhibition of the GH response to GHRP-1, Nle^{27}GHRH(1-29)NH_2, and GHRP-1 + Nle^{27}GHRH(1-29)NH_2.

	Dose (nmol, IV)	Serum GH (ng/ml ± SEM)	SRIF (IC_{50} nmol)
Pentobarbital-pretreated rat			
Control	1.0	152 ± 32	—
GHRP-1	1.0	2208 ± 304	9.5
GHRH*	0.3	1826 ± 260	13.5
Conscious intact rat			
Control	—	10 ± 0.6	—
GHRP-1	1.0	65 ± 19	9.8**
GHRH*	0.3	99 ± 17	—
GHRP-1 + GHRH	1.0 + 0.3	765 ± 188	169.0

Note: GHRP-1 = AlaHisDβNalAlaTrpDPheLysNH_2; values are mean of 6 ± SEM.
* Nle^{27}GHRH(1-29)NH_2.
** Value calculated from different experiment.

to represent a pharmacological action, rather than the physiological action of the GHRP-like ligand.

Summary

That GHRPs might interfere with, prevent, or reverse SRIF's pituitary inhibition of GH release in some unique way via a direct, as well as an indirect, pituitary action has encouraged us to consider a large number of theoretical possibilities of how this may occur and the mechanisms possibly involved. Some of them have been evaluated experimentally by in vitro and/or in vivo studies. Since SRIF appears to act via hyperpolarization and GHRP via depolarization of pituitary cell membranes, along with closing (SRIF) and opening (GHRP) of voltage-dependent calcium channels to inhibit and release GH, there are special biochemical reasons for carefully evaluating the possible interaction of the peptides (13, 17, 18). Unfortunately, neither the pattern nor the concentration of GHRH or SRIF in the hypothalamic-hypophyseal portal blood is well defined; if they were, they would possibly be of considerable value in helping to project the physiological role and possible action of the putative GHRP-like ligand. Projected portal GHRH and SRIF patterns have consisted of 180° out-of-phase secretion of the two peptides (32), as well as the tonic secretion of either GHRH or SRIF plus intermittent secretion of the opposite peptide (33). The actual concentration of SRIF and GHRH has varied from higher to lower levels of the peptides. Higher portal levels of SRIF would indicate SRIF is secreted intermittently because, unless GHRH had some way of reversing the inhibitory action of SRIF—that is, GHRP-like ligand + GHRH (see above)—GHRH alone would be unlikely to reverse the inhibitory action of the higher concentration of SRIF. Constant secretion of SRIF in contrast to constant secretion of GHRH is less likely to induce down-regulation of its own action on the pituitary; thus, this is another seemingly important possibility that may be relevant to projecting how these peptides are secreted.

Under physiological conditions, the mechanisms involved in the modulation of SRIF's pituitary inhibitory action and the degree to which this effect is mediated at the pituitary and/or hypothalamic level are still unclear and seemingly convoluted. It appears that different modulating mechanisms induced by GHRP(s) + GHRH are involved in vivo and in vitro. GHRPs are considered to act on the hypothalamus and this hypothalamic effect may very well mediate the synergistic GH response to GHRP(s) + GHRH via U-factor. Since synergism and U-factor may both be involved in the mechanism(s) that modulates SRIF's pituitary inhibitory action on GH release in vivo and neither are involved in vitro, in which GHRP(s) + GHRH modulates SRIF's inhibitory action, it is apparent that the mechanisms involved are convoluted. The degree to which the hypothalamic action of the GHRH-like ligand is required to inhibit SRIF's action in vivo is a very important and relevant point in proposing the physiological conceptual model for the GHRP-like ligand because if the hypothalamic action is not required, the pituitary conceptual model becomes the model of choice. The latter would mean that the GHRP-like ligand + GHRH would be released concomitantly from the median eminence of the hypothalamus and then would modulate the inhibitory action of SRIF on GH release. Release of the GHRP-like ligand from the hypothalamus with an immediate action on the pituitary is the classical way other hypothalamic-hypophysiotrophic hormones are secreted and function. Thus, without having to account for a hypothalamic action for the endogenous putative GHRP-like ligand, the model is greatly simplified, and the mechanisms involved in the physiological regulation of GH secretion by this regulator are much less convoluted. Although the above applies to the physiological aspect of the GHRP-like ligand, it does not invalidate the conclusion that exogenous GHRPs have an important and valuable hypothalamic action. The latter does appear to occur, but this may mainly represent a pharmacological action of the peptide and may be the reason why GHRPs release a much greater amount of GH in humans than GHRH.

Conclusion

Since it has been concluded that the GHRP-like ligand should, rather than could exist, the compelling guiding point seems to be that despite how irrational it may seem to be to include the GHRP-like ligand along with GHRH and SRIF as a physiological regulator of GH secretion, a variety of evidence indicates that the attempt to include this new putative regulator should be diligently pursued. Various aspects and conceptual models of the latter have been discussed and proposed. As a more detailed understanding of the GHRP(s) receptor, molecular intracellular mechanism, and endocrine mechanism by which it releases GH via both the hypothalamus and pituitary is elucidated, the physiological role of the GHRP-like ligand in the regulation of GH release will become more apparent. Evidence is strong that the actions of GHRP(s) and GHRH on release of GH can be dependent and independent, as well as complementary to each other. A seemingly important general conclusion that has evolved from these studies is that GH secretion can be markedly increased apparently without involving endogenous GHRH or involving it in only a permissive way. Thus, new results support that new dimensions will be revealed in the near future about the physiological regulation of GH secretion, as well as the pharmacological release of GH from the hypothalamic-pituitary unit.

Acknowledgments. This work was supported in part by National Institutes of Health Grants DK-40202 and 5 M01RR05096 (GCRC). Special appreciation is expressed to the technicians of the Endocrinology and Metabolism Section of the Department of Medicine, Patricia Giangrosso for graphic assistance, and Robin Alexander for typing the manuscript.

References

1. Momany F, Bowers CY, Reynolds GA, Hong A, Newlander K. Conformational energy studies and in vitro activity data on active GH releasing peptides. Endocrinology 1984;114:1531–6.
2. Bowers CY, Momany F, Reynolds GA, Hong A. On the in vitro and in vivo activity of a new synthetic hexapeptide that acts on the pituitary to specifically release growth hormone. Endocrinology 1984;14:1537–45.
3. Sartor O, Bowers CY, Reynolds GA, Momany F. Variables determining the GH response of His-D-Trp-Ala-Trp-D-Phe-Lys-NH_2 (GHRP-6) in the rat. Endocrinology 1985;117:1441–7.
4. Sartor O, Bowers CY, Chang D. Parallel studies of His-D-Trp-Ala-Trp-D-Phe-Lys-NH_2 and hpGRF-44NH_2 in rat primary pituitary cell monolayer culture. Endocrinology 1985;116:952–7.
5. Cheng K, Chan WWS, Barreto A, Convey EM, Smith RG. The synergistic effects of His-D-Trp-Ala-Trp-D-Phe-Lys-NH_2 on growth hormone (GH)-releasing factor-stimulated GH release and intracellular adenosine 3′,5′-monophosphate accumulation in rat primary pituitary cell culture. Endocrinology 1989;124:2791–8.
6. Codd EE, Shu AYL, Walker RF. Binding of a growth hormone releasing hexapeptide to specific hypothalamic and pituitary sites. Neuropharmacology 1989;28:1139–44.
7. Sethumadhavan K, Veeraragavan K, Bowers CY. Demonstration and characterization of the specific binding of growth hormone-releasing peptide (GHRP) to rat anterior pituitary and hypothalamic membranes. Biochem Biophys Res Commun 1991;178(1):31–7.
8. Bowers CY, Sartor AO, Reynolds GA, Badger TM. On the actions of the growth hormone-releasing hexapeptide, GHRP. Endocrinology 1991;128: 2027–35.
9. Clark RG, Carlsson LMS, Trohnar J, Robinson ICAF. The effects of a growth hormone-releasing peptide and growth hormone-releasing factor in conscious and anesthetized rats. J Neuroendocrinol 1989;1:249–55.
10. Goth MI, Lyons CE, Canny BJ, Thorner MO. Pituitary adenylate cyclase activating polypeptide, growth hormone (GH)-releasing peptide and GH-releasing hormone stimulate GH release through distinct pituitary receptors. Endocrinology 1992;130:939–44.
11. Cheng K, Chan WWS, Butler B, Barreto A, Smith RG. Evidence for a role of protein kinase-C in His-D-Trp-Ala-D-Phe-Lys-NH_2 induced growth hormone release from rat primary pituitary cells. Endocrinology 1991;129: 3337–42.
12. Walker RF, Yang SW, Bercu BB. Robust growth hormone (GH) secretion in aged female rats co-administered GH-releasing hexapeptide (GHRP-6) and GH-releasing hormone (GHRH). Life Sci 1991;49(20):1499–504.

13. Bercu BB, Weideman CA, Walker RF. Sex differences in growth hormone (GH) secretion by rats administered GH-releasing hexapeptide. Endocrinology 1991;129:2592–8.
14. Blake AD, Smith RG. Desensitization studies using perifused rat pituitary cells show that growth hormone-releasing hormone and His-D-Trp-Ala-Trp-D-Phe-Lys-NH_2 stimulate growth hormone release through distinct receptor sites. J Endocrinol 1991;129:11–9.
15. Bitar KG, Bowers CY, Coy DH. Effects of substance P/bombesin antagonists on the release of growth hormone by GHRP and GHRH. Biochem Biophys Res Commun 1991;180(1):156–61.
16. Veeraragavan K, Bowers CY. Evidence for the involvement of an unknown hypothalamic factor in the release of growth hormone (GH) by growth hormone-releasing peptide (GHRP-1) [Abstract]. FASEB annu meet, Anaheim, CA, April, 1992.
17. Pong SS, Chaung LY, Smith RG, Ertel E, Smith M, Cohen CJ. Role of calcium channels in growth hormone secretion induced by GHRP-6 (His-D-Trp-Ala-Trp-D-Phe-Lys-NH_2) and other secretagogues in rat somatotrophs [Abstract]. 74th annu meet Endocr Soc, San Antonio, TX, June, 1992:255.
18. Chik CL, Akman M, Girard M, Ho AK. Growth hormone releasing peptide stimulates GH via a calcium-dependent cAMP-independent mechanism [Abstract]. 74th annu meet Endocr Soc, San Antonio, TX, June, 1992:182.
19. Bowers CY, Reynolds GA, Durham D, Barrera CM, Pezzoli SS, Thorner MO. Growth hormone (GH)-releasing peptide stimulates GH release in normal men and acts synergistically with GH-releasing hormone. J Clin Endocrinol Metab 1990;70:975–82.
20. Ilson BE, Jorkasky DK, Curnow RT, Stote RM. Effect of a new synthetic hexapeptide to selectively stimulate growth hormone release in healthy human subjects. J Clin Endocrinol Metab 1989;69:212–4.
21. Bowers CY, Alster DK, Frentz JM. The GH releasing activity of a synthetic hexapeptide in normal men and short stature children after oral administration. J Clin Endocrinol Metab 1992;74:292–8.
22. DeBell WK, Pezzoli SS, Thorner MO. Growth hormone (GH) secretion during continuous infusion of GH-releasing peptide: partial response attenuation. J Clin Endocrinol Metab 1991;72:1312–6.
23. Bowers CY, Reynolds GA, Barrera CM. A second generation GH releasing heptapeptide [Abstract]. 73rd annu meet Endocr Soc, Washington, DC, June, 1991:422.
24. Bowers CY, Newell D, Granda-Ayala R, Garcia M, Barrera C. Comparative studies on GH release in younger and older men and women [Abstract]. 74th annu meet Endocr Soc, San Antonio, TX, June, 1992:172.
25. Mericq V, Cassorla F, Garcia H, Avila A, Merriam GR. Growth hormone responses to a second-generation growth hormone releasing peptide and to GH-releasing hormone (GHRH) in growth hormone deficient children [Abstract]. 74th annu meet Endocr Soc, San Antonio, TX, June, 1992:171.
26. Hartman ML, Farello G, Pezzoli SS, Thorner MO. Oral administration of growth hormone (GH)-releasing peptide stimulates GH secretion in normal men. J Clin Endocrinol Metab 1992;74:1378–84.
27. Hayashi S, Okimura Y, Yagi H, et al. Intra-nasal administration of His-D-Trp-Ala-Trp-D-Phe-Lys-NH_2 (growth hormone releasing peptide) increased

plasma growth hormone and insulin-like growth factor-I levels in normal men. Endocrinol Jpn 1991;38:15–21.

28. Jaffe CA, Ho PJ, Demott Friberg R, Bowers CY, Barkan AL. Effect of prolonged growth hormone releasing peptide infusion on GH secretion in normal men [Abstract]. 74th annu meet Endocr Soc, San Antonio, TX, June, 1992:169.
29. Huhn WC, Hartman ML, Pezzoli SS, Thorner MO. 24-hour continuous growth hormone releasing peptide (GHRP, SK&F 110679) augments pulsatile growth hormone (GH) secretion and selectively attenuates the response to a subsequent GHRP bolus [Abstract]. 74th annu meet Endocr Soc, San Antonio, TX, June, 1992:169.
30. Hoa EH, Malozowski S, Ren SG, Marin G, Southers J, Merriam G. A comparison of the effects of GH-releasing hormone (GHRH) and GH-releasing peptide (GHRP) on GH and somatostatin (SRIF) release [Abstract]. Endocr Soc, New Orleans, LA, June, 1988:123.
31. Koch BD, Blalock JB, Schonbrunn A. Characterization of the cyclic AMP-independent actions of somatostatin in GH cells, I. An increase in potassium conductance is responsible for both the hyperpolarization and the decrease in intracellular free calcium produced by somatostatin. J Biol Chem 1988;263: 216–25.
32. Tannenbaum GS, Ling N. The interrelation of growth hormone (GH)-releasing factor and somatostatin in generation of the ultradian rhythm of GH secretion. Endocrinology 1984;115:1952–7.
33. Thomas GB, Cummins JT, Francis H, Sudbury AW, McCloud PI, Clarke IJ. Effect of restricted feeding on the relationship between hypophysial portal concentration of growth hormone (GH)-releasing factor and somatostatin, and jugular concentration of GH in ovariectomizes ewes. Endocrinology 1991;128:1151–8.

16

Growth Hormone Releasing Hexapeptide: Basic Research and Clinical Application

KAZUO CHIHARA, HIDESUKE KAJI, SATOKO HAYASHI, HIROJI YAGI, YASUHIRO TAKESHIMA, MAKI MITANI, SHIN-ICHIRO OHASHI, AND HIROMI ABE

A synthetic hexapeptide, *His-D-Trp-Ala-Trp-D-Phe-Lys-*NH_2 (GHRP), was generated by a combination of conformational energy calculations and assessment of GH releasing activity (1). GHRP is reported to have remarkable GH releasing activity in a variety of species, especially in humans (2–6). First, we show clinical data regarding plasma GH responses to GHRP administration not only in normal young adults and elderly men but also in a variety of patients, including short stature and acromegaly. Subsequently, we performed animal experiments as well as in vitro studies using pituitary cells to elucidate the problems and mechanisms obtained in clinical studies.

GHRP Test in Normal Young Adults and Elderly Men

An intravenous (IV) administration of 1 μg/kg BW GHRP resulted in marked increases in plasma GH levels with mean (±SE) peaks of 50.9 ± 5.1 μg/l in normal young men aged 23–34 years and 41.8 ± 4.7 μg/l in normal young women aged 22–37 years, suggesting no sex difference in plasma GH responses to GHRP. GHRP was potent enough to increase plasma GH levels in elderly men in whom peak GH levels after GHRP injection, 36.8 ± 3.2 μg/l, were slightly lower than those in young men. However, this peak level is remarkable since plasma GH responses to *human GHRH(1-44)*NH_2 (hGHRH) were virtually absent in the same subjects. Intranasal administration of GHRP was also effective to stimulate GH release in normal young men, although a larger dose up to 30 μg/kg was required (6). When 15 μg/kg GHRP was repetitively administered every 8 h intranasally for 3 days, plasma IGF-I levels were significantly elevated after GHRP administration. These findings give a theoretical basis for therapeutic use of GHRP. Before a trial of chronic use of GHRP, however, we would evaluate the usefulness of a GHRP test as a provocative test for GH secretion.

Normal Short Stature

GHRP Test Versus Other GH-Provocative Tests

In all normal short children who failed to satisfy the criteria of GH deficiency, GHRP increased plasma GH with its peak of 26.6 ± 2.8 μg/l 15 min after IV injections of 1 μg/kg in a bolus dose. When the magnitude and pattern of GH response to GHRP were compared with those to hGHRH, L-dopa, and insulin-induced hypoglycemia, plasma GH peaks appeared earlier after GHRP injection than after other provocative stimuli. GHRP could stimulate GH release so rapidly that plasma GH peaks appeared 15 min after injection in more than half of normal short children and within 45 min after injection in the remaining subjects examined. The time until GH peaks appeared was highly varied after administration of hGHRH, L-dopa, or insulin. The averages of each plasma GH peak were 30.7 ± 2.8, 30.3 ± 3.0, 10.0 ± 1.2, and 10.2 ± 1.1 μg/l after GHRP, hGHRH, L-dopa, and insulin administration, respectively. The GH peaks after GHRP or hGHRH injection were significantly higher than those after L-dopa or insulin administration, whereas there was no difference in plasma GH peaks between GHRP and hGHRH injections. This finding is of interest since GHRP was always much more potent than hGHRH to stimulate GH release in adults as well as elderly subjects (5). The inhibitory control of GH release by hypothalamic *somatostatin* (SRIF) may be minimal during childhood but becomes stronger with age. As mentioned later, the action of hGHRH is more sensitive to SRIF inhibition than that of GHRP and, therefore, the discrepancy among generations in GH responses to GHRP and hGHRH may be explained by differences in the strength of hypothalamic somatostatinergic tone.

Reproducibility of GH Response in a GHRP Test and Other Tests

Next, we examined the reproducibility of GH responses to GHRP, hGHRH, L-dopa and insulin administration. Each provocative test was repeated after an interval of at least 1 week. Good reproducibility of plasma GH responses was obtained in the GHRP-provocative test. In contrast, plasma GH responses to hGHRH were highly variable. To elucidate the difference of GHRP and hGHRH in reproducibility of their GH responses, the effect of SRIF infusion on plasma GH responses to GHRP and hGHRH was examined in normal subjects. Our preliminary studies revealed that an IV infusion of 3 µg/kg BW/min SRIF resulted in complete suppression of plasma GH response to hGHRH, but not to GHRP, suggesting that GHRP may be less sensitive to SRIF inhibition in humans in vivo.

GH-Deficient Short Children with Pituitary Stalk Transection

GHRP Test

In some GH-deficient short children who have a past history of breech delivery and severe asphyxia at birth, the examination by *magnetic resonance imaging* (MRI) revealed that the pituitary stalk is transected with the formation of pseudoposterior pituitary in its proximal end (7). When hGHRH was injected IV in these short children with pituitary stalk transection, plasma GH levels were slightly but significantly elevated, confirming previous findings (8, 9). In contrast, plasma GH responses to GHRP injection were virtually absent in stalk-transected short children. These findings were not expected since we believed that GHRP was a very strong stimulator for GH release with a direct action at the pituitary. Therefore, we had an interest in the mechanisms by which GHRP fails to stimulate GH release in short children with pituitary stalk transection.

In Vivo Animal Studies

There is accumulating evidence in animal studies suggesting that endogenous GHRH plays a crucial role so that GHRP exerts its full GH releasing activity (10, 11). In conscious male rats, immunological neutralization of circulating GHRH with GHRH *antibodies* (ab) alone caused marked attenuation of plasma GH response to GHRP. To eliminate the modulation by endogenous SRIF of GH responses to GHRP, an additional experiment was performed using conscious male rats treated with SRIF-ab + GHRH-ab and those treated with SRIF-ab alone. The difference between the two groups of rats was only the presence or absence of circulating active endogenous GHRH. The plasma GH responses to GHRP in rats treated with SRIF-ab + GHRH-ab were markedly suppressed as compared with those in rats treated with SRIF-ab alone. Clark et al. (12) have reported that with GHRH infusions, GH release continued in the normal episodic pattern, whereas with GHRP infusion, GH secretion remained elevated over baseline and the normal pulsatile rhythm was disrupted, suggesting that GHRP disrupts cyclic SRIF release from the hypothalamus. To examine whether GHRP can stimulate hypothalamic SRIF release, we injected GHRP into the lateral ventricle of conscious male rats. However, intracerebroventricular injections of GHRP failed to increase, and instead suppressed, GH release. Rather, the synergistic action of GHRP and GHRH at the pituitary levels seemed much more important. Using a conscious animal model, we examined the synergistic action of GHRP and GHRH. All rats were treated with both SRIF-ab and GHRH-ab. The injection of GHRP alone resulted in significant but small increases of plasma GH. Also, a single injection of hGHRH analog, which does not cross-react with GHRH-ab (ab against rat GHRH), moderately stimulated GH release. However, a huge increase of plasma GH was observed when both agents were simultaneously injected. These findings in the conscious animal model are consistent with the previous observation in humans (5).

In Vitro Studies

There were conflicting data about the synergism of GHRP and GHRH on GH release in an in vitro system. Using a rat primary pituitary cell culture system, Sartor et al. (13) reported an additive instead of synergistic effect of GHRP and GHRH on GH release. On the other hand, Cheng et al. (14) demonstrated that when maximal concentrations of GHRH and GHRP were simultaneously added to the pituitary cells and incubated for 15 min, their effects on GH release were greater than additive. Subsequently, Bowers et al. (11) reported that synergism of GHRP and GHRH on GH secretion observed in the primary culture system of rat pituitary cells was less prominent than that observed in vivo and suggested that sites other than the pituitary level—that is, hypothalamus—may be involved in their synergistic interaction.

Using a similar primary culture system of rat pituitary cells, we also examined the effects of GHRP and GHRH on GH release in various combinations of different doses and incubation periods. During 15-min incubation with 100 nM GHRP or 100 nM hGHRH, GH release to the medium was significantly increased. The simultaneous incubation of GHRP and GHRH in considerably larger doses resulted in a synergistic increase of medium GH levels. During 2-h incubation, GHRP and hGHRH were both effective to stimulate GH release, but hGHRH was much stronger than GHRP in its potency. The synergistic effects of GHRP and hGHRH were no longer observed in the 2-h incubation experiment. After 24 h, only hGHRH increased GH levels in the medium. From these findings we agree that the synergistic effects of GHRP and GHRH on GH release can be demonstrated only under the limited conditions, using high doses of these peptides in a short incubation period. Next, to examine the effect of GHRP and GHRH on GH synthesis, GH mRNA levels on cultured pituitary cells were determined by

Northern blot analysis. The treatment with hGHRH for 3 days caused a significant increase of GH mRNA levels in pituitary cells, while GHRP had little effect on GH gene expression, suggesting that GHRP may not stimulate the synthesis of GH in pituitary.

Priming Effects of Human GHRH on Plasma GH Response to GHRP in Stalk-Transected Short Children

Since the interaction of GHRP and GHRH at the pituitary level was obvious from the above-mentioned animal studies, we examined the priming effect of repeated hGHRH administration on the GH response to GHRP in short patients with transection of pituitary stalk. The GHRP test was performed on the first and fifth days of the studies. The patients received daily 1-h IV infusions of 3 μg/kg BW hGHRH in the morning for 3 successive days following the day of the first GHRP test. We confirmed the previous observation that plasma GH response to a single IV bolus injection of GHRP was completely absent in these patients. After daily hGHRH infusions for 3 days, the plasma GH levels were significantly increased by the GHRP injection performed on the fifth day of the studies. However, the plasma GH peak levels after the latter GHRP injection were still lower than those after a single IV bolus injection of hGHRH tested on a separate day.

Although it is difficult to explain clearly the mechanism by which GH responsiveness to GHRP was improved after daily hGHRH infusions, one of the possibilities is that the releasable GH pool that could respond to GHRP was increased by hGHRH treatment. Alternatively, the number of receptors for GHRP might increase with hGHRH treatment. Apart from the interaction between GHRP and GHRH, the possibility cannot be ruled out that there may be functionally distinct subpopulations of the somatotrophs that respond to GHRP and/or GHRH in different ways. In addition, we have observed interesting findings in another clinical study in acromegaly.

Acromegaly

GHRP Test

In most patients with active acromegaly studied, a single bolus injection of 1 μg/kg GHRP caused a significant increase in plasma GH levels, the peaks of which appeared 15–45 min after GHRP injections. The magnitude of plasma GH peaks was highly variable among the patients and did not correlate with their basal GH levels. The percent increases of plasma GH above the basal level after GHRP injection varied from 146% to 953%, but were more than 200% in 12 of 14 patients. When the GH response to GHRP was compared with that to hGHRH, TRH, or LHRH, there were no correlations in terms of GH responsiveness. However, on closer examination of the GH responses to GHRP and hGHRH in individual patients, we noticed that one patient could respond to both GHRP and hGHRH, but another patient responded only to GHRP. To further clarify characteristics of GH-producing adenoma, GH responsiveness to GHRP and hGHRH was examined in vitro.

In Vitro Studies

Pituitary adenoma tissues, obtained at surgery from the patient who responded to both GHRP and hGHRH in vivo, were dispersed according to a standard method. These dispersed adenoma cells were responsive to both GHRP and GHRH in vitro as well. Incubation of adenoma cells with 1, 10 and 100 nM GHRP resulted in a dose-dependent increase of GH release into the medium. The GH release also enhanced with 10 nM hGHRH. The synergistic effects of GHRP and hGHRH on GH release were clearly observed when both agents were simultaneously incubated with dispersed pituitary cells. On the other hand, the GH-producing adenoma cells, derived from the acromegalic patient who could respond to only GHRP in vivo, were responsive only to GHRP but not to hGHRH in vitro. Since accumulating evidence indicates that GH-producing adenoma is a monoclonal proliferation of a single somatotroph cell, the heterogeneity of GH responsiveness to GHRP and hGHRH in acromegaly suggests the presence of subpopulations of somatotroph cells.

GHRP- and/or GHRH-Responsive Subpopulations of Rat Somatotroph Cells

To examine the responsiveness to GHRP and GHRH in individual rat pituitary cells, we used a *reverse hemolytic plaque assay* (RHPA) (15). Rat anterior pituitary cells and protein A-coupled ovine red blood cells were plated in a poly-L-lysine-coated Cunningham chamber. Then, anti-rat GH antibodies, GHRP, and/or GHRH and *SRIF analog* (SMS201995) were added to the chamber. Hemolysis of ovine red blood cells was triggered by adding complement. Both GHRH and GHRP caused significant increases in GH secretion when assessed by the percent of plaque-forming cells, the mean plaque area, and the total secretion index, which was calculated by multiplying percent of the plaque-forming cells and the mean plaque area. Addition of SMS201995 partially suppressed GH release induced by GHRH or GHRP. Next, a sequential double RHPA was performed to examine the responsiveness of individual cells to GHRP and GHRH according to the method reported by Jia et al. (16). Double RHPA is basically the same as single RHPA except that the first reaction is recorded in the videotape and then repeated using another secretagogue.

To examine the presence of subpopulations of somatotrophs, cells were initially challenged with GHRH or GHRP, and then with GHRP or GHRH in a reverse order. The GHRH- but not GHRP-responsive phenotype would form plaques during the incubation with GHRH, but not with GHRP. The GHRP but not GHRH phenotype would form plaque during the incubation with GHRP, but not with GHRH, whereas the GHRH or GHRP phenotype would form plaques in both incubation periods. Our preliminary data revealed that the GHRH but not GHRP phenotype, the GHRP but not GHRH phenotype, and the GHRH or GHRP phenotype corresponded to approximately 20%, 10% and 5%, respectively, of all anterior pituitary cells. Therefore, these findings suggest the presence of at least three functionally distinct subpopulations of somatotroph cell.

References

1. Momany FA, Bowers CY, Reynolds GA, Chang D, Hong A, Newlander K. Design, synthesis and biological activity of peptides which release growth hormone in vitro. Endocrinology 1981;108:31–9.
2. Bowers CY, Momany FA, Reynolds GA, Hong A. On the in vitro and in vivo activity of a new synthetic hexapeptide that acts on the pituitary to specifically release growth hormone. Endocrinology 1984;114:1537–45.
3. Momany FA, Bowers CY, Reynolds GA, Hong A, Newlander K. Conformational energy studies and in vivo and in vitro activity data on growth hormone-releasing peptides. Endocrinology 1984;114:1531–6.
4. Ilison BE, Jorkasky DK, Curnow RT, Stote RM. Effect of a new synthetic hexapeptide to selectively stimulate growth release in healthy human subjects. J Clin Endocrinol Metab 1989;69:212–4.
5. Bowers CR, Reynolds GA, Durham D, Barrera CM, Perroli SS, Thorner MO. Growth hormone (GH)-releasing peptide stimulates GH release in normal men and acts synergistically with GH-releasing hormone. J Clin Endocrinol Metab 1990;70:975–82.
6. Hayashi S, Okimura Y, Yagi H, et al. Intranasal administration of His-D-Trp-Ala-Trp-D-Phe-Lys-NH_2 (growth hormone releasing peptide) increased plasma GH and IGF-1 levels in normal men. Endocrinol Jpn 1991;38:15–21.
7. Fujisawa I, Kikuchi K, Nishimura K. Transection of the pituitary stalk: development of an ectopic posterior lobe assessed with MR imaging. Radiology 1987;165:487–9.
8. Chihara K, Kashio Y, Abe H, et al. Idiopathic growth hormone (GH) deficiency, and GH deficiency secondary to hypothalamic germinoma: effect of single and repeated administration of human GH-releasing factor (hGHR) on plasma GH level and endogenous hGRF-like immunoreactivity level in cerebrospinal fluid. J Clin Endocrinol Metab 1985;60:269–78.
9. Borges JLC, Blizzard RM, Gelato MC, et al. Effects of human pancreatic tumor growth hormone releasing factor on growth hormone and somatomedin

C levels in patients with idiopathic growth hormone deficiency. Lancet 1983;2:119–21.
10. Yagi H, Uchiyama T, Shakutsui S, Chihara K. Modulation by endogenous GHRH and SRIF of GH releasing hexapeptide (GHRP-6) induced GH release in conscious male rats [Abstract]. 72nd annu meet Endocr Soc, Atlanta, 1990:50.
11. Bowers CY, Sartor AO, Reynolds GA, Badger TM. On the actions of the growth hormone-releasing hexapeptide, GHRP. Endocrinology 1991;128: 2027–35.
12. Clark RG, Carlsson LMS, Trojnar J, Robinson ICAF. The effects of a growth hormone-releasing peptide and growth hormone-releasing factor in conscious and anesthetized rats. J Neuroendocrinol 1989;1:249–55.
13. Sartor AO, Bowers CY, Chang D. Parallel studies of His-D-Trp-Ala-Trp-D-Phe-Lys-NH_2 and human pancreatic growth hormone releasing factor-44-NH_2 in rat primary pituitary cell monolayer culture. Endocrinology 1985;116: 952–7.
14. Cheng K, Chan WWS, Barreto A, Convey EM, Smith RG. The synergistic effects of His-D-Trp-Ala-Trp-D-Phe-Lys-NH_2 on growth hormone (GH)-releasing factor-stimulated GH release and intracellular adenosine 3′,5′-monophosphate accumulation in rat primary pituitary cell culture. Endocrinology 1989;124:2791–8.
15. Neill JD, Smith PF, Luque M, Munoz de Toro M, Nagy G, Mulchahey JJ. Detection and measurement of hormone secretion from individual pituitary cells. Recent Prog Horm Res 1987;43:175–229.
16. Jia L-G, Canny BJ, Orth DN, Leong DA. Distinct classes of corticotropes mediate corticotropin releasing hormone- and arginine vasopressin-stimulated adrenocorticotropin release. Endocrinology 1991;128:197–203.

17

Molecular Endocrinology of Human Growth Hormone

James Wells, Brian Cunningham, Germaine Fuh,
Henry Lowman, Mark Ultsch, and Bart de Vos

The molecular basis for hormone-receptor interactions is fundamentally important for understanding how extracellular hormones bind their receptors and trigger intracellular responses. Such information is crucial to the rational design of protein hormones optimized for affinity, receptor selectivity, and agonistic or antagonistic activities. Such modified hormones are powerful tools that can permit pharmacologists to link the action of specific cellular receptors with their physiological effect(s). Moreover, a detailed molecular analysis facilitates the rational design of improved hormone therapeutics as engineered proteins and possibly small molecule mimics.

Human growth hormone (hGH) is a 191-amino acid pituitary hormone that elicits a wide variety of biological effects, including linear growth, lactation, nitrogen retention, and diabetagenic and insulin-like effects (reviewed in 1–2). To begin to understand the molecular basis for this complex array of pharmacological actions, we have elaborated the structural and functional details for the interaction between hGH and its two known receptors (the hGH and prolactin receptors). We have also established the mechanism by which hGH activates these receptors (reviewed in 3–4). Here, we summarize these studies and their applications to the design of receptor-selective agonists and antagonists of hGH.

Mutational Analysis of hGH-Receptor Interactions

Human growth hormone binds and activates the hGH and prolactin receptors. Cloning of these revealed that they contain an extracellular domain (about 210–250 residues in length) followed by a single transmembrane domain and an intracellular domain (5, 6). The extracellular domains of each receptor, called *binding proteins* (bps), have been expressed in *Escherichia coli* (7, 8) and bind hGH with high affinity (kds of 0.3 nM and 0.03 nM for the hGHbp and hPRLbp, respectively). These highly purified and soluble truncated receptors were used to assay binding of hGH mutants.

We utilized a two-tiered mutational strategy to identify residues in hGH important for modulating the affinity of the hormone for its receptor (reviewed in 9). In the first level of analysis, called homolog-scanning mutagenesis, we substituted segments of nonbinding homologs of hGH into the hormone and measured binding to the hGHbp (10) or hPRLbp (11). From the specific disruption in binding affinity caused by these substitutions, we could identify the general regions of hGH responsible for modulating receptor binding.

In the second level of analysis, called alanine-scanning mutagenesis, we mutated each residue to alanine within the disruptive segments. In this way we tested the importance of side-chain determinants larger than alanine for modulating binding. Each receptor binding site mapped to a discrete patch on the folded hormone and was made up of determinants from separated regions of the polypeptide sequence (a discontinuous epitope). For binding of the hGHbp, we found that hGH contains two separate binding sites (called sites 1 and 2), shown in Figure 17.1A (12, 13). Site 1 is a larger binding site and consists of weak determinants in the beginning of helix 1 and much stronger ones in the loop between residues 54 and 74 and the central portion of helix 4. Site 2 determinants are located in the extreme N-terminus through the central portion of helix 1 (opposite those employed in site 1) and the central region of helix 3 (13).

Biophysical studies showed that hGH binds two molecules of the hGHbp simultaneously to form an hGH·$(hGHbp)_2$ complex (13). Mutational studies showed that disruptive mutations in site 1 could not form 1:1 complexes, whereas mutations in site 2 could. These data suggested the complex forms sequentially—namely, hGH binds one molecule of the hGHbp to site 1, followed by binding of a second one to site 2 (Fig. 17.2). Addition of excess hGH to the hGH·$(hGHbp)_2$ complex could dissociate it into monomeric complexes. Moreover, site 2 mutants could dissociate the wild-type complex into monomeric ones. These data indicated that the corresponding site(s) on the receptor for binding to sites 1 and 2 overlapped.

The fact that hGH induced dimerization of the hGHbp in solution suggested it would also occur for the full-length receptor in the membrane. To test this, a cell-based receptor assay was constructed by transfecting a myeloid leukemia cell line, FDC-P1 cells, with a hybrid receptor containing the extracellular portion of the hGH receptor linked to a truncated portion of the *granulocyte colony-stimulating factor* (G-CSF) receptor (14). Addition of hGH to this cell line induces proliferation. The dose

A

B

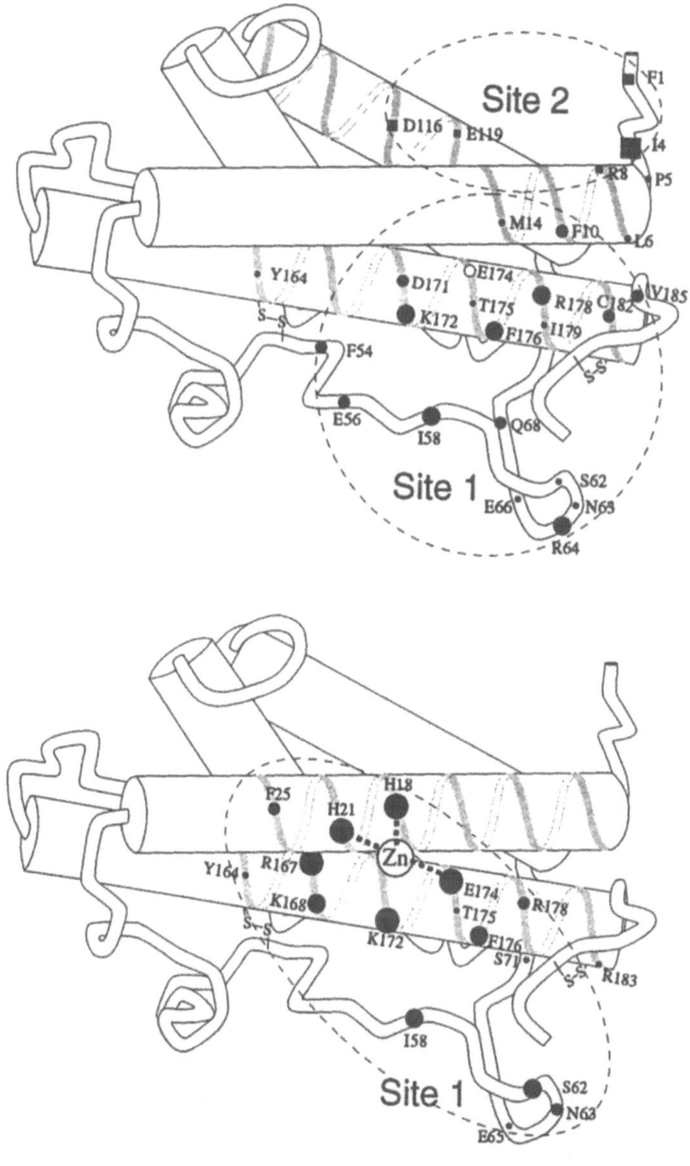

FIGURE 17.1. Functional epitopes for binding of hGH to the hGH receptor or the hPRL receptor. *A:* Positions of alanine replacements that reduce binding of hGH to the hGHbp by 2- to 4-fold, 4- to 10-fold, 10- to 50-fold, and more than 50-fold are shown by graduated symbols in either site 1 (solid circles) or site 2 (solid squares). The E174A mutant (open circle) enhances binding by 4-fold to site 1. Reprinted with permission from Fuh, Cunningham, Fukunaga, Nagata, Goeddel, and Wells (14). Copyright by the AAAS. *B:* Positions of alanine replacements causing a reduction in binding affinity to the hPRLbp of 2- to 4-fold, 4- to 10-fold, 10- to 80-fold, or >80-fold to site 1 are shown by graduated symbols (solid circles). The ligands that coordinate Zn^{2+} are shown. Reprinted with permission from Cunningham, Ultsch, de Vos, Mulkerrin, Clauser, and Wells (13). Copyright by the AAAS.

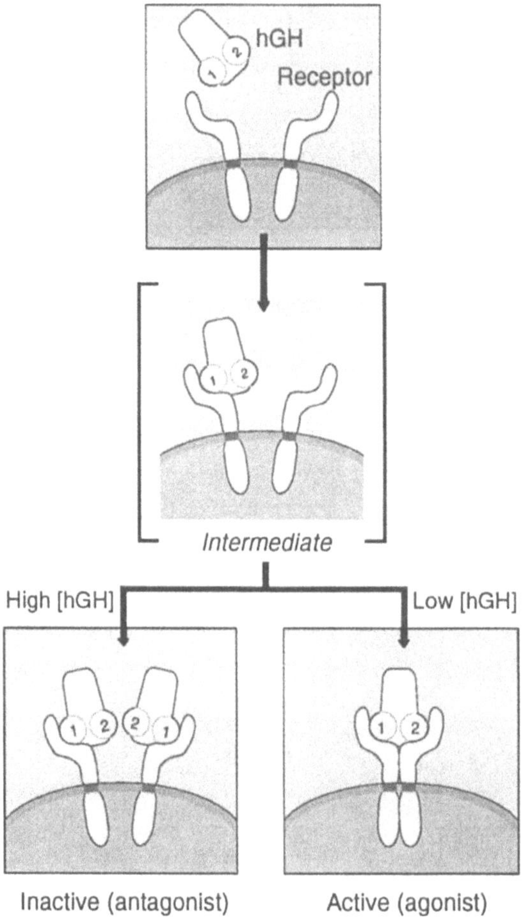

FIGURE 17.2. Sequential dimerization model for the activation of the hGH receptor. At low concentrations, hGH binds first at site 1 and subsequently at site 2 to produce an active hGH·(receptor)$_2$ complex. At high concentrations, hGH saturates the receptor through site 1 interactions and acts as an antagonist by preventing receptor dimerization. Reprinted with permission from Fuh, Cunningham, Fukunaga, Nagata, Goeddel, and Wells (14). Copyright by the AAAS.

response curve showed a striking bell shape—that is, cells showed increased rates of DNA synthesis at low concentrations of hGH, a maximum rate at intermediate concentrations, and then inhibition at very high concentrations of hGH. This is consistent with the mechanism shown in Figure 17.2. Moreover, most bivalent antibodies to the hGHbp showed similar bell-shaped dose response curves, whereas their corresponding FAb fragments were inactive. This further indicated that dimerization was critical for receptor activation. Similar effects have been found for the full-length hGH receptor transfected into FDC-P1 cells (Colosi, Wood, Fuh, Wells, unpublished results) and for phosphorylation of proteins induced by hGH in IM-9 cells (15).

The site 1 epitope for binding of hGH to the hPRLbp has also been characterized (8, 11). This functional epitope overlaps the site 1 epitope for binding the hGHbp, but is not identical (Fig. 17.1B). One of the striking features of this epitope is the unique requirement for binding a Zn^{2+} in the central portion of this epitope. Mutational studies showed that hGH contributes three of the coordinating ligands, and the fourth is donated from His 188 of the prolactin receptor (8).

Studies in cells containing prolactin receptors suggest that hGH activates the prolactin receptor by sequential dimerization, as seen for the hGH receptor (Fig. 17.2). For example, dose response curves for stimulating cell proliferation by either hGH or prolactin show similar bell shapes (16). Furthermore, mutations in site 1 are inactive as agonists, and lowering the exogenous zinc concentration reduces the potency of hGH. The putative site 2 epitope for binding the prolactin receptor has not been fully characterized, but a mutation (Gly120Arg) that disrupts binding of hGH via site 2 to the hGH receptor is inactive in stimulating the prolactin receptor. This suggests that hGH has two sites for binding the prolactin receptor and that its site 2 determinants overlap those used for binding to the hGH receptor. Also, some antibodies to the prolactin receptor are weak agonists, whereas their corresponding FAb fragments are not (17), further indicating that dimerization is important for activation of the prolactin receptor.

Zinc binding studies carried out separately on hGH showed that it bound a single zinc ion weakly (kd $\sim 1\,\mu M$) and in a positive cooperative fashion (18). It was further shown that physiological concentrations of zinc induced dimerization of hGH at hormone concentrations exceeding $\sim 1\,\mu M$ to form a $(Zn^{2+} \cdot hGH)_2$ complex. The maximal concentration of hGH ($\sim$1–5 nM) is too low in serum for this to form as a stable complex. However, hGH is stored in somatotrophs in the pituitary in the presence of stoichiometric quantities of zinc at concentrations where most would exist in this complex. Moreover, hGH is more stable to denaturation as a $(Zn^{2+} \cdot hGH)_2$ complex (18). We believe that hGH is stored as a $(Zn^{2+} \cdot hGH)_2$ complex and dissociates to the free hormone after its release from somatotrophic vesicles to interact with cellular receptors or the hGHbp.

The natural hGHbp appears to be a shed form of the hGH receptor (5). Its role may be to prolong the clearance of hGH in circulation (19). Gel filtration studies have shown that the major form of the bound hGH is a 1:1 complex with the hGHbp (20, 21). This is also consistent with calculations based on the estimated concentrations of hGH and the hGHbp in serum and the known dissociation constants for the monomeric and dimeric complexes of hGH with the hGHbp (Baumann, Lowman, Wells, unpublished results).

Structure of the hGH·$(hGHbp)_2$ Complex

After the site 1 and site 2 functional epitopes were established by mutational analysis, a 2.7-Å resolution X-ray crystal structure of the hGH·$(hGHbp)_2$ complex was solved (22) and provided the atomic details of the protein folds and molecular interactions (Fig. 17.3). The folding of the bound hGH was nearly identical to that reported for unbound porcine

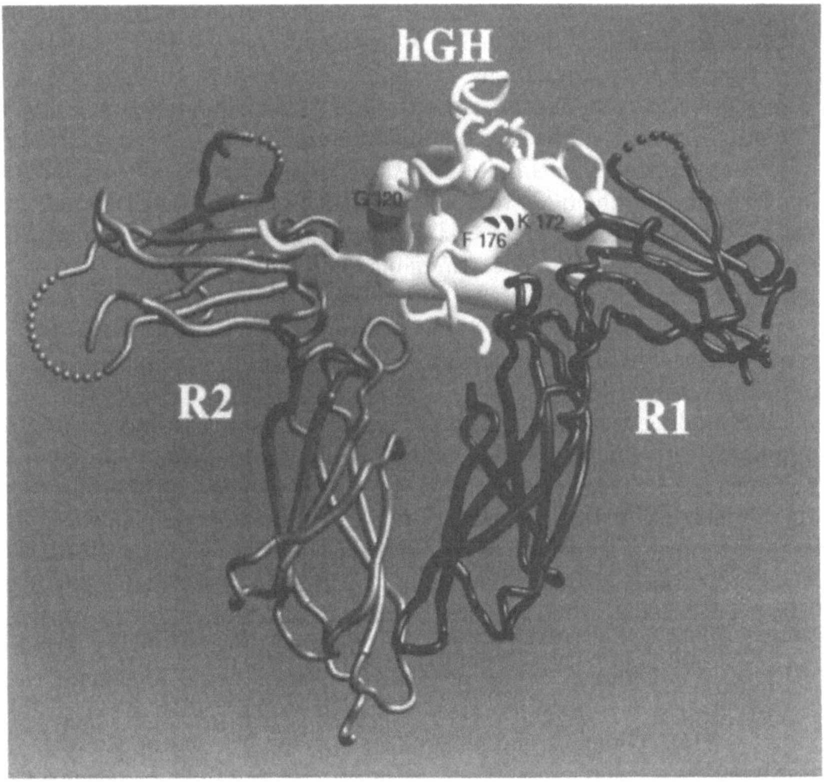

FIGURE 17.3. Ribbon diagram of the hGH·$(HGHbp)_2$ complex taken from a 2.8 Å X-ray structure (22). The two molecules of the hGHbp are shown complexed to site 1 (left) and site 2 (right) of hGH. The carboxyltermini of the hGHbp are shown to penetrate a model lipid bilayer at foreground. This illustration is based on X-ray coordinates of the hGH·$(hGHbp)_2$ complex. Reprinted with permission from Fuh, Cunningham, Fukunaga, Nagata, Goeddel, and Wells (14). Copyright by the AAAS.

GH (23). Both hormones have a basic 4-helix bundle motif that contains two sets of parallel helices linked by a short loop connection. Thus, helices 1 and 2 are parallel to each other and anti-parallel to helices 3 and 4 (reviewed in 4). One major difference between the hormone structures is that the long crossover loop in site 1 that connects helices 1 and 2 contains two ordered mini-helices in the hGH·$(hGHbp)_2$ complex. These mini-helices are unstructured loops in the pGH structure. It is possible this difference is due to the low level of refinement of the pGH structure or that these mini-helices assume their structure only upon binding to receptor 1.

The folding of each hGHbp is similar to immunoglobulins. It contains two beta-sandwich domains connected by a short linker. Each domain contains seven strands of alternating beta-sheet. The aminoterminus is in the domain farthest from the membrane, while the carboxyterminus is nearest the membrane. The binding of hGH brings the two hGHbps together so that they are parallel to each other and their carboxytermini are close together. This is consistent with the view that dimerization of the extracellular domain brings the intracellular domains into proximity to signal by some as yet unknown mechanism.

The structure shows there are three contact epitopes: two between each hGHbp and hGH (sites 1 and 2) and one between the two receptors. The contact epitopes for sites 1 and 2 generally match those functional epitopes defined by the alanine scan. However, there are some fundamental differences. First, both sites 1 and 2 contact epitopes are smaller than their corresponding functional epitopes. We find that some of the contacts near the periphery of the contact epitope are functionally silent; that is, when these side chains are converted to alanine we do not detect a 2-fold or greater change in binding affinity. In fact only about 7 of the 31 contact side chains in site 1 can account for about 70% of the binding energy (Cunningham, Lowman, Wells, unpublished results). Thus, one cannot easily interpret from the contact surface alone the functionality of the residues. Second, not all the side-chain mutations in site 1 that affected binding by alanine scanning are in contact—about 5 of them are mostly buried in the hormone just underneath the site 1 contact epitope. This indicates it is possible to engineer the binding affinity by mutating side chains that do not make direct contact but influence those that do.

Perhaps the most striking aspect of the hGH·$(hGHbp)_2$ complex structure (22) is that while the sites on hGH that bind the two hGHbp's are very different and have no obvious internal symmetry, the sites on the receptor that bind them are virtually identical. Thus the receptor uses essentially the same set of residues to bind totally different sites on the hormone. That the binding determinants on the receptor that bind sites 1 and 2 overlap is consistent with the observation that excess hGH will dissociate the dimeric complex (13). Mutational studies have defined many, but not all, of the important determinants used by the receptor in binding (24). The importance of the receptor-receptor contacts near the receptor stems has not been analyzed by mutational studies.

The structural analysis (22) also provided data consistent with the sequential dimerization mechanism (13). There is a general correlation between the binding constant of an interface and the surface area that becomes buried. The X-ray structure shows that the site 1 interface buries about 1250 $Å^2$, while site 2 buries only 850 $Å^2$. However, if the second receptor binds to the preformed hGH·hGHbp complex, then 1350 $Å^2$ of surface would become buried because of the additional 500 $Å^2$ buried at the receptor-receptor interface. We have estimated that the binding constant (kd) for formation of the hGH·hGHbp site 1 complex is 0.3 nM (7), and that for formation of the hGH·$(hGHbp)_2$ complex from the monomeric intermediate complex is 0.5 nM (13). Thus, the structure provides a plausible explanation for the sequential binding mechanism and for the similarities of these binding constants.

Protein Engineering of hGH

There are numerous applications of the functional and structural data base for designing useful variants of hGH. The fact that the functional epitopes in site 1 for binding the hGH and hPRL receptors are not identical has allowed us to design receptor-selective analogs of hGH (11). By mutating shared or nonshared residues, it has been possible to produce variants that bind over 100-fold weaker to either or both receptors. Such analogs have been useful for sorting out which of the large number of pharmacological effects of hGH can be attributed to one, both, or neither of the hGH or hPRL receptors. For example, the use of receptor-selective analogs and metal ion specificity studies has shown that activation of macrophages by hGH is mediated by the prolactin receptor (25).

Based on the functional data alone, it was possible to "recruit" nonbinding or weakly binding homologs of hGH—namely, hPRL and human placental lactogen (hPL)—to bind to the hGHbp via site 1 only 6-fold and 2-fold more weakly than hGH, respectively (26, 27). This was accomplished by changing nonconserved residues in each homolog to hGH residues in site 1 that were known to be important for binding to the hGHbp by alanine scanning. These recruited analogs were biologically inactive and provided the first clues that simple binding via site 1 was insufficient for biological activity.

From the functional, structural, and mechanistic data, it has been possible to build potent antagonists of hGH for the hGH and hPRL receptors (14, 16). A mutation (Gly120Arg) introduced a large side chain in site 2 epitope that prevented binding of the second hGH receptor. This analog is inactive as an agonist, but is a potent antagonist of the hybrid hGH-G-CSF receptor in FDC-P1 cells (14), the hGH receptor in IM-9 cells (15), and endogenous prolactin receptors in rat Nb2 cells or transfected into FDC-P1 cells (16). A Gly119Arg variant of bovine GH was independently found by Kopchick and coworkers to antagonize the action of GH in transgenic animals (28).

We have shown it is possible to improve the potency of the Gly120Arg antagonist by introducing mutations that enhance the affinity of the analog for site 1 either by alanine scanning (14) or by monovalent phage display (29, Lowman and Wells, unpublished results). These antagonists are powerful probes to specify the action of hGH. Moreover, these are potentially useful therapeutics for treating pathological conditions of growth hormone excess (as in acromegly) or hyperprolactinemia. In conclusion, an understanding of the molecular details of hormone-receptor interactions can greatly facilitate both the unraveling of their complex biology and the design of better pharmaceuticals.

Acknowledgments. We thank our colleagues at Genentech and coauthors for stimulating discussions, sharing unpublished data, and providing an extremely supportive environment.

References

1. Chawla RK, Parks JS, Rudman D. Structural variants of human growth hormone: biochemical, genetic and clinical aspects. Annu Rev Med 1983; 34:519–47.
2. Isaksson OGP, Lindahl A, Nilsson A, Isgaard. Action of growth hormone: current views. Acta Paediatr Scand 1988;343:12–8.
3. Wells JA, Cunningham BC, Fuh G, et al. The molecular basis for growth hormone-receptor interactions. Recent Prog Horm Res 1992;48:253–75.
4. Wells JA, de Vos AM. Structure and function of human growth hormone: implications for hematopoetins. Annu Rev Biophys Biomol Struct 1993.
5. Leung DW, Spencer SA, Cachianes G, et al. Growth hormone receptor and serum binding protein: purification, cloning and expression. Nature 1987; 330:537–43.
6. Boutin JM, Edery M, Shirota M, et al. Identification of a cDNA encoding a long form of prolactin receptor in human hepatoma and breast cancer cells. Mol Endocinol 1989;3:1455–61.
7. Fuh G, Mulkerrin MG, Bass S, et al. The human growth hormone receptor: secretion from *Escherchia coli* and disulfide bonding pattern of the extracellular binding domain. J Biol Chem 1990;265:3111–5.
8. Cunningham BC, Bass S, Fuh G, Wells JA. Zinc mediation of the binding of human growth hormone to the human prolactin receptor. Science 1990; 250:1709–12.
9. Wells JA. Systematic mutational analyses of protein-protein interfaces. Methods Enzymol 1991;202:390–411.
10. Cunningham BC, Jhurani P, Ng P, Wells JA. Receptor and antibody epitopes in human growth hormone identified by homolog-scanning mutagenesis. Science 1989;243:1330–6.

11. Cunningham BC, Wells JA. Rational design of receptor-specific variants of human growth hormone. Proc Natl Acad Sci USA 1991;88:3407–11.
12. Cunningham BC, Wells JA. High-resolution epitope mapping of hGH-receptor interactions by alanine-scanning mutagenesis. Science 1989;244:1081–5.
13. Cunningham BC, Ultsch M, de Vos AM, Mulkerrin MG, Clauser KR, Wells JA. Dimerization of the extracellular domain of the human growth hormone receptor by a single hormone molecule. Science 1991;254:821–5.
14. Fuh G, Cunningham BC, Fukunaga R, Nagata S, Goeddel DV, Wells JA. Rational design of potent antagonist to the human growth hormone receptor. Science 1992;256:1677–80.
15. Silva CM, Weber MJ, Thorner MO. Stimulation of tyrosine phosphorylation in IM-9 cells by human growth hormone receptor activation. Mol Endocrinol 1993.
16. Fuh G, Colosi P, Wood WI, Wells JA. Mechanism-based design of prolactin receptor antagonists. J Biol Chem 1993;268.
17. Elberg G, Kelly PA, Djiane J, Binder L, Gertler A. Mitogenic and binding properties of monoclonal antibodies to the prolactin receptor in Nb2 rat lymphoma cells: selective enhancement by anti-mouse IgG. J Biol Chem 1990;265:14770–6.
18. Cunningham BC, Mulkerrin MG, Wells JA. Dimerization of human growth hormone by zinc. Science 1991;253:545–8.
19. Moore J, Celnicker A, Light D, Spencer S. Cloned human growth hormone binding protein effects on disposition of hGH in rats [Abstract]. 71st annu meet Endocr Soc, Seattle, 1989.
20. Baumann G, Stolar MW, Ambarn K, Barsano CP, DeVries BC. A specific growth hormone-binding protein in human plasma: initial characterization. J Clin Endocrinol Metab 1986;62:134–41.
21. Herington AC, Ymer SI, Stevenson JL. Identification and characterization of specific binding proteins for growth hormone in normal human sera. J Clin Invest 1986;77:1817–23.
22. de Vos AM, Ultsch M, Kossiakoff AA. Human growth hormone and extracellular domain of its receptor: crystal structure of the complex. Science 1992;255:306–12.
23. Abdel-Meguid SS, Shieh HS, Smith WW, Dayringer ME, Violand BN, Bentle LA. Three-dimensional structure of a genetically engineered variant of porcine growth hormone. Proc Natl Acad Sci USA 1987;84:6434–7.
24. Bass SH, Mulkerrin MG, Wells JA. A systematic mutational analysis of hormone-binding determinants in the human growth hormone receptor. Proc Natl Acad Sci USA 1991;88:4498–502.
25. Fu YK, Arkins S, Fuh G, et al. Growth hormone augments superoxide anion secretion of human neutrophils by binding to the prolactin receptor. J Clin Invest 1992;89:451–7.
26. Cunningham BC, Henner DJ, Wells JA. Engineering human prolactin to bind to the human growth hormone receptor. Science 1990;247:1461–5.
27. Lowman HB, Cunningham BC, Wells JA. Mutational analysis and protein engineering of receptor-binding determinants in human placental lactogen. J Biol Chem 1991;266:10982–8.

28. Chen WY, Wright DC, Wagner TE, Kopchick JJ. Expression of a mutated bovine growth hormone gene suppresses growth of transgenic mice. Proc Natl Acad Sci USA 1990;87:5061–5.
29. Lowman HB, Bass SH, Wells JA. Selecting high-affinity binding proteins by monovalent phage display. Biochemistry 1991;30:10832–8.

18

Growth Hormone Binding Protein: A Modulator of Growth Hormone Activity

Ross Clark

Growth hormone (GH) is present in the blood in many forms. A recent review (1) counts over 100 naturally occurring variant forms of *human GH* (hGH). Many of these forms of hGH are also probably present in blood bound, in varying amounts, to a specific high-affinity *GH binding protein* (GHBP). The activity and physiological relevance of all these hGH variants in their bound and unbound forms are ill defined. It is therefore somewhat surprising that when the biological activities of hGH are considered they are usually restricted to the actions of the predominant pituitary form of hGH, the 22-kd form. However, it has been estimated (1) that only about 20% of the GH in the blood circulates "free" as 22-kd hGH. The same reviewer estimates that another 20% of the hGH in blood is also 22-kd hGH, but this circulates bound to the GHBP. The older literature contains evidence that forms of GH in plasma other than 22-kd GH possess the majority of the bioactivity of plasma, as measured in the classical rat in vivo GH bioassays. There is a body of older work (2) indicating that there is much more bioassayable GH-like activity in plasma in high molecular weight fractions than in the lower molecular weight fractions containing free GH. We now know that this high molecular weight material would contain GH bound to the GHBP. These findings suggest that the GH-GHBP complex might maintain or even enhance GH bioactivity.

History of the Growth Hormone Binding Protein (GHBP)

Discovery in the 1960s

It was shown first by Hadden (3, 4) nearly thirty years ago that there are proteins in the blood of mammals that specifically bind GH. At this time methods for the immunological measurement of peptide hormones were being developed. The discovery of the GHBP came about when it was found that GH was especially difficult to measure in the presence of serum. Initial studies (reviewed in 4) showed that if radiolabeled human GH was incubated with serum, a high molecular weight complex was formed, due to the labeled GH binding to a protein.

In subsequent years doubts were cast on both the significance, specificity, and nature of this binding activity. It was generally believed that the binding activity was an artifact, due to GH being denatured by being radiolabeled, or that the GH was aggregating to produce high molecular weight oligomeric forms. This view was supported by the dogma of the time that all protein hormones circulated in the blood in a "free" or unbound form (1). In the following decade there was a body of literature showing biochemical data that GH did aggregate, especially in preparations of pituitary-derived GH (5). Similar high molecular weight forms of GH were also reported to be present in plasma. Such molecules were given names, such as "big" and "big-big" forms of GH (6). The relationship between these species of GH and the binding protein described by Hadden and others was unclear.

Further Characterization in the 1970s and 1980s

Interest in this area of GH research was rekindled in 1977 by the discovery (7) of a GHBP in the blood of pregnant mice. If serum from pregnant, but not from nonpregnant, mice was incubated with radiolabeled hGH and fractionated on a sizing column, a high molecular weight form of hGH of approximately 60 kd was eluted. This binding was specific to GH and was displaceable with cold GH, clearly demonstrating the presence of a GHBP in serum. The failure of previous investigators to identify the GHBP may have been due to its relatively low concentration in the blood and to its dissociation during chromatography. GH oligomers migrate in the same size range, which may have also contributed to the presence of bound GH going undetected (8).

Clinical endocrinologists became more aware of this research when Herington (9) and Baumann (8) separately described a high-affinity GHBP in human sera. They found that adding radiolabeled hGH to blood, followed by running a sizing column, gave two peaks of radiolabel, due to a higher molecular weight peak eluting before the main peak of free hGH. A second peak of binding activity, which is of high molecular weight, low affinity, and high capacity, as yet has been poorly characterized (1).

Relationship of the GHBP to the GH Receptor

Baumann (8) and Herington (9) both considered the possibility that the GHBP was related to the GH receptor. Baumann leaned away from thinking that the GHBP was a shed receptor for three reasons: (i) The affinity of the GHBP for GH was lower than that of the receptor; (ii) the molecular weight was different; and (iii) the concentration of the GHBP in blood was too high relative to the low density of GH receptors in tissues. On the other hand Herington speculated that the GHBP and GH receptor were related based on their antigenic similarity. At this time Barnard and Waters showed (10) that the GHBP and liver GH receptor in the rabbit were antigenically indistinguishable using a panel of monoclonal antibodies. Humans with Laron dwarfism were found to be deficient in both GH receptors in the liver (11) and GHBP in serum (12), suggesting that the GH receptor and the GHBP might be closely related.

Therefore, there was clear evidence of similarities between the GHBP and the GH receptor. This implied that perhaps the hormone binding domain of the GH receptor was cleaved to circulate in the blood. At that time the idea that protein hormones circulated bound to specific binding proteins was not widely accepted (1). However, it had already been shown that the insulin-like growth factors circulated bound to serum binding proteins (13, 14), so the concept of binding proteins for protein hormones was not new to GH endocrinology.

Purification and Cloning

GHBPs and GH Receptors in Rabbits and Humans

The purification of the GHBP from rabbit serum, and of the GH receptor from detergent-solubilized rabbit liver membranes, was finally accomplished by Spencer and colleagues (15) at Genentech, chiefly by using hGH affinity columns. Partial amino acid sequences of the rabbit GHBP and the GH receptor were then obtained. The evidence of similarity between the rabbit GHBP and GH receptor was confirmed; there was clear sequence identity between the serum GHBP and the liver GH receptor. Using an oligonucleotide based on the partial amino acid sequence, the full-length rabbit GH receptor was then cloned in the Wood Laboratory at Genentech (16). The same group cloned the human GH receptor and showed that the human GHBP shared sequence identity to the extracellular domain of the human GH receptor. The extracellular domain of the cloned receptor, equivalent to the serum GHBP, was then expressed and was shown to bind GH with high affinity (17). GH receptors have since been cloned from several species, including rodents (see below), and GHBPs unequivocally demonstrated in some of these species.

The cloning of the GH receptor made it clear how the GHBP could be generated. The GH receptor, of 620 amino acids in the human and the rabbit, has one intracellular domain of 350 amino acids, one transmembrane domain of approximately 24 amino acids, and at the aminoterminal one extracellular domain of 246 amino acids (16). Thus, the GHBP could be produced by the intact extracellular domain of the GH receptor being shed due to the receptor being cleaved near its transmembrane domain. This would give an intact hormone binding domain capable of binding GH. There has been surprisingly little evidence to prove directly that the GHBP can be generated by the proteolytic cleavage of the extracellular domain of the GH receptor. The only published data come from one cell culture study (18), using cultured human IM-9 lymphocytes, which showed that sulfhydryl group inactivators caused the cleavage and appearance of the GHBP in the medium.

GHBP and GH Receptor in Rodents

The subsequent cloning of rodent GH receptors gave an unexpected finding. In the rat (19) and mouse (20) a separate mRNA, coding for a truncated GH receptor, is produced by alternative splicing. This truncated GH receptor, of 261 amino acids in the rat, is essentially the extracellular domain of the full-length receptor with a short C-terminal hydrophilic extension of 17 amino acids. This truncated receptor lacks the transmembrane and intracellular domains, which one would assume to be important for signal transduction, so it would appear to be incapable of producing an intracellular signal.

It remains unclear whether proteolytic cleavage is the only mechanism for the generation of the GHBP in primates. As yet no alternatively spliced GH receptors have been characterized in humans. The prolactin receptor is highly homologous to the GH receptor, and a variety of alternatively spliced prolactin receptor messages have been detected in rodents (21), but not in humans.

Why a Soluble Receptor?

Why should rodents produce a truncated receptor that may be incapable of functioning as a traditional transmembrane signaling receptor? The truncated alternate mRNA codes for the soluble binding protein that appears as a GHBP in serum. Subsequent investigation in the rat of the distribution of the expression of the mRNAs of the full-length receptor and the binding protein showed that the receptor and GHBP were expressed at differing times during development and expressed in the adult in different tissues (22). Some tissues appeared to produce more GHBP than full-length GH receptor. This could be taken as evidence that the GHBP might fulfill some functional role in GH physiology.

However, in the cytokine receptor superfamily, of which the GH receptor is a member, there are other receptors that are truncated. Some of these truncated receptors—for example, the interleukin-6 receptor—have been shown to be capable of participating in signal transduction and inducing a cellular response (23). There are reports (24, 25) that the GHBP is present associated with intracellular membrane fractions and, therefore, speculation that it not only circulates as a GHBP but also participates actively in the cellular response to GH.

Measurement of the GHBP

The regulation and function of a blood-borne protein can sometimes be inferred by its measurement in different physiological states. When a new protein or hormone is discovered, it is usual in initial experiments to attempt to remove the source of the molecule and then to attempt to measure changes in body function. With the GHBP such experiments do not seem possible, except perhaps using molecular biology to produce targeted gene knockouts. Discovering the function of the GHBP by administering it to animals is a possibility. However, there are relatively low concentrations of the GHBP in blood, making it difficult to purify large amounts of GHBP. This lack of material has restricted such experimental approaches. Therefore, the measurement of the GHBP in blood has been a major approach to studying GHBP function.

Column Chromatography

The first assays for the GHBP were based on the addition of radiolabeled GH to a serum sample, followed by column chromatography and the measurement of the amount of radiolabel in the high molecular weight peak where the GHBP eluted (8, 9). These assays have identified changes in serum GHBP concentrations during development and in various hormonal states in experimental animals and humans in health and disease. Such assays are relatively slow to run, and there is significant interference if GH concentrations in the sample are high since the GH in the sample competes for binding with the labeled GH. Despite their limitations chromatographic assays have provided much information on GHBP concentrations in blood.

Radioimmunoassay

More rapid and specific methods for the measurement of the GHBP are now available. With the production of recombinant rodent GHBP it has been possible to produce suitable polyclonal antibodies suitable for use in immunoassays. In addition, the pure GHBP can be iodinated and also used as an assay standard. For these reasons, specific and sensitive radioimmunoassays for rodent GHBP have now been developed (26, 27). The GHBP can now be measured in small blood samples in the rat, so GHBP profiles can be obtained and the regulation of the GHBP studied in more detail.

Ligand-Mediated Immuno-Functional Assay (LIFA)

The LIFA is a new type of "sandwich" assay that uses both hGH binding and immunological recognition of the hGHBP for the measurement of the hGHBP (28). In the LIFA a coat antibody is used to capture the hGHBP, and an antibody against hGH is used to detect the amount of hGH bound to the hGHBP. If the assay is flooded with excess hGH before the antibody against hGH is used, then total hGHBP is measured. The LIFA, therefore, combines the advantages of the traditional method of detecting the GHBP, measuring the amount of GH binding in a sample, with the advantage of using antibodies to specifically measure the GHBP.

The LIFA assay has been used to study the regulation of the GHBP in children with growth disorders. As measured by the LIFA, children with GH deficiency do not have a lowered serum GHBP concentration, and hGH replacement does not seem to alter GHBP concentrations. However, many children previously diagnosed as being of idiopathic short stature do have very low serum GHBP concentrations compared to age-matched normal children. Their lowered GHBP concentration may reflect a low level of GH receptors and GHBP in the body or, more interestingly, a relationship between their low GHBP levels and their growth failure.

Regulation of GHBP Production by GH

Short-Term Effects of GH

Little is known about the physiological regulation of the concentration of the GHBP in the blood. For the reliable measurement of the GHBP it has been important to establish the normal variation in GHBP concentrations throughout the day. For example, the pulsatile pattern of GH production might be expected to impact GHBP concentrations. Both GH and GHBP concentrations have now been measured in serial blood samples, and it seems that there is very little variation in total serum GHBP concentrations throughout the day in the rat (27), measured using a GHBP radioimmunoassay, or in humans (29), measured using the GHBP LIFA. The concentration of GHBP is very stable despite large changes in GH concentrations in the same samples. Using other methods of measurement, it had been reported that there is an inverse relationship between serum GH and GHBP concentrations (30). There now seems to be a consensus that serum GHBP concentrations change little throughout the day and that a single blood sample drawn at any time of the day can provide a reliable estimate of total serum GHBP.

Effect of Long-Term GH Exposure

Most research on the regulation of the GHBP has been in the rat, where the GH receptor and the GHBP are produced by separate mRNA and thus have the potential to be independently regulated. Therefore, in rodents there is not necessarily a positive correlation between GH receptor number in target tissues and GHBP concentrations in blood. The regulation of GH receptors in the rat is at first sight confusing; for example, hypophysectomy reduces liver GH receptor mRNA and serum GHBP concentrations in the female, but not in the male rat (31). This sex difference reflects the importance of the pattern of GH exposure in the rat.

Effect of Pattern of GH Administration in Rodents

In mammals, both the total amount of GH that is secreted by the pituitary and the pattern of GH secretion are believed to be important for the growth-promoting actions of GH (32). In the rat, frequent injections of GH give larger growth responses than infrequent injections or continuous infusions of GH (32). Some responses to GH, particularly of some enzymes in the rat liver, depend more on the pattern of GH exposure (continuous or intermittent) than on the amount of GH exposure (33).

The female rat has higher serum GHBP concentrations than the male rat (26, 27). In the GH-deficient dwarf rat (dw/dw) there is no sex difference in serum GHBP concentrations, suggesting that the higher level of GHBP in normal females is due to a higher basal GH secretion (27). This appears to be the case since giving male dw/dw rats (27) or hypophysectomized rats (31) a continuous GH infusion, to mimic the female pattern of GH secretion, leads to a large increase in serum GHBP concentrations. In contrast, giving injections of GH to a hypophysectomized female rat does not increase serum GHBP concentrations (31). Therefore, in the rat, continuous GH exposure increases serum GHBP concentrations, while GH pulses from bolus injections have little effect. This might help explain why GHBP concentrations rise in rodents in some situations where GH exposure is continuous—for example, in pregnancy (7), where GH is present in a nearly continuous manner.

Pattern-Dependent Effects of GH in Humans?

Pattern-dependent effects of GH have not been as actively studied in humans as they have been in rodents. This has in part been due to a lack of blood markers that could be identified as truly dependent on the pattern of GH exposure. As the serum GHBP concentration is a clear marker of the pattern of GH exposure in the rodent, it will be of great interest to discover if the serum GHBP concentration is also dependent on the pattern of GH exposure in humans.

Regulation of GH Activity by GHBP

GHBP Inhibits GH Activity In Vitro

The concentration of the GHBP in blood varies with the physiological state of an animal, demonstrating that the level of circulating GHBP is regulated. However, this does not supply direct evidence of the GHBP modulating the in vivo activities of GH. A large amount of the GH in the blood is present in a bound form, complexed to the GHBP (34). What is the physiological significance of this relatively large amount of bound GH in the blood? Few experiments have directly addressed the functional role in vivo of the GH bound to the GHBP. However, some in vitro experiments, using either natural or recombinant GHBP, have attempted to test the bioactivity of GH-GHBP complex (35, 36). Such experiments have shown that the GHBP can inhibit GH binding to cellular GH receptors. In addition, the GHBP can inhibit the actions of GH in cell culture: in the case of 3T3-F433A pre-adipocytes, by inhibiting their conversion into adipocytes. Raising the concentration of GH to saturate the GHBP, and thus elevating free GH, removed the inhibitory effect of the GHBP on GH binding and activity.

Such in vitro evidence from cell-based assays suggests that the role of the GHBP in vivo may be to inhibit GH activity. This in vitro evidence may not be directly applicable to the in vivo situation and should be regarded with some caution. In such static cell-based assay systems the GHBP and the cellular GH receptors are competing for the same binding sites on the GH molecule, so the demonstrations of inhibition are the most likely and expected result.

It may be of value to compare studies with GH and the GHBP with similar studies with IGF-I and the IGFBPs. For example, there is some evidence that IGFBP-3 can enhance or inhibit the activity of IGF-I under different conditions of in vitro cell culture (37). When IGF-I is coadministered with recombinant IGFBP-3, the efficacy of IGF-I can be enhanced or reduced depending on the form of IGFBP-3 given and the response measured (38). It seems that similar experimental approaches could be used to dissect the GH-GHBP system.

Clearance of Exogenous GH-GHBP Complex

Compared to the static situation of an in vitro culture system, there is a very dynamic situation in vivo, where GH concentrations change rapidly and widely and free GH is cleared relatively quickly from the circulation. In contrast to this rapid clearance of free GH, it has been shown that the clearance of GH is greatly slowed when either natural GHBP or recombinant GHBP is administered to rats along with GH (39, 40). This is not surprising, considering that the complex of GH and the GHBP has a much higher molecular weight than free GH and is therefore not as easily filtered by the kidney.

Multiple Species of the GH-GHBP Complex

Another explanation for the slowed clearance of GH following the formation of the GH-GHBP complex has come from recent studies (41) using recombinant GHBP, where the nature of the hGH-hGHBP complex has been elucidated. These studies have produced the surprising finding that the hGH-hGHBP complex can exist in two states, with either one or two molecules of hGHBP being bound to one molecule of hGH. The abundance of these two forms of the hGH-hGHBP complex depends on their relative concentrations. At high hGH concentrations relative to hGHBP concentrations, a one-to-one complex of hGH-hGHBP is the preferred form, while at low hGH concentrations and relatively high hGHBP concentrations, each hGH molecule binds to two hGHBP molecules.

The cocrystallization of hGH and the hGHBP has elegantly confirmed these findings. The crystal structure of the complex gave a surprising result, as each GH molecule was bound to two GHBP molecules to form a trimer (42). It is likely that native GHBP also occurs as both forms of the GH-GHBP complex. Smith and Talamantes (43) have reported that in mouse serum the GH-GHBP complex can occur as a dimer of two GHBP molecules bound to one labeled GH molecule. The conversion of free GH into this high molecular weight species of the GH-GHBP complex obviously would slow the clearance of GH and thereby cause GH to be retained in the body.

Clearance of Exogenous and Endogenous GH-GHBP Complex

Therefore, GH administered prebound to natural or recombinant GHBP shows a slowed clearance. However, the evidence from measuring endogenous GH and GHBP in serially drawn blood samples indicates that free and bound GH disappear with similar time courses (27). It seems that despite the maintained presence of relatively high concentrations of

GHBP in the blood and the binding of endogenous GH to this GHBP (28, 29), circulating GH is cleared from the blood as if it were unbound. This can be measured directly in the LIFA, where in serially drawn blood samples the fraction of GH bound to the GHBP disappears in parallel with the total amount of GH. The similar clearance of bound and unbound GH is reflected in the very low basal GH concentrations between endogenous pulses of GH, despite relatively high concentrations of GHBP. This area is in need of much further research.

hGHBP Enhances hGH Activity in the Rat

Recent studies at Genentech (44) have indicated that recombinant nonglycosylated hGHBP can enhance the activity of coadministered hGH. In hypophysectomized and in dwarf rats the enhancing activity of the hGH-hGHBP complex, compared to hGH alone, can be seen on many measures of growth, including body weight gain, organ weight, long bone growth, and blood IGF-I concentrations.

Increasing the half-life of a hormone might be expected to increase activity in vivo. This has been shown previously for hGH, as the growth-promoting activity of GH can be enhanced if it is given with some anti-GH monoclonal antibodies (45, 46). As noted above, the GH receptor and the GHBP compete for the same binding sites on the GH molecule, so although co-administering the GHBP may slow the clearance of GH, it also "hides" the active sites of GH from the receptor, especially when two GHBPs are bound to each GH molecule. There may therefore be a balance between the GHBP enhancing the activity of the GH by maintaining its presence in the body and the GHBP inhibiting activity by binding the active sites on the GH.

The GHBP might also influence the activity of GH by acting as a "trap" for GH in the extracellular compartment. One might expect that in the immediate surroundings of cells producing GHBP there would be a higher GHBP concentration than in the bloodstream. There should therefore be a concentration gradient for the GHBP from such tissues into the bloodstream. In addition, the recombinant nonglycosylated hGHBP has a lower affinity for hGH than the full-length hGH receptor (17). If the same is true for natural GHBP and GH receptors, then when GH dissociates from the GH-GHBP complex it would have a higher affinity for the GH receptor than for the GHBP. It may be that the GHBP can therefore "hand off" GH to the GH receptor.

Further studies are needed to define how the many activities of GH are affected by GHBP. For example, the metabolic activities of GH, including lipolytic and diabetogenic activities, should also be studied. In the above experiments, recombinant hGHBP was used, which does not bind to rat GH so that it is very unlikely that in the rat this hGHBP would show activity when given by itself. It is possible that in a homologous system (e.g., if rat GHBP were given by itself to a rat whose pituitary was capable of secreting GH) the rat GHBP alone may be able to enhance the activity of the endogenous rat GH and thereby stimulate body growth.

As the *E. coli*-produced hGHBP has a clearance rate similar to natural GHBP in the rat, it is likely that the growth-enhancing activity of this form of the recombinant GHBP also reflects the activity of natural GHBP. Although hGHBP can enhance hGH activity when they are coadministered in the rat, it does not follow necessarily that the role of endogenous GHBP is to enhance the growth-promoting activity of GH. Recombinant hGHBP can enhance hGH activity in the rat, but GHBPs in other species may have different effects on GH activity. In fact, some recent in vivo data (31) on the effect of rat GH administration in the rat were interpreted as showing that endogenous rat GHBP inhibited the activity of the administered rat GH.

The most likely reason for the GHBP enhancing the activity of GH is the pharmacokinetic explanation outlined above. Another possibility is that the GH-GHBP complex can directly participate in signal transduction, as described above for the soluble IL-6 receptor (23). Perhaps when the signal transduction mechanisms for GH are resolved, the role of the GHBP in mediating GH activity will be revealed.

Summary

Doubts expressed within the last ten years (48) regarding the existence of binding proteins for protein hormones have now disappeared. With the availability of specific assays, mRNA probes, and recombinant GHBP, much will be learned in the next ten years regarding the endocrinology of binding proteins. Understanding these molecules will lead to a greater understanding of GH action on growth and metabolism and should assist in the development of treatments for human disorders involving the GH system.

It seems likely that the GHBPs, like the IGFBPs, modulate the activities of these hormones. The use of GH and IGF-I in medicine is set to expand greatly in the future, and it may be that their BPs, already useful for diagnosis, will one day prove useful as therapeutics. GHBP administered in combination with GH can slow the clearance of GH from the blood and enhance the growth-promoting activity of GH. In the future, the GHBP-GH complex may be of use as a more potent and more long-acting form of GH therapy.

References

1. Baumann G. Growth hormone heterogeneity: genes, isohormone variants, and binding proteins. Endocr Rev 1991;12:424–49.

2. Ellis S, Vodian MA, Grindeland RE. Studies on the bioassayable growth hormone-like activity in plasma. Recent Prog Horm Res 1978;34:213–38.
3. Hadden DR, Prout TE. A growth hormone binding protein in normal human serum. Nature 1964;202:1342–3.
4. Hadden DR. Historical review of early studies on protein binding of growth hormone. Acta Endocrinol (Copenh) 1991;124:7–9.
5. Berson SA, Yalow RS. State of human growth hormone in plasma and changes in stored solutions of pituitary growth hormone. J Biol Chem 1966; 241:5745–9.
6. Gorden P, Hendricks C, Roth J. Evidence for "big" and "little" components of human plasma and pituitary growth hormone. J Clin Endocrinol Metab 1973;36:178–84.
7. Peeters S, Friesen HG. A growth hormone-binding factor in the serum of pregnant mice. Endocrinology 1977;101:1164–79.
8. Baumann G, Stolar MW, Amburn K, Barsano CP, DeVries BC. A specific growth hormone-binding protein in human plasma: initial characterization. J Clin Endocrinol Metab 1986;62:134–41.
9. Herington AC, Ymer SI, Stevenson JL. Identification and characterization of specific binding proteins for growth hormone in normal human sera. J Clin Invest 1986;77:1817–23.
10. Barnard R, Waters MJ. Serum and liver cytosolic growth-hormone binding proteins are antigenically identical with liver membrane "receptor" types 1 and 2. Biochem J 1986;237:885–92.
11. Eshet R, Laron Z, Pertzelan A, Arnon R, Dintzman M. Defect of human growth hormone receptors in the liver of two patients with Laron-type dwarfism. Isr J Med Sci 1984;20:8–11.
12. Daughaday WH, Trivedi B. Absence of serum growth hormone binding protein in patients with growth hormone receptor deficiency (Laron dwarfism). Proc Natl Acad Sci USA 1987;84:4636–40.
13. Zapf J, Waldvogel M, Froesch ER. Binding of non-suppressable insulin-like activity to human serum. Evidence for a carrier protein. Arch Biochem Biophys 1975;168:638–45.
14. Hintz RL, Liu F. Demonstration of specific plasma protein binding sites for somatomedin. J Clin Endocrinol Metab 1977;45:988–95.
15. Spencer SA, Hammonds RG, Henzel WJ, Rodriguez H, Waters MJ, Wood WI. Rabbit liver growth hormone receptor and serum binding protein: purification, characterization and sequence. J Biol Chem 1988;263:7862–7.
16. Leung DW, Spencer SA, Cachianes G, et al. Growth hormone receptor and serum binding protein: purification, cloning and expression. Nature 1987; 330:537–43.
17. Fuh G, Mulkerrin MG, Bass S, et al. The human growth hormone receptor. Secretion from *Escherichia Coli* and disulfide bonding pattern of the extracellular domains. J Biol Chem 1990;265:3111–5.
18. Trivedi B, Daughaday WH. Release of growth hormone binding protein from IM-9 lymphocytes by endopeptidase is dependent on sulfydryl group inactivation. Endocrinology 1988;123:2201–6.
19. Baumbach WR, Horner DL, Logan JS. The growth hormone-binding protein in rat serum is an alternatively spliced form of the rat growth hormone receptor. Genes Dev 1989;3:1199–205.

20. Smith WC, Kuniyoshi J, Talamantes F. Mouse serum growth hormone binding protein has growth hormone receptor extracellular and substituted transmembrane domains. Mol Endocrinol 1989;6:984–90.
21. Kelly PA, Dijane J, Postel-Vinay M-C, Edery M. The prolactin/growth hormone receptor family. Endocr Rev 1991;12:235–51.
22. Carlsson B, Billig H, Rymo L, Isaksson OGP. Expression of the growth hormone binding protein messenger RNA in the liver and extra-hepatic tissues in the rat; co-expression with the growth hormone receptor. Mol Cell Endocrinol 1990;73:R1–6.
23. Hibi M, Murakami M, Saito M, Hirano T, Taga T, Kishimoto T. Molecular cloning and expression of an IL-6 signal transducer, gp130. Cell 1990;63: 1149–57.
24. Lobie PE, Garcia-Aragon J, Wang BS, Baumbach WR, Waters MJ. Cellular localization of the growth hormone binding protein in the rat. Endocrinology 1992;130:3057–65.
25. Frick GP, Goodman HM. Characterization of the short isoform of the growth hormone receptor synthesized by rat adipocytes. Endocrinology 1992;131: 3083–90.
26. Sadeghi H, Wang BS, Lumanglas AL, Logan JS, Baumbach WR. Identification of the origin of the growth hormone binding protein in rat serum. Mol Endocrinol 1990;4:1799–1805.
27. Carmignac DR, Wells T, Carlsson LMS, Clark RG, Robinson ICAF. Growth hormone (GH)-binding protein in normal and GH-deficient dwarf rats. J Endocrinol 1992;135:447–57.
28. Carlsson LMS, Rowland A, Clark RG, Gesundheit N, Wong WT. Ligand-mediated immuno-functional assay for quantitation of growth hormone binding protein in human blood. J Clin Endocrinol Metab 1991;73:1216–23.
29. Carlsson LMS, Rosberg S, Wong WL, Albertsson-Wikland K. Analysis of 24-hour profiles of growth hormone (GH)-binding protein in healthy children [Abstract 63]. Horm Res 1991;35(suppl 2).
30. Bick T, Amit T, Barkey R, Hertz P, Youdim MBH, Hochberg Z. The interrelationship of growth hormone (GH), liver membrane GH receptor, serum GH-binding protein activity, and insulin-like growth factor-I in the male rat. Endocrinology 1990;126:1914–20.
31. Maiter D, Walker JL, Adam E, et al. Differential regulation by growth hormone (GH) of insulin-like growth factor I and GH receptor/binding protein gene expression in rat liver. Endocrinology 1992;130:3257–64.
32. Clark RG, Jansson J-O, Isaksson O, Robinson ICAF. Intravenous growth hormone: growth responses to patterned infusions. J Endocrinol 1985;104: 53–61.
33. Jeffrey S, Carter ND, Clark RG, Robinson ICAF. The episodic secretory pattern of growth hormone regulates liver carbonic anhydrase III—studies in normal and mutant growth-hormone-deficient dwarf rats. Biochem J 1990; 266:69–74.
34. Baumann G, Vance ML, Shaw MA, Thorner MO. Plasma transport of human growth hormone in vivo. J Clin Endocrinol Metab 1990;71:470–3.
35. Mannor DA, Winer LM, Shaw MA, Baumann G. Plasma growth hormone (GH)-binding proteins: effect of GH binding to receptors and GH action. J Clin Endocrinol Metab 1991;73:30–4.

36. Lim L, Spencer SA, McKay P, Waters MJ. Regulation of growth hormone (GH) bioactivity by a recombinant GH-binding protein. Endocrinology 1990;127:1287–91.
37. Conover CA. Potentiation of insulin-like growth factor (IGF) action by IGF-binding protein-3: studies of underlying mechanism. Endocrinology 1992; 130:3191–9.
38. Clark RG, Mortensen D, Reifsynder D, Mohler M, Etcheverry T, Mukku V. Recombinant human insulin-like growth factor binding protein-3 (rhIGFBP-3): effects on the glycemic and growth promoting activities of rhIGF-1 in the rat. Growth Regul 1993;3:48–50.
39. Baumann G, Amburn KD, Buchanan TA. The effect of circulating growth hormone binding protein on metabolic clearance, distribution and degradation of human growth hormone. J Clin Endocrinol Metab 1987;64:657–60.
40. Moore J, Celniker A, Fuh G, Light D, McKay P, Spencer S. Cloned human growth hormone binding protein effects on disposition of human growth hormone in rats [Abstract 1652]. 71st annu meet Endocr Soc, Seattle, 1989.
41. Cunningham B, Ultsch M, de Vos A, Mulkerrin M, Clauser K, Wells J. Dimerization of the extracellular domain of the human growth hormone receptor by a single hormone molecule. Science 1991;254:821–5.
42. de Vos AM, Ultsch M, Kossiakoff AA. Human growth hormone and extracellular domain of its receptor; crystal structure of the complex. Science 1992;255:307–12.
43. Smith WC, Talamantes F. Gestational profile and affinity cross-linking of the mouse serum growth hormone-binding protein. Endocrinology 1988;123: 1489–94.
44. Clark R, Cunningham B, Moore J, et al. Growth hormone binding protein enhances the growth promoting activity of GH in the rat [Abstract]. 73rd annu meet Endocr Soc, Washington, DC, 1991:433.
45. Holder AT, Aston R, Preece MA, Ivanyi JJ. Monoclonal antibody-mediated enhancement of growth hormone activity in vivo. J Endocrinol 1985;107: R9–12.
46. Thomas HM, Green IC, Wallis M, Aston RJ. Induction of lactogenic receptors in liver of hypophysectomized rats treated with bovine growth hormone-monoclonal antibody complexes. J Endocrinol 1990;124:469–74.
47. Lobie PE, Barnard R, Waters MJ. The nuclear growth hormone receptor binding protein—antigenic and physiochemical characterization. J Biol Chem 1991;266:22645–52.
48. Berson SA, Yalow RS. Peptide hormones in plasma. Harvey Lect 1986;62: 107–63.

Part V

Clinical Application of Growth Hormone Treatment

19

Growth Hormone Treatment, Acromegaly, and Relationship to Cancer and Leukemia

LOUIS E. UNDERWOOD

Use of *growth hormone* (GH) as a therapeutic agent has expanded over the past several years to include children whose short stature is due to conditions other than GH deficiency. Larger numbers of children are being treated and are receiving doses of GH that are generally higher than those used when only pituitary GH was available. One of the potential side effects of GH about which there is concern is that of a tumorigenic action of this peptide. This concern came into focus in 1988 when several children developed leukemia after receiving GH therapeutically (1). In a previous report we discussed some of the factors that should be considered in assessing the influence of GH on tumor formation and tumor growth (2). We observed that several studies indicate that the production rates for GH in normal adults range between 5.5 and 17.4 μg/kg/day and that acromegalic patients produce 50–>1000 μg/kg/day. Most children treated with GH received 25–50 μg GH/kg/day. Therefore, the amounts of GH given therapeutically are slightly above daily GH production by normal subjects, but lower than the amount produced by most patients with acromegaly. The relevance of this information to the possible tumorigenic actions of GH is problematic because these data do not account for any differences that might result from the secretion of GH in multiple daily bursts, compared to once-daily injections. Likewise, they do not take into account whether injections of exogenous GH cause the suppression of endogenous secretion (2).

In this chapter we consider the possible role of therapy with GH in the genesis and growth of tumors from the perspective of the effect of GH or IGF-I on tumor cells in vitro; the influence of GH or IGF-I on the growth of experimental tumors in vivo; the occurrence of tumors in patients with GH excess; and occurrence (or recurrence) of neoplasms during or after GH therapy.

Effects of GH or IGF-I on Tumor Cells In Vitro

A variety of cultured tumor cell lines have receptors for IGF-I, a major mediator of GH action (3–8), and some cell lines exhibit more IGF-I binding than cells from normal surrounding tissues (4, 7, 9). Breast cancer cells were targeted for study by Furlanetto and DiCarlo (3), who showed that IGF-I receptors were present in the four lines studied; that specific binding of radiolabeled IGF-I ranged between 2% and 8%; that the kds estimated from curves of competitive binding were between 0.5 and 4 nM; and that, as expected for the type-I IGF receptor, insulin competed with a potency of only 1/10 to 1/100 that of IGF-I. In another study, Pekonen et al. (4) compared the IGF-I binding of membrane preparations from breast cancers removed from 15 patients with normal breast tissue from the same patients. Cancerous tissue consistently bound more IGF-I/mg of membrane protein (usually about two times more binding) than normal tissue. Receptors for IGF-I have also been demonstrated in cell membranes prepared from colon cancers (5) and in cultured human erythroleukemia cells (9). As with breast cancer cells, the receptors for IGF-I in these tumors have the characteristics expected for the type-I IGF receptor. In erythroleukemia cells, the IGF-I binding subunit of the type-I IGF receptor has been shown to exhibit the 130-kd size expected for this receptor (9).

IGF-I is also produced by breast cancers (10, 11). Huff et al. (10) presented evidence that IGF-I-related activity is present in conditioned media from cultured lines of breast cancer cells (Table 19.1) and that the activity measured is not the result of interference from IGF binding proteins. They showed that dose response curves of acid-ethanol purified conditioned medium is parallel to IGF-I in the IGF-I RIA; that the IGF-I from conditioned media is similar in size to authentic IGF-I; and that the expected species of mRNA hybridize to an IGF-I cDNA probe (10).

Evidence for involvement of IGF-I in tumor growth comes from several reports showing that addition of IGF-I to cultured tumor cells stimulates

TABLE 19.1. Immunoreactive IGF-I produced by human breast cancer cells in culture.

Cell line	IGF-I in conditioned media (ng/ml)
MCF-7	63.9 ± 3.4
ZR-75-1	40.7 ± 6.7
MDA-231	158.0 ± 34
Hs 578T	173.0 ± 44

Note: Values are expressed as mean ± 1 SD. Conditioned media were collected for 48 h and subjected to acid-ethanol extraction before assay.
Source: Data from Huff, Kaufman, Gabbay, Spencer, Lippman, and Dickson (10).

DNA synthesis and cell proliferation (3, 8, 10, 12–14). Furlanetto and DiCarlo (3) reported that IGF-I stimulated the incorporation of 3H-thymidine into DNA in the four breast cancer cell lines they studied. Half-maximal stimulation occurred at IGF-I concentrations of 0.03–0.6 nM (0.2–4.5 ng/ml). Similar effects were produced by porcine insulin at concentrations 10–1000 times higher than IGF-I.

Vetter et al. (8), using an established Burkitt-type *acute lymphoblastic leukemia* (ALL) cell line (X 308), showed that IGF-I caused 3H-thymidine incorporation to increase 170% over control. A similar effect of IGF-I was observed in a Hodgkin-derived cell line (L-428KSA). IGF-I added in vitro also stimulates DNA synthesis in cultured human leukemia cells (HL 60) and cells derived from a patient with Burkitt's lymphoma (Daudi, 15). IGF-I at concentrations of 0.5 ng/ml increased 3H-thymidine incorporation by 50% in both cell lines and stimulated cell proliferation, as measured by a colony growth assay. This effect was inhibited by addition of antibodies to IGF-I.

Similar phenomena were observed when IGF-I was used to stimulate growth of leukemic cells from the marrow of children with ALL and *acute myelocytic leukemia* (AML) (15). Of particular interest in the studies by Estrov et al. (15) was the observation that GH itself had effects that were similar to IGF-I. Whether these effects of GH are direct or are mediated through IGF-I that is produced by leukemic cells and that acts through autocrine/paracrine mechanisms was not addressed.

A study by Murphy et al. (16) focused on the question of whether tumorigenic effects of GH might represent direct actions of GH rather than, or in addition to, actions mediated through IGF-I. These investigators showed that injection of GH into hypophysectomized rats caused a 12- and 3.4-fold increase in the expression of the c-*myc* proto-oncogene in liver and kidney, respectively, within 1 h. On the other hand, IGF-I mRNA did not reach maximal values until 6–12 h after GH, at a time when c-*myc* had returned to basal values. This suggests that GH itself might possess a mitogenic effect that does not require the induction of IGF-I. In addition to the important question of whether GH has direct IGF-I-independent mitogenic actions, the in vitro studies described above do not deal with the question of whether the mitogenic effects of IGF-I are dependent on the induction of IGF-I by GH or whether they can result from IGF that is produced constitutively. The latter question is addressed in part in the section that follows.

Influence of GH or IGF-I on Growth of Experimental Tumors In Vivo

Few reports are available on the effects of GH injections on in vivo growth of experimental tumors, and the results of such experiments are in conflict. Akaza et al. (17) examined the effect of rGH on rat bladder cancers that had been induced by *N-butyl-N-(4-hydroxybutyl) nitrosamine* (BBN). In animals exposed to BBN in their drinking water for 9 weeks (controls), the mean number of tumors per bladder was 11.26 ± 5.21 (SD) and the mean volume of tumor per bladder was 126.1 ± 212.7 μl (Table 19.2). When GH (0.5 units sc, once weekly) was administered during weeks 7–12 of the experiment, the number of tumors per bladder was increased to 16.95 ± 1.07 ($P < 0.01$ vs. controls) and tumor volume was 204.5 ± 317.7 μl (P = NS). These investigators concluded that GH enhances the growth of this variety of chemically induced tumor. At the other extreme are the results of Torosian and Donoway (18), who examined the effects of GH on the growth of a subcutaneously implanted metastasizing prostate adenocarcinoma (PA-II) in rats. Implanted rats received either placebo or 1 IU/kg/day of GH for 14 days. The GH-treated animals had no increase in tumor volume (11.3 ± 1.5 in GH-treated vs. 12.8 ± 2.2 in controls), but exhibited fewer pulmonary metastases (27 ± 10 in GH-treated vs. 70 ± 20 in controls; $P < 0.05$) (Table 19.3). At the same time, the GH-treated animals had better weight gain (final weight for GH treatment = 209 ± 8 g vs. 186 ± 11 g for controls; $P < 0.05$). In this experimental model, therefore, GH supports host growth and well-being while inhibiting pulmonary metastases. A similar beneficial effect of IGF-I in rats who are suffering from cancer cachexia has been observed by Eng-Hen et al. (19). These investigators observed that tumor-related cachexia of rats bearing an implanted fibrosarcoma was greatly impeded by administration of exogenous IGF-I, without stimulation of growth of the malignancy.

One explanation for the diverse and, at times, opposite in vivo effects of GH on experimental tumors is related to the type of tumor being

TABLE 19.2. Evidence that GH injections stimulate chemically induced rat bladder carcinogenesis.

Experimental group	Volume of tumor per bladder (mean ± SD)	Number of tumors per bladder (mean ± SD)
I BBN	126.1 ± 212.7	11.26 ± 5.21
II BBN + GH (1)	206.6 ± 318.0	12.15 ± 6.59
III BBN + GH (2)	204.5 ± 317.7	16.95 ± 7.07
IV BBN + GH (3)	213.4 ± 274.6	16.79 ± 10.75
V GH (1)	—	—

Note: Animals had BBN (0.05%) in their drinking water for 9 weeks. GH was given once weekly at a dosage of 0.51 IU sc. BBN: N-butyl-N-(4-hydroxybutyl) nitrosamine; GH (1): growth hormone administered 1–6 weeks; GH (2): growth hormone administered 7–12 weeks; GH (3): growth hormone administered 13–18 weeks.
Source: Data from Akaza, Matsuki, Matsushima, Koiso, and Aso (17).

TABLE 19.3. Evidence that GH inhibits metastasis of subcutaneous implants of the prostate tumor PA-III in rats.

	Saline Rx	GH Rx
Volume of tumor (cm^3)	12.8 ± 2.2	11.3 ± 1.5
Tumor weight (g)	19.1 ± 2.0	20.9 ± 1.7
Tumor metastases	70 ± 20	27 ± 10*

Note: Tumors were implanted on day 0; animals were begun on GH (1 U/kg/day) or placebo at 28 days and sacrificed at 42 days. Values are expressed as mean ± SE.
* $P < 0.05$.
Source: Data from Torosian and Donoway (18).

studied. Some tumors may be GH/IGF-I dependent, while others might not.

The work of Moon et al. as much as three decades ago lends credence to the concept that GH may enhance susceptibility to tumors. From a series of studies, these and other investigators presented evidence that hypophysectomized rats have a lower rate of spontaneous tumors than nonhypophysectomized animals (20–22). Further, when hypophysectomized animals were treated with pituitary GH preparations, the tendency to develop tumors was restored (20–22). These early studies, however, were flawed by the obvious fact that hypophysectomy causes reduction or elimination of a variety of hormones in addition to GH. Also, the GH preparations with which test animals were injected were impure by today's standards and undoubtedly contained a variety of hormones and growth factors.

Occurrence of Tumors in Patients with GH Excess

The relationship between acromegaly and increased risk of malignancy has recently been reviewed (23). Alexander et al. reported that malignancy was the second most common cause of death, after cardiovascular disease, among men with acromegaly (24). This increased incidence, however, did not hold for women. Overall, the death rate from malignancy was observed to be two times greater than the general population (24). Bengtsson et al. (25), reporting on 166 Swedish patients, noted a 3-fold increase in cancer-related deaths in women with acromegaly. While a variety of tumors are reported in patients with acromegaly, neoplasms of the gastrointestinal tract appear to be the most common. These include an increased rate of occurrence of both colonic polyps (26–28) and colonic carcinoma (23, 27–30). Summing the experience reported by Ezzat et al. in 323 acromegalic patients (23), there were 31 instances of colon polyps and 13 patients reported to have colon cancer. These same investigators performed colonoscopy on 23 patients with acromegaly and found that 8 (35%) had adenomatous polyps of the colon. They point out that this is a higher prevalence rate (19%) than the adenomatous polyps reported in relatives of patients with cancer of the colon or rectum.

The seminal question with regard to the prevalence of malignancy in patients with acromegaly is whether these observations are relevant to individuals who received GH therapy. Some of the differences in these two groups of patients are probably important: (i) As indicated earlier, the quantities of GH secreted by patients with acromegaly are almost always several times greater than the amounts of exogenous GH administered therapeutically; (ii) the duration of acromegaly is often longer than the period of GH therapy; and (iii) the secretion of GH in acromegalic patients is usually either continuous or occurs in multiple daily bursts, while GH given therapeutically is not administered more than once daily. Unfortunately, it is not possible to quantify the importance of each of these factors in relation to a possible tumorigenic effect of GH.

Occurrence (or Recurrence) of Neoplasms After GH Therapy

The principle reason for concern about the possible role of GH in the production of neoplasms are the reports of leukemia occurring in children and young adults following GH therapy for short stature (1). GH appears to play an important function in maintenance of lymphoid tissue in normal individuals. For example, it prevents involution of splenic follicles and of the thymus following hypophysectomy (31). Similarly 3H-thymidine incorporation by thymocytes and splenocytes from hypophysectomized rats is restored by GH treatment (32), and thymic size in mice is reduced when these animals are injected with antibody to GH (33). The role of GH in immune function is discussed in more detail by Kelley and Blalock elsewhere in these proceedings.

Relating GH and pituitary status to leukemia, hypophysectomy in rats decreases susceptibility to infection-induced leukemia (34) and the occurrence of leukemia following the injection of chemical carcinogens (35, 36) or radiation (37). Additionally, GH in concentrations as low as 0.1 ng/ml stimulates in vitro proliferation of Friend virus-infected erythroleukemia cells, and at concentrations of 200 ng/ml this stimulation reaches 2.4 times control (38). GH has also been shown to stimulate human leukemia cells to undergo colony growth in vitro (39, 40) and to take up thymidine (41), albeit at supraphysiologic concentrations. Possibly of greater relevance to the potential role of GH in the genesis of leukemia are the observations detailed earlier on the effect of IGF-I in vitro on cells from patients with Burkitt-type ALL (8), a Hodgkin-derived cell line (8), and cells from children with ALL and AML (15).

As of September 1992, 33 cases of hematologic malignancy were known to have occurred in GH-treated children and young adults. Of these, 32 patients had been treated because of GH deficiency. Fourteen of the 32 had possible risk factors: 11 patients had previous radio- or chemotherapy for tumors; 2 had Fanconi anemia, and 1 had Bloom syndrome. The geographic distribution is focused in Japan (12 cases) and the United States (10 cases). While overall incidence of leukemia in all GH-treated patients (1:13,000) is greater than that in the general population of 0–15-year-old children (1:30,000), the incidence in the 19 patients without additional risk factors is only 1:22,400 (Table 19.4). Therefore, there is no clear evidence that links GH treatment with the occurrence of leukemia. In addition to the 33 GH-treated children who have had leukemia, 7 children with GH deficiency have developed leukemia, but have not received GH therapy. This, of course, raises the question of whether hypopituitarism itself might predispose to the development of leukemia. In this regard, Dequesnoy and Pedersen (42) reported that the injection of oGH into mice delayed the onset and lowered the incidence of virus-induced leukemia. They speculate that GH delayed thymic involution, a step that may ordinarily precede leukemic changes.

The most extensive study on the role of GH in tumor recurrence is that of Ogilvy-Stuart et al. (43). These investigators reported that of 161 children with ALL, the relapse rate among those not treated with GH (n = 146) was 7% (11 patients). However, there were no cases of leukemia relapse among the 15 patients treated with GH (Table 19.5). Among 207

TABLE 19.4. Incidences of leukemia as of September 1992.

	Patient-years
Incidence of acute leukemia (0–15 years) in general population	1:30,000
Incidence of leukemia in all GH-treated patients	1:13,000
Incidence of leukemia in 19 patients without additional possible risk factors	1:22,400

Note: Estimated patient-years of GH therapy: 425,000; 7 cases of leukemia have been reported in GH-deficient children *not* treated with GH (5 had radiotherapy).

TABLE 19.5. Influence of GH treatment on relapse of ALL.

161 children with ALL

15 GH Rx — 146 no GH Rx

No relapses — 11 relapses (7%)

Source: Data from Ogilvy-Stuart, Ryder, Gattamaneni, Clayton, and Shalet (43).

TABLE 19.6. Evidence that GH does not increase the risk of recurrence of brain tumors.

	207 with brain tumor	
	↙ ↘	
47 GH Rx		160 no GH Rx
↙		↘
5 relapses (11%)		42 relapses (26%)

Note: Relative risk of GH on tumor recurrence after adjustment for diagnosis, sex, age, and use of chemotherapy = 0.82 (confidence interval 0.28–2.37); adjusted relative risk minus chemotherapy = 1.01 (confidence interval 0.36–2.83).
Source: Data from Ogilvy-Stuart, Ryder, Gattamaneni, Clayton, and Shalet (43).

children previously treated for brain tumors (Table 19.6), there was a 26% relapse rate among the 160 not treated with GH and an 11% relapse rate among the 47 patients that received GH therapeutically. The relative risk of GH on tumor recurrence after the adjustment for diagnosis, sex, age, and use of chemotherapy was 0.82. These authors, therefore, concluded that GH does not increase the likelihood of recurrence of leukemia or other tumors.

References

1. Fisher DA, Job J-C, Preece M, Underwood LE. Growth hormone deficiency, human growth hormone and the occurrence of leukemia. Lancet 1988;1: 1159–60.
2. Underwood LE. The risks of growth hormone therapy: does growth hormone influence tumor formation or tumor growth? In: Frisch H, Thorner M, eds. Hormonal regulation of growth. New York: Raven Press, 1989:231–9.
3. Furlanetto RW, DiCarlo JN. Somatomedin-C receptors and growth effects in human breast cells maintained in long-term tissue culture. Cancer Res 1984;44:2122–8.
4. Pekonen F, Partenen S, Makinen T, Rutanen EM. Receptors for epidermal growth factor and insulin-like growth factor I and their relationship to steroid receptors in human breast cancer. Cancer Res 1988;48:1343–7.
5. Pollak MN, Perdue F, Margolese RG, Baer K, Richard M. Presence of somatomedin receptors on primary breast and colon carcinomas. Cancer Lett 1987;38:223–30.
6. Ritvos O, Rutanen E-M, Pekonen F, Jalkanen J, Suikkari A-M, Ranta T. Characterization of functional type I insulin-like growth factor receptors from human choriocarcinoma cells. Endocrinology 1988;122:395–401.
7. Gansler T, Allen KD, Burant CF, et al. Detection of type 1 insulin-like growth factor (IGF) receptors in Wilms' tumors. Am J Pathol 1988;130:431–5.
8. Vetter U, Schlickenrieder JHM, Zapf J, et al. Human leukemic cells: receptor binding and biological effects of insulin and insulin-like growth factors. Leukemia Res 1986;10:1201–7.
9. Hizuka N, Sukegawa I, Takano K, et al. Characterization of insulin-like growth factor I receptors on human erythroleukemia cell line (K-562 cells). Endocrinol Jpn 1987;34:81–8.

10. Huff K, Kaufman D, Gabbay KH, Spencer EM, Lippman ME, Dickson RB. Secretion of an insulin-like growth factor I related protein by human breast cancer cells. Cancer Res 1986;46:4613–9.
11. Minuto F, Del Monte P, Barreca A, et al. Evidence for an increased somatomedin-C/insulin-like growth factor I content in primary lung tumors. Cancer Res 1986;46:985–8.
12. Myal Y, Shiu RPC, Bhaumick B, Bala M. Receptor binding and growth-promoting activity of insulin-like growth factors in human breast cancer cells (T-47D) in culture. Cancer Res 1984;44:5486–90.
13. Macauly VM, Teale JD, Everard MJ, Joshi GP, Smith IE, Millar JL. Somatomedin-C/insulin-like growth factor I is a mitogen for human small cell lung cancer. Br J Cancer 1988;57:91–3.
14. Binnerts A, Uitterlinden P, Hofland LJ, Koetsveld P, Lamberts SWJ. The in vitro and in vivo effects of human growth hormone administration on tumor growth of rats bearing a translatable rat pituitary tumor (7315b). Eur J Cancer 1990;26:269–76.
15. Estrov Z, Meir R, Barak Y, Zaizov R, Zadik Z. Human growth hormone and insulin-like growth factor-I enhance the proliferation of human leukemic blasts. J Clin Oncol 1991;9:394–9.
16. Murphy LJ, Bell GI, Friesen HG. Growth hormone stimulates sequential induction of *c-myc* and insulin like growth factor I expression in vivo. Endocrinology 1987;120:1806–12.
17. Akaza H, Matsuki K, Matsushima H, Koiso K, Aso Y. Stimulatory effects of growth hormone on rat bladder carcinogenesis. Cancer 1991;68:2418–21.
18. Torosian MH, Donoway RB. Growth hormone inhibits tumor metastasis. Cancer 1991;67:2280–3.
19. Eng-Hen NG, Rock CS, Lazarus DD, Stiaino-Coico L, Moldawer LL, Lowry SF. Insulin-like growth factor I preserves host lean tissue mass in cancer cachexia. Am J Physiol 1992;262:R426–31.
20. Moon HD, Li CH, Simpson ME. Effect of pituitary hormones on carcinogenesis with 9,10-dimethyl-1,2 dibenzanthacene in hypophysectomized rats. Cancer Res 1956;16:111–6.
21. Moon HD, Simpson ME, Evans HM. Inhibition of methylchlolanthrene carcinogenesis by hypophysectomy. Science 1952;116:331.
22. Rose DP, Noonan JJ. Influence of prolactin and growth hormone on rat mammary tumors induced by N-nitrosomethylurea. Cancer Res 1982;42:35–8.
23. Ezzat S, Melmed S. Clinical review 18: are patients with acromegaly at increased risk for neoplasia? J Clin Endocrinol Metab 1991;72:245–9.
24. Alexander L, Appleton D, Hall R, Ross WM, Wilkinson R. Epidemiology of acromegaly in the Newcastle region. Clin Endocrinol (Oxf) 1980;12:71–9.
25. Bengtsson B-A, Eden S, Ernest I, Oden A, Sjogren B. Epidemiology and long-term survival in acromegaly. Acta Med Scand 1988;223:327–35.
26. Klein I, Parveen G, Gavaler JS, Van Thiel DH. Colonic polyps in patients with acromegaly. Ann Intern Med 1982;97:27–30.
27. Pines A, Rozen P, Ron F, Gilat T. Gastrointestinal tumors in acromegalic-patients. Am J Gastroenterol 1985;80:266–9.
28. Brunner JE, Johnson CC, Zafar S, Peterson EL, Brunner JF, Mellinger RC. Colon cancer and polyps in acromegaly: increased risk associated with family history of colon cancer. Clin Endocrinol (Oxf) 1990;32:65–71.

29. Ituarte EA, Petrini J, Hershmann JM. Acromegaly and colon cancer. Ann Intern Med 1984;101:627–8.
30. Ritter MM, Richter WO, Schwandt P. Acromegaly and colon cancer. Ann Intern Med 1987;106:636–7.
31. Talwar GP, Pandian MR, Kumar N, et al. Mechanism of action of pituitary growth hormone. Recent Prog Horm Res 1975;31:141–70.
32. Talwar GP, Hanjan SNS, Kidwai Z, et al. Growth hormone action on thymus and lymphoid cells. In: Pecile A, Muller EE, eds. Growth hormones and related peptides. New York: North American Elsevier, 1976:104–15.
33. Pierpaoli W, Sorkin E. Effect of heterologous anti-growth hormone (ASTH) antiserum on thymus and peripheral lymphatic tissue in mice. Induction of a wasting syndrome. J Immunol 1968;101:1036–43.
34. Bentley HP, Hughes ER, Peterson RDA. Effect of hypophysectomy on a virus-induced T-cell leukemia. Nature 1974;252:747–8.
35. Huggins CB, Oka H. Regression of stem-cell erythroblastic leukemia after hypophysectomy. Cancer Res 1972;32:239–42.
36. Huggins CB, Yoshida H, Bird CC. Hormone-dependent stem-cell rat leukemia evoked by a series of feedings of 7,12 dimethylbenz[a]anthracene. J Natl Cancer Inst 1974;52:1301–3.
37. Kaplan HS, Nagareda CS, Brown MB. Endocrine factors and radiation-induced lymphoid tumors in mice. Recent Prog Horm Res 1954;10:293–333.
38. Golde DW, Bersch N, Li CH. Growth hormone modulation of murine erythroleukemia cell growth in vitro. Proc Natl Acad Sci USA 1978;75: 3437–9.
39. Mercola KE, Cline MJ, Golde DW. Growth hormone stimulation of normal and leukemic human T-lymphocyte proliferation in vitro. Blood 1981;58: 337–40.
40. Blatt J, Wenger S, Stitely S, Lee PA. Lack of mitogenic effects of growth hormone on human leukemic lymphoblasts. Eur J Pediatr 1987;146:257–60.
41. Desai LS, Lazarus H, Li CH, Foley GE. Human leukemia cells: effect of human growth hormones. Exp Cell Res 1973;81:330–2.
42. Dequesnoy RJ, Pedersen GM. Inhibitory effect of growth hormone on Gross passage A virus-induced thymus leukemia in C3Hf mice. Cancer Res 1979; 39:415–8.
43. Oglivy-Stuart AL, Ryder WDJ, Gattamaneni HR, Clayton PE, Shalet SM. Growth hormone and tumour recurrence. Br Med J 1992;304:1601–5.

20

Clinical Applications of Growth Hormone Treatment in Childhood

M.A. Preece

Before the advent of *recombinant human growth hormone* (rhGH) the supply of hormone manufactured from cadaveric human pituitaries was always less than that required to treat those with the classical indication of *growth hormone* (GH) deficiency and, therefore, there was relatively little exploration of higher dosage regimens or wider indications (1). Now that there are supplies of rhGH that are limited only by cost, the situation has changed dramatically, and over the past seven years many new avenues have been explored. The most important new indications are *idiopathic short stature* (ISS), *intrauterine growth retardation* (IUGR), and *Turner's syndrome* (TS), which is now a licensed indication in many countries. Before discussing these, however, there still remain important issues concerning *growth hormone insufficiency* (GHI).

A General Issue: Dosage

Before addressing the more specific issues raised by the diagnostic groups defined above, it is important to discuss the dose dependence of response since this influences the interpretation of growth responses to rhGH in any clinical situation. That there is a dosage effect is hardly surprising and has been described over several years (2, 3), but these studies restricted the patients to those with GHI of a relatively tightly defined degree. Also, their patients were receiving the GH two or three times weekly, which does not accord with most current practice.

A more recent investigation (4) has widened the area of study and attempted to separate the effect of dosage from the degree of GHI, including a group of children with normal GH secretion and a normal growth velocity prior to treatment. The rhGH was always given six or seven times weekly. The key results are shown in Figure 20.1, which is redrawn from the original publication (4), with the x-axis plotted

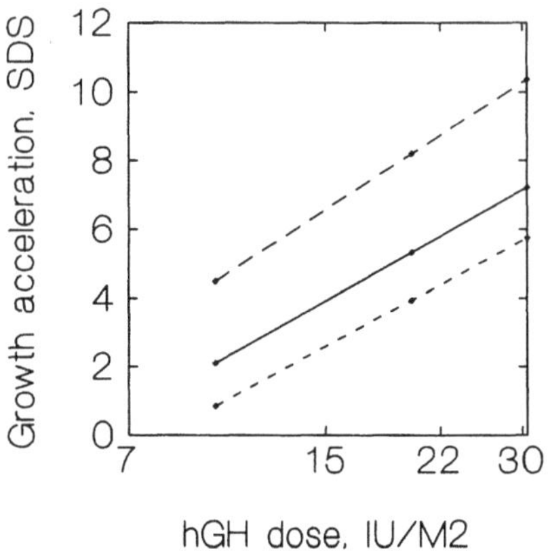

FIGURE 20.1. Dose response curves for growth response to various doses of rhGH in three patient groups. The upper line refers to patients with severe GHI, the middle line to patients with partial GHI, and the lower line to children with normal GH secretion. Growth acceleration is defined as the change in height velocity from the year prior to treatment to the end of the first treatment year. Reproduced from Darendeliler, Hindmarsh, and Brook (4) by permission of the Journal of Endocrinology Ltd.

logarithmically. The upper line represents patients with severe GHI; the middle, those with moderate GHI; and the lower, those with normal GH secretion and a normal growth velocity. The curves clearly show a dose dependence in each group and demonstrate that the more severe the GHI, then the more dramatic is the growth response to a given dose of rhGH. Therefore, when assessing the growth response of a given patient, it is vital to take into account both the diagnostic group and the dosage used.

Growth Hormone Insufficiency

The principal issues concern the long-term prognosis for mature height and how this may be optimized, the management of puberty, and dosage and frequency of rhGH administration.

Prognosis for Mature Height

In 1981, we published the long-term results for mature height in 55 patients with idiopathic GHI attending the Growth Disorder Clinic at the Hospital for Sick Children, London (5), who had reached mature height by 1980; the key figures are shown in Table 1.1 (1980 cohort). Final height in those with isolated GHI was on average −2.5 SDS, whereas

TABLE 20.1. Height SDS for children starting on hGH treatment and at mature height.

	1980 cohort			1990 cohort		
	n	Start of treatment	End of treatment	n	Start of treatment	End of treatment
Isolated GH insufficiency						
Boys	30	−4.5 ± 0.3	−2.2 ± 0.2	16	−3.1 ± 0.3	−1.5 ± 0.3
Girls	9	−5.0 ± 0.4	−2.9 ± 0.4	6	−4.1 ± 0.4	−2.4 ± 0.5
GH and gonadotropin insufficiency						
Boys	11	−5.1 ± 0.4	−1.4 ± 0.3	18	−4.0 ± 0.2	−1.7 ± 0.2
Girls	5	−5.5 ± 0.7	−1.3 ± 0.4	6	−4.9 ± 0.4	−2.1 ± 0.5

Note: The results for two cohorts are presented: Those finishing treatment by 1980 and previously reported (5) and a more recent cohort finishing treatment between 1980 and 1990 reported in 1992 (8). The data are presented as mean ± SEM.

in those patients with *multiple pituitary hormone deficiencies* (MPHD) (which included gonadotropin deficiency in all cases) the overall result was about −1.5 SDS. This apparent advantage was counterbalanced by a significant degree of trunk/limb disproportion in the gonadotropin-deficient group with a very eunuchoid body appearance (5); this was also seen in a second group of patients with GH and gonadotropin deficiency secondary to a craniopharyngioma treated over the same period of time (6). The most likely explanation is that the gonadotropin deficiency led to a failure of spontaneous puberty that, coupled with a clinical policy of delaying the introduction of sex steroids for as long as possible, allowed for more prepubertal growth, mostly in the limbs. However, this was at the expense of a very poor adolescent growth spurt and a diminished adult sitting height (7).

In 1992, a second review of a further 46 similar patients who reached mature height between 1980 and 1990 was reported (8), and these data are also shown in Table 20.1. The main differences between the two cohorts are: The mean age of starting GH in the first was, for boys and girls respectively, 13.7 and 12.0 years as compared to 10.3 and 10.0 years in the second, and sex steroids were introduced at a more physiologically appropriate time. It is clear that all groups in the second cohort are about 1.0 SDS taller at the start of treatment, with a gain in mature height of about 0.5 SDS in mature height for the isolated GHI patients, but with a slightly poorer prognosis for mature height (compared to the 1980 cohort) for the MPHD group. This presumably reflects the earlier induction of puberty.

Management of Puberty

This remains one of the most vexing problems in management of GHI, whether isolated or part of a complex of multiple pituitary hormone insufficiency. The principle questions are: What dosage of rhGH should be used, and when should pubertal induction take place in the presence of gonadotrophin insufficiency?

The question of GH dosage is still largely undecided, and there are a number of clinical trials currently addressing this question. The main point is that in normal adolescents the peak amplitude of spontaneous GH pulses increases markedly (9), and it seems sensible to ask whether rhGH replacement should attempt to mimic this. Answers are not yet available, but other data suggest a need for caution. In a retrospective study of a large group of GH-treated children we have previously reported an acceleration of progression through puberty relative to normal children, although the age of onset of puberty (as measured by the age of entry to breast or genitalia stage 2) was no younger than normal (10).

The right age to induce puberty is also problematic, although probably more easily resolved. In children whose GHI is diagnosed and treated at an early age and who therefore achieve good statural growth in mid-childhood, it is seldom an issue. Since their height deficit by the age when their peers are entering puberty will usually be small, there is no reason why they should not commence pubertal induction at a socially appropriate time. The difficulties arise in those less fortunate who still have a significant height deficit and where there is therefore something of a conflict between the pressures of achieving maximum height benefit and those of delaying puberty excessively. The data in Table 20.1 suggest that earlier induction does reduce final height in such individuals. Clearly, this is a decision that needs a degree of individual consideration and with the current absence of unequivocal data, is largely a clinical decision. It is also important to recognize that the synergy between GH and the sex steroids is an important part of the adolescent growth spurt (11, 12), which probably should bias the clinical decision toward earlier induction if other considerations allow.

Frequency of Growth Hormone Administration

This is an easier point to decide. There is now clear evidence that daily administration of rhGH is superior to alternate daily or less frequent regimens, given an equal total weekly dose (13). Whether more frequent administration would further increase the height gain per unit of rhGH is as yet unclear, although early evidence suggests that it would not (13).

Idiopathic Short Stature

This is the most heterogeneous group of short stature disorders and a term that often means different things to different clinicians and research groups. In the present context ISS includes patients whose stature is more than 2.0 SDS below the mean, whose growth velocity is normal or low, and who have no dysmorphic features, no as yet defined endocrinological abnormalities, and no evidence of underlying chronic pediatric disease.

Many early studies have suggested that either pituitary hGH or rhGH could induce an acceleration in growth velocity in children with ISS, at least in the short term (14–18). It is also clear that generally larger doses of GH are required for an acceleration equivalent to that seen in GHI (4). Randomized, controlled clinical trials including a placebo-treated limb have been far fewer, but have confirmed that while there may be a small placebo effect, it is far smaller than that seen with active rhGH treatment and biologically insignificant (19).

The more difficult issue pertains to the long-term benefits. Clearly definitive conclusions must await the attainment of adult height, for which data are very scanty at present. Conclusions based on predicted mature height calculated after three or more years of therapy suggest little long-term benefit (20, 21), but firm conclusions must wait for several years yet.

Intrauterine Growth Retardation

Compared to the previous clinical group these children are relatively less common, but still form a significant group among short children. The definition is also rather likely to embrace a varied group of specific diagnoses, but is usually taken to mean children born light for their gestational age and who are generally short, strikingly lean, sometimes with some minor dysmorphic features, and with a characteristic growth pattern. At the severe end of the clinical picture lies the Silver-Russell syndrome, whose pattern of growth is most clearly defined (22, 23). These children all show failure of catch-up growth in early life, a rather blunted adolescent growth spurt that may start a little early, and a poor prognosis for adult height.

Rather like the ISS group, there are no data on long-term growth outcome following rhGH treatment, but data to three years are now available (24) and are summarized in Figure 20.2. With a dose of 30 IU/M^2/week of rhGH, growth velocity accelerated from −0.75 SDS to +3.6 SDS in the first year and then continued at 2.4 and 1.1 SDS in years two and three, respectively. Clearly the third-year velocity is still significantly greater than that prior to treatment, but shows a significant decline from the first year of treatment. A dose of 15 IU/M^2/week showed a similar, but less dramatic, acceleration in the first year (−0.77 to +1.4 SDS). When the data are studied in relationship to skeletal maturity and predicted mature height, the situation is much like the ISS group in that skeletal maturation proceeds at an enhanced pace such that long-term height prognosis is unchanged. As with ISS, many years must elapse before the effect on adult height can be evaluated, and until then both

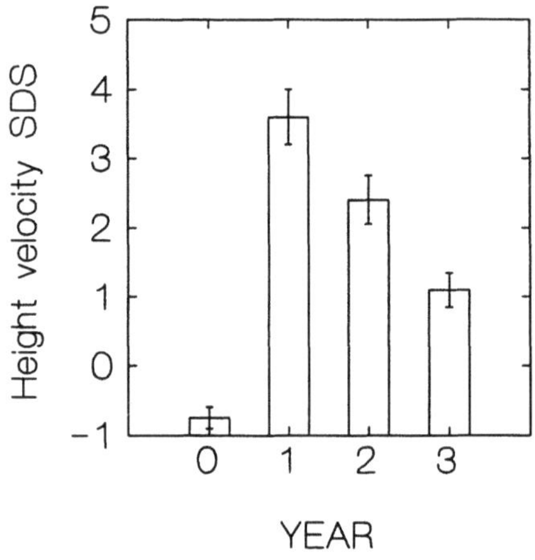

FIGURE 20.2. Height velocity response in 13 patients with IUGR. Bars represent the mean velocities in SDS units and the error bars are 1 SEM. Data from Stanhope, Preece, and Hamill (24).

these indications must be considered in the realm of research and treatment only instituted within formal clinical trials.

Turner's Syndrome

This is a much more homogeneous population of a well-defined syndrome, at least from a growth point of view. Before the advent of rhGH, limited randomized controlled clinical trials of pituitary hGH had started and had suggested at least short-term benefit (14). At this time, the nearly definitive trial is the multicenter study in the USA, which has been the subject of several interim reports and which now is beginning to report final or near-final height data (25–27). A short-term benefit of rhGH, alone or in combination with the mild anabolic steroid oxandrolone, was rapidly demonstrated (25), and subsequent reports have confirmed these observations. The most recent published data are summarized in Figure 20.3, which shows the current heights of the girls in the study according to age groups. Those nearest to adult height (age >17 years) are on average 8 cm taller than the mean of a comparable untreated adult population of women with Turner's syndrome (28). In the same publication, details are also given of each girl's currently achieved height compared to her individually predicted adult height (28): Of the 17 treated with rhGH alone, 82% now exceed their predicted adult heights; and for the 45 girls treated with combination therapy, 93% have achieved the same. In all, 32 are still growing so that the expectation is that they will further exceed their original potential adult height.

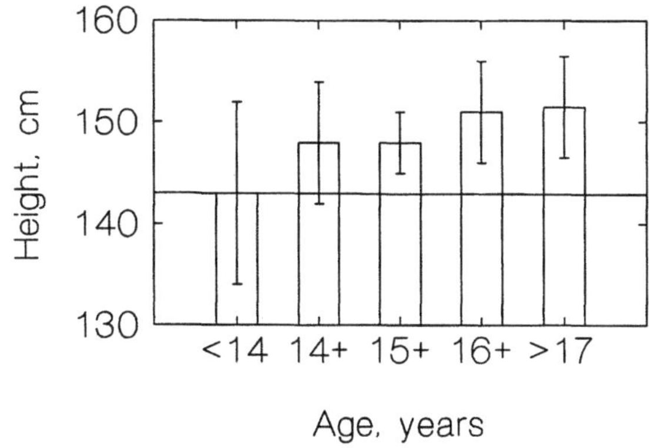

FIGURE 20.3. Long-term height outcome in girls with Turner's syndrome treated with rhGH, either alone or in combination with oxandrolone (27). The horizontal line represents the mean height of untreated Turner women taken from reference 28. Reprinted with permission from Rosenfeld (27).

Summary

The ready availability of rhGH has made it possible to explore many more potential indications in the last seven years. We are now in a position to evaluate these applications critically and to address the following clinical situations.

1. Growth hormone insufficiency: The effectiveness of rhGH is not in doubt, but there are still questions concerning the best regimen. It seems clear that daily administration of subcutaneous rhGH is required, in doses between 15 and 20 IU/M^2/week. More frequent injection does not seem to improve growth response.

2. Turner's syndrome: There is now extensive data showing short- and middle-term benefit of rhGH administered to girls with Turner's syndrome. In one major study, results from patients followed for up to six years strongly suggest an actual increase in mature height in the majority of patients. These seems most marked when the rhGH is combined with the anabolic steriod oxandrolone.

3. Idiopathic short stature: There are now a large number of clinical trials studying the effect of rhGH on children who are short, with no identifiable endocrine or dysmorphic disorder. They vary as to whether they are growing at normal or sub-normal velocities. The short-term results are encouraging, with significant increases in growth velocity that are markedly greater than the small placebo effect. However, the longer-term benefit is less clear, with a tendency for the growth response to lessen greatly after two to three years and also for skeletal maturation to accelerate so that predicted mature height is unlikely to be significantly

changed. This requires much more research before clear recommendations can be made.

4. Intrauterine growth retardation: This group of children is characterized by a birth weight that is inappropriately low for gestational age, often associated with some minor dysmorphic features. They show marked increases in growth velocity following doses of rhGH of about 30 IU/M^2/week, but there is associated acceleration of skeletal maturation, and it seems likely that the long-term benefits may be small and similar to those children with idiopathic short stature.

Conclusions

The short- to medium-term safety of rhGH now seems well established in a wide number of clinical situations. In the case of GHI the issues in the 1990s are ones of relative detail, of which the management of puberty is probably the most vexing. The issues of dose frequency, beyond the clear advantage of daily over less frequent regimens, are not likely to make great differences in height outcome, although it is possible that more frequent injection may have influence over the metabolic effects of the therapy. In many ways the more important issues center around early diagnosis and the greater flexibility that is possible in the management of puberty as the height deficit is minimized as adolescence approaches.

Turner's syndrome seems to be a clear indication for rhGH therapy, although the magnitude of the benefit at adult height is still unclear in many cases. Doubtless, further trials that are less well advanced than the one described above will modify conclusions over the next few years, but it seems unlikely that the main message will alter.

Finally the situation with ISS and IUGR remains the least clear. It is certain that short-term growth acceleration can be achieved in the vast majority of individuals if sufficient rhGH is administered. What is far less clear is whether the patients will feel that a significant gain in height during childhood with little or no gain in adult height justifies the long period of a relatively invasive treatment. This requires considerably greater clinical experience and substantial new data as to the final heights achieved.

References

1. Milner R. Which children should have growth hormone therapy? Lancet 1986;i:483–5.
2. Preece M, Tanner J, Whitehouse R, Cameron N. Dose dependence of growth response to hGH in "isolated" growth hormone deficiency. J Clin Endocrinol Metab 1976;42:477–83.
3. Frasier S, Costin G, Lippe B, Aceto T, Bunger P. A dose-response curve for human growth hormone. J Clin Endocrinol Metab 1981;53:1213–7.

4. Darendeliler F, Hindmarsh PC, Brook CGD. Dose-response curves for treatment with biosynthetic human growth hormone. J Endocrinol 1990;125:311–6.
5. Burns E, Tanner J, Preece M, Cameron N. Final height and pubertal development in 55 children with idiopathic growth hormone deficiency, treated for between 2 and 15 years with human growth hormone. Eur J Pediatr 1981;137:155–64.
6. Burns E, Tanner J, Preece M, Cameron N. Growth hormone treatment in children with craniopharyngioma: final growth status. Clin Endocrinol (Oxf) 1981;14:587–95.
7. Preece MA, Pan H, Ratcliffe SG. Auxological aspects of male and female puberty. Acta Paediatr Scand [Suppl] 1992;383:11–3.
8. Michalacos S, Buchanan CR, Adlard P, Preece MA. Final height and endocrine status after treatment of "idiopathic" growth hormone deficiency. Horm Res 1992;37(suppl 4):2.
9. Maurus N, Blizzard RM, Link K, Johnson ML, Rogol AD, Veldhuis JD. Augmentation of growth hormone secretion during puberty: evidence for a pulse amplitude-modulated phenomenon. J Clin Endocrinol Metab 1987; 64:596–601.
10. Darendeliler F, Hindmarsh PC, Preece MA, Cox L, Brook CGD. Growth hormone increases rate of pubertal maturation. Acta Endocrinol (Copenh) 1990;122:414–6.
11. Tanner J, Whitehouse R, Hughes P, Carter B. Relative importance of growth hormone and sex steroids for the growth at puberty of trunk length, limb length, and muscle width in growth hormone-deficient children. J Pediatr 1976;89:1000–8.
12. Aynsley-Green A, Zachmann M, Prader A. Interrelation of the therapeutic effects of growth hormone and testosterone on growth in hypopituitarism. J Pediatr 1976;89:992–9.
13. Smith P, Hindmarsh P, Brook C. Contribution of dose and frequency of administration to the therapeutic effect of growth hormone. Arch Dis Child 1988;63:491–4.
14. Buchanan C, Law C, Milner R. Growth hormone in short, slowly growing children and those with Turner's syndrome. Arch Dis Child 1987;62:912–6.
15. Gertner J, Genel M, Gianfredi S, et al. Prospective clinical trial of human growth hormone in short children without growth hormone deficiency. J Pediatr 1984;104:172–6.
16. Van Vliet G, Styne D, Kaplan S, Grumbach M. Growth hormone treatment for short stature. N Engl J Med 1983;309:1016–22.
17. Buzi F, Adlard P, Jones J, Stanhope R, Preece MA. Short children with normal growth hormone response to pharmacological stimuli: physiologic growth hormone secretion and response to growth hormone treatment. J Pediatr Endocrinol 1991;4:61–6.
18. Hindmarsh P, Brook C. Effect of growth hormone on short normal children. Br Med J 1987;295:573–7.
19. Ackland FM, Jones J, Buckler J, Dunger DB, Rayner PHW, Preece MA. Growth hormone treatment in non-growth hormone deficient children: effects of stopping treatment. Acta Paediatr Scand [Suppl] 1990;366:32–7.
20. Wit J, Fokker M, Keizerschrama S, et al. Effects of 2 years of methionyl growth hormone therapy in 2 dosage regimens in prepubertal children with

short stature, subnormal growth rate, and normal growth hormone response to secretagogues. J Pediatr 1989;115:720–5.
21. Wit JM, Rikken B, De Muinck Keizer-Schramer SMPF, et al. Growth hormone therapy for short, slowly growing children without classical growth hormone deficiency: 3 year results. Acta Paediatr Scand [Suppl] 1990;370:186.
22. Tanner J, Lejarraga H, Cameron N. The natural history of the Silver-Russell syndrome: a longitudinal study of thirty-nine cases. Pediatr Res 1975;9: 611–23.
23. Davies P, Valley R, Preece M. Adolescent growth and pubertal progression in the Silver-Russell syndrome. Arch Dis Child 1988;63:130–5.
24. Stanhope R, Preece MA, Hamill G. Does growth hormone therapy increase final height in children with intra-uterine growth retardation? Arch Dis Child 1991;66:1180–3.
25. Rosenfeld RG, Hintz RL, Johanson AJ, Sherman B. The Genentech Collaborative Group results from the first 2 years of a clinical trial with recombinant DNA-derived human growth hormone (somatrem) in Turner's syndrome. Acta Paediatr Scand [Suppl] 1987;331:59–66.
26. Rosenfeld R, Hintz R, Johanson A, et al. Three-year results of a randomized prospective trial of methionyl human growth hormone and oxandrolone in Turner syndrome. J Pediatr 1988;113:393–400.
27. Rosenfeld RG, Genentech National Cooperative Group. Growth hormone therapy in Turner's syndrome: an update on final height. Acta Paediatr Scand [Suppl] 1992;383:3–6.
28. Lyon A, Preece M, Grant D. Growth curve for girls with Turner syndrome. Arch Dis Child 1985;60:932–5.

21

Medical and Scientific Aspects of Growth Hormone Receptor Deficiency (Laron Syndrome) in Ecuador

PINCHAS COHEN, JAIME GUEVARA-AGUIRRE, PAUL J. FIELDER, KRISTIN WILSON, SHARRON E. GARGOSKY, MARY A. VACCARELLO, FRANK B. DIAMOND, JR., ARLAN L. ROSENBLOOM, AND RON G. ROSENFELD

In 1966, Laron et al., in Israel, described a genetic form of dwarfism in three siblings with physical features similar to *growth hormone* (GH) deficiency, but with abnormally high serum levels of immunoreactive GH (1). In subsequent reports, they described 29 patients in Israel, primarily of Oriental Jewish ancestry from North Africa and the Persian Gulf (2, 3). An additional 70 subjects have been reported (4), many of Mediterranean ancestry, including individuals from Iran, Iraq, Yemen, Tunisia, Lebanon, Saudi Arabia, Algeria, France, Italy, Sardinia, Spain, Brazil, Argentina, Mexico, Pakistan, Bangladesh, northern Europe, Japan, South Africa, Canada, and the United States.

Evaluation of these patients confirmed that despite the presence of severe short stature, serum levels of GH were normal or elevated. Additional features included low serum levels of *insulin-like growth factor I* (IGF-I), failure to accelerate growth or raise serum IGF-I levels following treatment with exogenous GH, and decreased GH binding capacity in tissues and in serum (1–4). Originally named "Laron-type dwarfism," this condition is better termed "Laron syndrome" to avoid the pejorative connotations of "dwarfism." An alternative term currently used to describe this syndrome is *growth hormone receptor deficiency* (GHRD). GHRD is now recognized to constitute one form of GH insensitivity and shares many phenotypic features with GH deficiency type 1A, particularly when the latter is associated with anti-GH antibodies. Over the last three years, we have identified a constellation of over 60 patients with GH insensitivity in Ecuador, derived from an inbred white population of purported pure Spanish descent (5, 6).

Clinical Features of GHRD

The mean height standard deviation score of the Ecuadorian GHRD population (relative to American standards) was −8.4 for prepubertal subjects and −8.6 for adults. Bone age was significantly delayed, but not nearly to the extent of height age, so that the mean bone age/height age ratio in preadolescent patients was 3:2. In most newborn males, a small penis was noted. However, penile growth during puberty was normal. Sexual maturation in both males and females was normal, although puberty was delayed in half of the patients by 3–5 years. In adults, arm span minus height difference averaged −7.8 cm, consistent with the appearance of short limbs. Similarly, upper to lower segment ratios were >1.10 in most adults. Hand and/or foot length was <10th percentile for height in most patients. Elbow limitation, a finding not previously noted in this condition, was observed in almost half of the children and the majority of adults. Blue scleras, also a newly reported finding in the syndrome, was seen in all patients. Contrary to other reports, intelligence of all patients was within the normal range. However, physical developmental parameters, such as walking, were delayed in the first years of life due to hypomuscularity. Evidence of premature osteoporosis, including hip, femur, and acetabulum degeneration, was present in some of the adult patients (5, 6). Table 21.1 summarizes the clinical features of GHRD.

Biochemical Features of GHRD

The major biochemical features of patients with GHRD are summarized in Table 21.2. Basal serum GH levels were markedly elevated in prepubertal subjects; GH levels in adults, while generally elevated, were significantly lower than in prepubertal subjects. Approximately half of the subjects had serum *GH binding protein* (GHBP) levels below the limits of detection; the remainder had measurable, but low levels, with several having values within the lower limits of the normal adult range. Serum IGF-I, IGF-II, and IGFBP-3 levels were dramatically reduced, especially in prepubertal subjects. It is interesting to note that serum levels of IGF-I, IGF-II, and IGFBP-3 all rose following puberty, although GHBP levels remained low, and random serum GH levels actually fell. These results suggest that the IGF axis is regulated to some extent by sex steroids, independent of GH secretion and binding. IGFBP-3 levels correlated with IGF-I levels ($r = 0.48$, $P < 0.01$). Serum IGF-I levels correlated with statural SDS in adults ($r = 0.64$, $P < 0.001$); serum IGFBP-3 levels (by RIA) correlated with statural SDS in both adults ($r = 0.61$, $P < 0.001$) and children ($r = 0.58$, $P < 0.05$). In the majority of children with GHRD, but in none of the adults, IGFBP-3 serum protease activity was

TABLE 21.1. Common clinical characteristics of Ecuadorian patients with Laron syndrome.

Clinical feature	% of patients
Extreme short stature	100
Small hands or feet	67
Abnormal body proportions	75 (adults)
Limited elbow extension	80 (older children and adults)
Blue sclera	100
High-pitched voice	90
Slow hair growth	90 (early childhood)
Nasal bridge hypoplasia	100
Prominent forehead	100
Vertically short facies	100
Decay of primary teeth	75
Truncal obesity	90 (adults)
Osteopenia	50 (adult women)
Neonatal hypoglycemia	50
Micropenis	100 (prepubertal males)

TABLE 21.2. Biochemical characteristics of Ecuadorian patients with Laron syndrome.

	Prepubertal*	Adult*	Normal adult range
Random GH (ng/ml)	32 (6–99)	11 (ND–40)	<5
GH binding protein (pmol/L)	23 (ND–62)	21 (ND–106)	79–746
IGF-I (ng/ml)	3 (1–7)	26 (6–84)	96–270
IGF-II (ng/ml)	72 (24–186)	150 (65–389)	388–792
IGFBP-3 (ng/ml)	274 (67–1135)	428 (174–744)	2200–3700
IGFBP-2 (% of normal)	290 (85–500)	173 (45–480)	50–150

* Values are given as mean with range in parentheses.
ND = below limits of detection.

observed (7). Compatible with this finding, IGFBP-3 levels assayed by RIA (measuring intact IGFBP-3 as well as IGFBP-3 proteolyzed fragments) correlated with those assayed by *Western ligand blotting* (WLB) (measuring only intact IGFBP-3) only in adult patients. Serum IGFBP-2 levels in GHRD patients were twofold elevated and negatively correlated with IGF-I and IGFBP-3 levels.

Genetics of GHRD

Recent advancements in the molecular biology of the human GH receptor gene have made possible the identification of the molecular defects resulting in some cases of GH insensitivity (8, 9). The gene spans 87 kilobases, including nine exons that encode the receptor. Exon 2 encodes the signal peptide; exons 3–7 encode the extracellular binding domain; exon 8 encodes the transmembrane domain; and exons 9–10 encode the intracellular domain. The GH receptor belongs to a family of transmembrane receptors that includes the receptors for prolactin, erythropoietin, interleukins-2, -4, and -6, and colony-stimulating factor. Several mutations and deletions of this gene have been reported to cause GHRD in non-Ecuadorian patients. Noncontiguous deletions of exons 3–6, which normally encode a portion of the extracellular binding domain of the receptor (8, 9), have been detected in five patients, and point mutations in exons encoding the extracellular domain of the GH receptor in four patients (10, 11). The latter included a codon 96 Phe-to-Ser substitution resulting from a TTT to TCT mutation. Additionally, three independent patients born of consanguineous parents were found to be homozygous for nonsense mutations resulting from CG to TG transitions at codons 38 or 43.

Obligate heterozygotes from the Ecuadorean population were evaluated by Berg et al. using polymerase chain reaction amplification of GH receptor gene sequences, followed by denaturing gradient gel electrophoresis (12). Homo- and heteroduplexes were identified only for exon 6; sequencing revealed a substitution of guanine for adenine in codon 180. However, both the original GAA and mutated GAG codons code for the same amino acid, glutamic acid. Subsequent sequencing of the exon 6-exon 7 splice junction from RNA-polymerase chain-reaction-amplified cellular RNA of an affected individual indicated that the A-to-G substitution at codon 180 activates a 5′ splice site 24 nucleotides upstream from the usual exon 6-intron 6 splice site. As a result of this splicing abnormality, exon 6 is missing its final 24 nucleotides, and the resulting protein is missing 8 amino acids, residues 181–188 of the extracellular domain. This is a region that is highly conserved across species, with at least 75% homology among human, pig, cow, sheep, rabbit, rat, mouse and chicken, and is likely to be critical to the ability of the GH receptor to bind GH. Berg et al. (12) subsequently developed an allele-specific oligonucleotide hybridization assay capable of detecting codon 180 substitutions. This mutation was identified in 37 of 38 Ecuadorean probands; all 26 parents tested proved to be heterozygotes. It appears likely that Laron syndrome in this Ecuadorean population is the result of a new mutation that has been propogated in a highly inbred population of Spanish descent. The codon 180 substitution has, to date, not been identified outside of Ecuador in any normal control subjects or in other patients with Laron syndrome.

Therapy of GHRD

As would be expected, attempts to treat GHRD patients with *human growth hormone* (hGH) resulted in a lack of consistent growth response

(4). Insensitivity to GH in GHRD was also documented in vitro by documenting lack of response of fibroblasts and other cells from GHRD patients to exogenously added GH (13). Those studies, however, demonstrated a normal response of GHRD cells to IGF-I.

The availability of recombinant human IGF-I for investigative clinical use has prompted several trials in Laron syndrome. Intravenous studies have shown that IGF-I has hypoglycemic effects and that it suppresses insulin secretion. Laron et al. have reported that in five prepubertal and five adult GHRD patients, daily subcutaneous injections of 120–150 μg/kg/d for a week led to a marked drop in serum GH levels and a marked rise in serum type III procollagen (14). Additionally, Walker et al. (15) infused IGF-I over 11 days to a child with GHRD. The infusion resulted in a marked rise in serum IGF-I levels, modest asymptomatic hypoglycemia, decreased BUN and elevated urinary nitrogen, increased urinary calcium excretion, decreased urinary phosphate and sodium excretion, and suppression of serum GH, insulin, and IGF-II.

In six Ecuadorian GHRD adult patients (16), we demonstrated that subcutaneous injections of IGF-I at a dose of 40 μg/kg BID for 6 days was safe and increased serum IGF-I levels into the normal range. No hypoglycemia was observed in that study; GH, insulin, and IGF-II were suppressed, while IGFBP-2 levels rose. Urinary calcium excretion was increased without a change in serum calcium levels. Serum levels of IGFBP-3, measured by either WLB or *radioimmunoassay* (RIA), were unaffected by six days of therapy with IGF-I and remained well below the normal range.

It is not yet known whether IGF-I therapy will improve final height of children with GHRD, although preliminary studies from several sources are encouraging. Laron et al. (17) reported that three patients treated with IGF-I at a dose of 120 μg/kg/day doubled their growth velocity. Walker et al. reported similar results on a single patient (18). Wilton et al. (19) demonstrated an increase in growth velocity in 22 out of 24 patients enrolled in a European study of the use of IGF-I in GHRD. The increase in growth velocity of the children in the above studies varied from 2 to 10 cm/year. The heterogeneity of the patient population in these studies may explain the variability in the response observed. Transient side effects of IGF-I observed by these authors included hypokalemia, papilloedema, and renal colic, but overall, the drug seems to be well tolerated. Additional studies with IGF-I are in progress in the US and in Ecuador. Preliminary findings in those patients support the growth-promoting effects of IGF-I in GHRD.

Summary

We have identified in Ecuador the largest population of GHRD patients in the world. We are now able to biochemically and genetically diagnose this condition and identify the carrier state in this population. Future population screening with a recently developed oligonucleotide hybridization assay will allow genetic counseling for families with this disorder. Treatment with IGF-I appears to be safe and beneficial to patients with this condition, although longer-term studies of the efficacy and potential morbidity of such therapy are necessary. In the future, we expect to improve the therapeutic management of these patients by optimizing IGF-I therapy, possibly replacing other potentially important factors (such as IGF-II, IGFBP-3, and GHBP) and, eventually, by replacing the GH receptor through gene therapy.

Acknowledgments. Supported in part by grants from the NORD Foundation, March of Dimes Birth Defects Foundation, HG-00298 from the National Institutes of Health, and a contract for a Regional Diabetes and Endocrine Program with the Department of Health and Rehabilitative Services of the State of Florida.

References

1. Laron Z, Pertzelan A, Mannheimer S. Genetic pituitary dwarfism and high serum concentration of growth hormone: a new inborn error of metabolism. Isr J Med Sci 1966;2:152–5.
2. Pertzelan A, Adam A, Laron Z. Genetic aspects of pituitary dwarfism due to absence or biological inactivity of growth hormone. Isr J Med Sci 1968; 4:895–900.
3. Laron Z, Pertzelan A, Karp M. Pituitary dwarfism with high serum levels of growth hormone. Isr J Med Sci 1968;4:883–94.
4. Laron Z. The syndrome of familial dwarfism and high plasma immunoreactive human growth hormone. Birth Defects 1974;10:231–8.
5. Rosenbloom AL, Guevara-Aguirre J, Rosenfeld RG, Fielder PJ. The little women of Loja: growth hormone receptor deficiency in an inbred population of southern Ecuador. N Engl J Med 1900;323:1367–74.
6. Guevara-Aguirre J, Rosenbloom AL, Fielder PJ, Diamond FB, Rosenfeld RG. Growth hormone receptor deficiency in Ecuador: clinical and biochemical phenotype in two populations. J Clin Endocrinol Metab 1993.
7. Fielder PJ, Guevara Aguirre J, Rosenbloom AL, Carlsson L, Hintz RL, Rosenfeld RG. Expression of serum insulin-like growth factors (IGFs), IGF-binding proteins (IGFBPs) and the growth hormone binding protein (GHBP) in heterozygote relatives of Ecuadorian GHR-deficient patients. J Clin Endocrinol Metab 1992;74:743–50.
8. Leung DW, Spencer SA, Cachianes G, et al. Growth hormone receptor and serum binding protein: purification, cloning and expression. Nature 1987;330:537–43.
9. Godowski PJ, Leung DW, Meacham LR, et al. Characterization of the human growth hormone receptor gene and demonstration of a partial gene deletion in two patients with Laron-type dwarfism. Proc Natl Acad Sci USA 1989;86:8083–7.

10. Amselem S, Duquesnoy P, Attree O, et al. Laron dwarfism and mutations of the growth hormone receptor. N Engl J Med 1989;321:989–95.
11. Amselem S, Sobrier ML, Duquesnoy P, et al. Recurrent nonsense mutations in the growth hormone receptor from patients with Laron dwarfism. J Clin Invest 1991;87:1098–102.
12. Berg MA, Guevara-Aguirre J, Rosenbloom AL, Rosenfeld RG, Francke U. Mutation causing a new splice site in the growth hormone receptor gene of 37 Ecuadorean patients with Laron syndrome. Hum Mut 1992;1:24–34.
13. Geffner ME, Golde DW, Lippe BM, Kaplan SA, Bersch N, Li CH. Tissues of the Laron dwarf are sensitive to insulin-like growth factor I but not to growth hormone. J Clin Endocrinol Metab 1987;64:1042–6.
14. Laron Z, Klinger B, Jensen LT, Erster B. Biochemical and hormonal changes induced by one week of administration of rIGF-I to patients with Laron type dwarfism. Clin Endocrinol (Oxf) 1991;35:145–50.
15. Walker JL, Ginalska-Malinoska M, Romer TE, Pucilowska JB, Underwood LE. Effects of infusion of insulin-like growth factor-I in a child with growth hormone insensitivity syndrome (Laron dwarfism). N Engl J Med 1991;324: 1483–8.
16. Vaccarello MA, Diamond FB, Guevara-Aguirre J, et al. Hormonal and metabolic effects and pharmacokinetics of recombinant insulin-like growth factor-I in growth hormone receptor deficiency/Laron syndrome. J Clin Endocrinol Metab 1993.
17. Laron Z, Anin S, Klipper-Aurbach Y, Klinger B. Effects of insulin-like growth factor-I on linear growth, head circumference, and body fat in patients with Laron type dwarfism. Lancet 1992;339:1258–61.
18. Walker JL, Van Wyk JJ, Underwood LE. Stimulation of statural growth by insulin-like growth factor-I in a child with growth hormone insensitivity syndrome (Laron type). J Pediatr 1992;121:641–6.
19. Wilton J, Kabi-Pharmacia study group on insulin-like growth factor-I in growth hormone insensitivity syndrome. Treatment with recombinant human insulin-like growth factor-I of children with growth hormone receptor deficiency (Laron syndrome). Acta Paediatr Scand [Suppl] 1992;383:137–41.

22

Spontaneous Growth Hormone Secretion in Children With Normal and Abnormal Growth

E. Verena Jorgensen, Dorothy I. Shulman,
Frank B. Diamond, Jr., Allen W. Root, and Barry B. Bercu

We have previously described the use of spontaneous *growth hormone* (GH) release studies in evaluating poor growth in children (1, 2). A group of children was defined as having GH neurosecretory dysfunction if they had short stature, slow growth, delayed bone age, normal responses to provocative GH testing, low *insulin-like growth factor I* (IGF-I) levels, and abnormal 24-h GH secretory patterns (3). These children were found to respond well to GH replacement therapy, and therefore spontaneous GH studies have been used diagnostically in many centers.

Spontaneous GH studies can also be used to study normal GH physiology. Computer analysis has allowed detailed evaluation of GH pulse characteristics and exploration of the mechanisms by which GH secretion is altered in different clinical states. In the past seven years, we have done 24- and 12-h spontaneous GH release studies of children with growth disorders and also as part of research studies. We have analyzed these studies by clinical groups to examine the relationship of endogenous GH secretion to normal growth and puberty and to investigate how GH secretion is altered in several clinical conditions affecting growth.

Methods

Clinical Evaluation

Three hundred studies of spontaneous GH release were performed on 272 children from July 1984 to December 1991. Children who were being evaluated for growth disorders or were enrolled in a clinical research protocol were admitted to All Children's Hospital, St. Petersburg, FL.

All research studies were approved by the University of South Florida and All Children's Hospital Institutional Review Boards, and written informed consent was obtained from the patient and/or parent(s).

Clinical assessments included careful measurement of height using a stadiometer, weight, and pubertal staging using breast or genital development according to the method of Tanner (4). *Body mass index* (BMI) was calculated as weight (kg)/(height [m])^2. Percentiles for BMI by sex and age were obtained from Hammer et al. (5). Growth velocity was determined by two height measurements at least six months apart, which encompassed the spontaneous GH study. Skeletal age was estimated according to the atlas of Greulich and Pyle (6).

Based on clinical criteria, children were classified into groups. Control children were defined as having a growth velocity *standard deviation score* (SDS) of ≥ -1.0 and a height SDS of ≥ -3.0 of the mean for chronologic age, without a recognizable syndrome, obesity, cranial lesion, or precocious puberty. Separate groups were analyzed for children with Turner's syndrome, a history of cranial irradiation for brain tumors or leukemia (radiation doses 12–56 Grey), obesity (BMI >99th percentile for age), and precocious puberty (pubertal development before the age of 8 years for girls and 10 years for boys). A group of children was identified who were growing slowly (growth velocity SDS <-1.0), short (height SDS <3.0), and not GH deficient by provocative testing.

GH Testing

An indwelling intravenous catheter was inserted, and 1.0 ml of blood was withdrawn every 20 min from 0800 to 0800 h for 24-h studies, and from 2000 to 0800 h for 12-h studies, as previously described (1). The children ate regular meals and were encouraged to continue normal activity. Awake vs. sleep state was assessed at each 20-min blood sampling.

An equal volume of serum from each of the sample points from the 24- or 12-h study was combined to create timed pooled specimens. The 24-h pool contained aliquots from 0800 to 0800 h. The A.M. pool contained aliquots from 0800 to 1600 h, and the P.M. pool from 2200 to 0600 h. Extra blood for IGF-I determinations were drawn at 0800 and 2000 h.

Two or more provocative tests of GH secretion were performed on 249 children. The highest GH response to any of the provocative tests was recorded as the peak GH concentration. Provocative tests included insulin (0.05–0.15 U/kg), arginine HCl infusion (0.5 g/kg over 30 min), L-dopa (500 mg), clonidine (0.15 mg/m^2), and propranolol (0.75 mg/kg). Blood samples were obtained at 10- to 30-min intervals after the stimulus.

GH releasing hormone (GHRH) was administered to 122 children. Human pancreatic GHRH-40 was generously provided by Dr. M. Thorner of the University of Virginia, Charlottesville, VA, and Drs. W. Vale and J. Rivier at the Salk Institute, La Jolla, CA. An intravenous bolus of

1.0 μg/kg was given, and blood samples were obtained at frequent intervals for 2 h.

Hormone Assays

Serum GH and prolactin concentrations were measured with double antibody polyclonal radioimmunoassays according to published methods (7, 8). All GH samples from an individual patient were measured in one assay. The intra- and interassay coefficients of variation for the GH assay were 6.2% and 8.1%, and for the prolactin assay 5.5% and 7.5%. The sensitivities of these assays were 0.5 and 1.0 μg/L, respectively. Testosterone was measured by radioimmunoassay, with intra- and interassay coefficients of variation of 8.0% and 12.2%. IGF-I was measured by radioimmunoassay without acid extraction, with intra- and interassay coefficients of variation of 5.1% and 9.2%.

Statistical Analysis

Results are presented as mean ± 1 *standard deviation* (SD). Mean 24-h GH concentration (M24GH) was determined by calculating the average of all 73 samples from each 24-h study. The mean 12-h GH concentration (M12GH) was calculated as the average of all 37 samples from a 12-h study or of samples from 2000 to 0800 h in a 24-h study. Those samples that were below the sensitivity of the assay were assigned the measurement at the sensitivity of the assay, thus giving a maximum estimation of the averaged GH secretion. Analysis of GH pulse frequency, pulse amplitude, and area under the curve was performed using Cluster analysis (9). Peak GH after onset of sleep was defined as a peak occurring within one hour of falling asleep. An average IGF-I value was obtained from the mean of the 0800- and 2000-h samples.

Correlations were done by Pearsons correlation. Groups were compared using analysis of variance for multiple comparisons and two-tailed Student's t-test for single comparisons. Diurnal variation was analyzed by the paired Student's t-test.

Results

Control Population

Sixty children met the control population criteria of a growth velocity SDS of ⩾−1.0 and a height SDS of ⩾−3.0, without a recognized syndrome, cranial irradiation, precocious puberty, or obesity (BMI >95th percentile for age). There were 33 males and 27 females, ranging in age from 3.8 to 17.8 years. Thirty-six of the children were prepubertal at the time of the study; nine were siblings of patients and were enrolled in a study of normal volunteers.

The control children had a M24GH of 3.8 ± 2.1 μg/L and a M12GH of 5.6 ± 3.4 μg/L. Mean GH concentrations were therefore 47% higher during the nocturnal 12-h period than during the whole 24-h period. Control children had 8.1 ± 2.3 peaks in 24 h and 4.2 ± 1.2 peaks in 12 h, indicating that the total number of peaks was evenly spaced during the two types of studies. The number of large peaks increased at night, since the proportion of peaks that were ≥5.0 μg/L increased from 58% during the 24-h period to 71% when only the 12-h period was examined. The mean peak amplitude was 10.2 ± 5.2 μg/L, and the highest GH peak was 24.3 ± 12.3 μg/L. The highest peak occurred within an hour of the onset of sleep in 59% of the studies. Using a definition of a normal mean GH concentration being within one SD of the mean, the lower limit of M24GH is 1.7 μg/L and that of M12GH is 2.2 μg/L.

As a group, growth velocity correlated significantly with M12GH ($r = 0.37$, $P < 0.01$, Fig. 22.1A), area under the curve ($r = 0.35$, $P = 0.02$), number of peaks ≥5.0 μg/L ($r = 0.39$, $P < 0.01$), and peak GH concentration after onset of sleep ($r = 0.32$, $P = 0.02$). Growth velocity did not correlate with M24GH, total number of GH peaks, or mean peak amplitude. The correlation of growth velocity and M12GH was strengthened if only the males were considered ($r = 0.59$, $P < 0.001$, Fig. 22.1B) and was not significant if the females or the prepubertal subjects

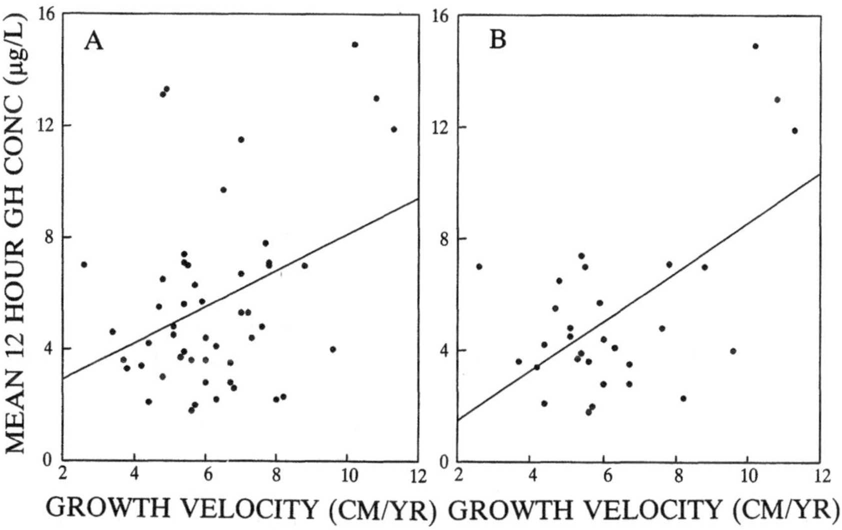

FIGURE 22.1. Relationship of growth velocity to mean 12-h GH secretion in control children (*A*) and control males only (*B*). Regression analysis shows for control children $r = 0.37$, $P < 0.01$, and for control males only $r = 0.59$, $P < 0.001$.

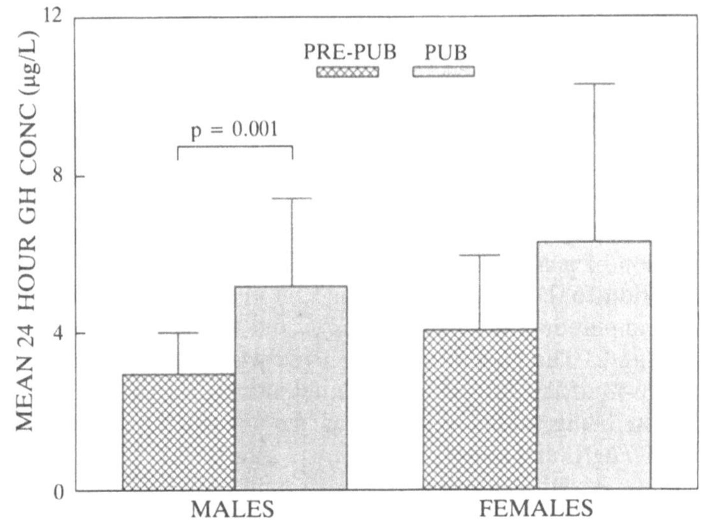

FIGURE 22.2. Changes in the control group mean 24-h GH secretion with attainment of a pubertal bone age. In males, a prepubertal bone age was defined as <12.5 years, and a pubertal bone age as 13–15 years. In females, a prepubertal bone age was defined as <10.5 years, and a pubertal bone age as 11–13 years. (PUB = pubertal.)

were considered alone. In the males, the testosterone P.M. pool correlated strongly with both growth velocity (r = 0.66, $P < 0.0001$) and M12GH (r = 0.51, $P < 0.005$).

M24GH correlated with average IGF-I concentrations in the whole group (r = 0.53, $P < 0.0001$), males only (r = 0.41, $P < 0.05$), and females only (r = 0.58, $P < 0.05$). Similar results were obtained when M12GH was correlated with average IGF-I concentrations, with the exception that only M12GH, not M24GH, correlated with average IGF-I concentrations in prepubertal subjects (r = 0.39, $P < 0.05$). Height SDS did not correlate with M24GH, M12GH, or IGF-I concentrations. Similarly, BMI did not correlate with M24GH or M12GH in this control group, which excluded obese subjects.

M24GH varied by Tanner stage (P = 0.01), with Tanner stage 4 (7.6 ± 3.4 µg/L) being significantly higher than stages 1 (3.4 ± 1.7 µg/L) or 2 (3.3 ± 2.5 µg/L). Males and females are combined for this analysis because few subjects in late puberty were studied. In order to look at males and females separately, each sex was divided into a prepubertal and pubertal group based on bone age (Fig. 22.2). Males had a significant increase in M24GH at a bone age of 13–15 years compared to a bone age of <12.5 years. Females also had an increase in M24GH at a bone age of 11–13

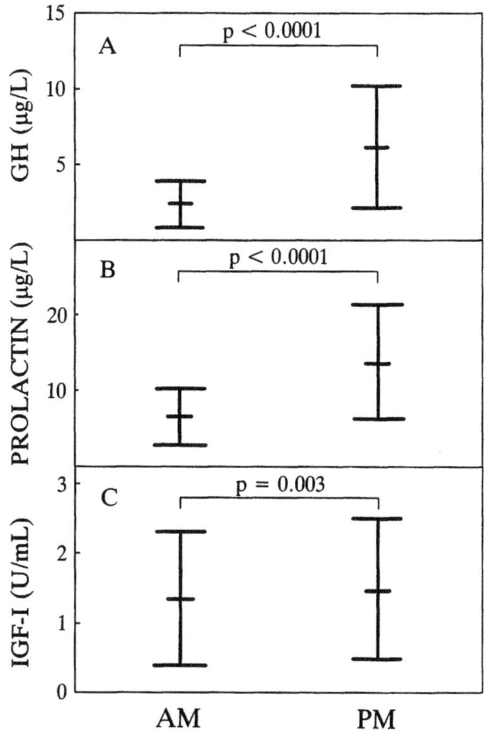

FIGURE 22.3. Diurnal variation of growth hormone (*A*) and prolactin (*B*) pools and IGF-I (*C*) measurements. The GH and prolactin A.M. pools contained aliquots sampled from 0800 to 1600 h, and the P.M. pools from 2200 to 0600 h. Single sample IGF-I determinations were drawn at 0800 and 2000 h.

years compared to a bone age of <10.5 years, but the difference did not reach statistical significance.

Diurnal variation was examined by comparing A.M. (0800–1600 h) and P.M. (2200–0600 h) pools in the same child (Fig. 22.3). GH and prolactin showed clear increases in secretion at night. IGF-I concentrations were also significantly higher at 2000 h than at 0800 h, although the difference is quite small (1.3 ± 1.0 vs. 1.5 ± 1.0 U/ml).

Whole Group

In order to look at the effects of body weight across a wider spectrum, studies from a larger population of children were examined. The only children excluded were those with GH or cortisol excess, or a recognizable syndrome that affects growth, such as Turner's, Down's, or Noonan's syndrome. This population includes 260 children with a mixture of di-

TABLE 22.1. Spontaneous growth hormone characteristics in obese, normal-weight, and underweight children.

	Obese	Normal weight	Underweight
Mean 12-h GH (μg/L)	2.4 ± 1.5^{a}	4.1 ± 2.9	4.9 ± 2.3^{a}
Mean GH peak amplitude (μg/L)	4.5 ± 3.9^{a}	7.9 ± 5.4^{b}	$10.5 \pm 6.5^{a,b}$
Number of total pulses in 12 h	$2.7 \pm 2.3^{a,b}$	4.0 ± 1.7^{a}	4.3 ± 1.4^{b}
Number of pulses ≥5 μg/L in 12 h	1.5 ± 1.7^{a}	2.5 ± 1.5^{b}	$3.4 \pm 1.5^{a,b}$

Note: Values with the same superscript are significantly different from each other.

agnoses, including tall stature, obesity, short stature, precocious puberty, and previous cranial irradiation, and normal volunteers.

BMI correlated negatively with M24GH and M12GH ($r = -0.35$, $P < 0.0001$), peak GH concentration after provocative testing ($r = -0.43$, $P < 0.0001$), and peak GH concentration after GHRH administration ($r = -0.47$, $P < 0.0001$). The GH values were log transformed to facilitate the correlations.

Children in this population were classified as underweight if their BMIs were ≤1st percentile for age ($n = 35$), normal if their BMIs were between the 1st and 99th percentile for age ($n = 208$), and obese if their BMIs were ≥99th percentile for age ($n = 16$). Body weight had significant effects on spontaneous GH secretion characteristics (Table 22.1). M12GH was significantly increased in the underweight patients compared with the obese patients. Mean GH peak amplitude was also significantly increased in the underweight patients. The total number of peaks in 12 h was decreased in the obese, but not different from normal in the underweight, whereas the number of GH peaks ≥5.0 μg/L was increased in the underweight and decreased in the obese.

Selected Groups

Studies from five diagnostic groups were selected for further evaluation: the control population as defined above ($n = 60$), obese subjects with BMI ≥99th percentile for age and no other diagnoses ($n = 9$), children with precocious puberty ($n = 15$), children with a previous history of cranial irradiation ($n = 30$), and girls with Turner's syndrome ($n = 12$). The obese and irradiated groups had low M24GH when compared to the control or precocious puberty groups, and the magnitudes of these changes were similar (Fig. 22.4A). The same pattern was seen in the mean GH peak amplitude, number of GH peaks ≥5.0 μg/L, and peak GH concentration after sleep onset. The total number of peaks in 24 h was decreased only in the obese (4.3 ± 4.4 peaks) compared to the control group (8.1 ± 2.4 peaks); the irradiated group (7.2 ± 4.7 peaks) was not different from control. Conversely, the average IGF-I concentration was decreased in the irradiated group and increased in the pre-

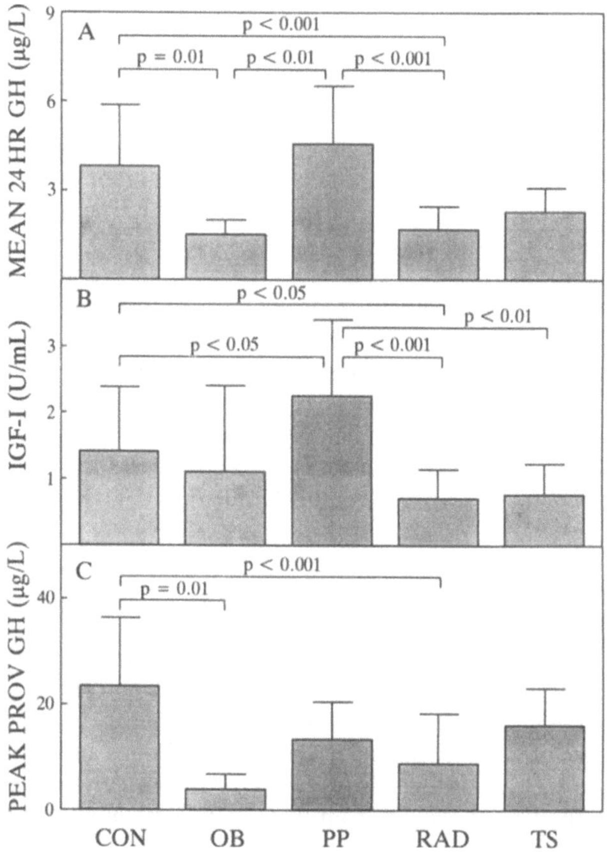

FIGURE 22.4. Variation in mean 24-h GH secretion (*A*), average IGF-I concentration (*B*), and peak GH concentration after provocative stimuli (*C*) in five clinical groups. (CON = control group, OB = obese subjects, PP = precocious puberty, RAD = history of cranial irradiation, TS = Turner's syndrome, and PEAK PROV GH = peak GH concentration after provocative stimuli.)

cocious puberty group compared with control (Fig. 22.4B); the obese group was not different from control. Peak GH concentration after provocative testing showed a similar pattern to M24GH, but the obese appeared even more decreased than the irradiated group compared with controls (Fig. 22.4C). The groups did not differ in their GH responses to GHRH infusion.

Short, Slow-Growing, Non-GH-Deficient Children

Finally, studies from a group of children with unexplained slow growth were examined. Thirty-six children had a height SDS <−3.0, a growth velocity SDS <−1.0, and a peak GH concentration after provocative testing ≥10.0 μg/L, with no systemic illness. The studies in these patients were compared with those of 60 children in the control population defined above.

M24GH was not significantly different between these two groups, but the M12GH was slightly lower in the slow-growing group (4.1 ± 1.9 μg/L) compared with controls (5.6 ± 3.4 μg/L, $P < 0.05$). The total number of GH peaks in 24 h was slightly lower in the slow-growing group (6.8 ± 2.0 peaks) vs. controls (8.1 ± 2.4 peaks, $P < 0.05$), an average of 1.3 fewer peaks, but the total number of peaks in 12 h was exactly the same between the two groups (4.2 ± 1.6 vs. 4.3 ± 1.2 peaks, respectively). Similarly, the number of GH peaks ≥5.0 μg/L and the mean GH peak amplitude were no different between groups. The peak GH concentration after sleep onset was significantly lower in the slow-growing group (16.5 ± 10.5 vs. 21.2 ± 11.3 μg/L, $P < 0.05$), as was the average IGF-I concentration (0.7 ± 0.6 vs. 1.4 ± 0.9 U/ml, $P < 0.001$).

Discussion

We have taken our seven-year clinical and research experience with spontaneous GH studies and compared the studies by clinical groups that encompass both normal and abnormal growth states. Our studies have the advantage of being done under uniform conditions and were analyzed with a single GH assay. Because GH assays vary widely between laboratories (10), the trends elicited from this analysis are more important than the specific numbers.

In our analysis of 60 children with normal stature and growth rate, we found a modest correlation between nighttime GH secretion and growth velocity. Other investigations that have found a correlation of spontaneous GH secretion and growth velocity have included *GH-deficient* (GHD) children (11, 12). Studies looking at growth velocity using non-GHD children have found no relationship with spontaneous GH secretion (13–15). Albertsson-Wikland and Rosberg (16) found a correlation between height and 24-h GH secretion, whereas we and other investigators did not (13, 15, 17).

The relationship of growth velocity and nighttime GH secretion was probably due to the increase in GH secretion seen with puberty, since the correlation was not present when only the prepubertal controls were considered. Other studies have shown a rise in 24-h GH concentrations in pubertal subjects (18, 19), and we noted an increase in 24-h GH secretion in our control group at Tanner stage 4 and in boys with pubertal bone ages. In our study, testosterone secretion in the males correlated more strongly with growth velocity than did GH secretion, suggesting an important action of testosterone on growth velocity. Martha et al. (18) also found a positive correlation of testosterone with mean 24-h GH secretion in boys before epiphyseal fusion.

In several cases, nocturnal GH secretion was more informative than total daily secretion. In the control group, growth velocity correlated significantly with M12GH, but did not correlate with M24GH. In prepubertal controls, M12GH correlated with IGF-I levels, whereas M24GH did not. In the group of short, slow-growing, non-GHD children, M12GH was decreased compared to controls, whereas M24GH was not. These relationships suggest that nocturnal GH secretion has more physiologic impact than total daily secretion.

Changes in total GH secretion could theoretically be mediated by modification of pulse amplitude and/or frequency. In most situations, we found pulse amplitude to be the parameter that varied: It increased with sleep in the control population and with undernutrition, and it decreased with obesity and after irradiation. In our studies, obesity was unique in producing a change in pulse frequency. Veldhuis et al. also found that obesity in adult men causes a clear decrease in pulse frequency (20).

BMI was seen to have an effect on spontaneous GH secretion only when the extremes of weight were considered; there was no correlation of BMI with M24GH within the control group. When obese and underweight patients were also considered, BMI correlated negatively with M24GH, peak GH concentration after provocative testing, and peak GH concentration after GHRH administration. This would imply a threshold must be reached before BMI affects GH secretion. The r values were higher for the relationship of BMI with provocative and GHRH testing, suggesting they are more directly affected by weight than is spontaneous GH secretion.

Analysis of the five clinical groups revealed divergences between 24-h GH secretion and IGF-I concentrations. The obese group had decreased M24GH, but normal IGF-I levels compared with the controls. Conversely, the precocious puberty group had no significant change in GH secretion compared to control, but showed an increase in IGF-I concentration. Clearly, other factors affect IGF-I levels, such as insulin and sex hormone concentrations.

The irradiated group showed a decrease in GH pulse amplitude without a change in pulse frequency. Others have also noted decreased GH pulse amplitude in cranially irradiated children (21–23), suggesting that the "pulse generator" is intact, but that GHRH secretion may be impaired. The Turner's syndrome group did not vary significantly from controls in any of the parameters tested, but appeared to have a mild decrease in 24-h GH secretion. Other studies have shown that girls with Turner's syndrome secrete normal amounts of GH at a young age, but they do not have the expected pubertal increase in GH secretion (24–26).

The short, slow-growing, non-GHD children had a subtle difference in their GH secretion compared to controls. Only the nocturnal GH secretion was decreased, and this appeared to be mediated largely by a decrease in sleep-associated GH release. GH pulse amplitude was not different from controls, and the mean GH pulse frequency decrease of 1.3 peaks during the day is of doubtful physiologic significance. Nonetheless, these factors, and potentially others, resulted in lower average IGF-I concentrations, suggesting these children may have a mild GH dysregulation.

In summary, we have analyzed the 300 24- or 12-h spontaneous GH studies done in our clinical and research facility over the last seven years. Sixty children who had a normal height and growth velocity served as a control group and showed a significant correlation of M12GH with growth velocity, but most of the variation in GH secretion was related to puberty. M24GH increased significantly at Tanner stage 4. Diurnal variation was seen in GH, prolactin, and IGF-I secretion. Cranially irradiated and obese subjects had significant decreases in M24GH and mean peak GH amplitude. Only the obese children had a change in peak frequency; they had approximately half the number of GH peaks as controls. A group of 36 slowly growing, short, non-GHD children had mildly decreased M12GH values, peak frequency, sleep-associated peak GH concentration, and IGF-I concentration compared to controls, suggesting a subtle GH dysregulation.

References

1. Bercu BB, Shulman D, Root AW, Spiliotis BE. Growth hormone (GH) provocative testing frequently does not reflect endogenous GH secretion. J Clin Endocrinol Metab 1986;63:709–15.
2. Shulman DI, Bercu BB. Evaluation of growth hormone secretion: provocative testing vs. endogenous 24-hour growth hormone profile. Acta Paediatr Scand [Suppl] 1987;337:61–71.
3. Spiliotis BE, August GP, Hung W, Sonis W, Mendelson W, Bercu BB. Growth hormone neurosecretory dysfunction: a treatable cause of short stature. JAMA 1984;251:2223–30.
4. Tanner JM. Growth at adolescence. Springfield, IL: Charles C. Thomas, 1972.
5. Hammer LD, Kraemer HC, Wilson DM, Ritter PL, Dornbusch SM. Standardized percentile curves of body-mass index for children and adolescents. Am J Dis Child 1991;145:259–63.
6. Greulich WW, Pyle SJ. Radiographic atlas of skeletal development of the hand and wrist. Palo Alto, CA: Stanford University Press, 1973.
7. Root AW, Rosenfield RL, Bongiovanni AM, Eberlein WR. The plasma growth hormone response to insulin-induced hypoglycemia in children with retardation of growth. Pediatrics 1967;39:844–52.
8. Reiter EO, Root AW. The effect of pyridoxine in pituitary release of growth hormone and prolactin in childhood and adolescence. J Clin Endocrinol Metab 1978;47:689–90.

9. Veldhuis JD, Johnson ML. Cluster analysis: a simple, versatile, and robust algorithm for endocrine pulse detection. Am J Physiol 1986;250:E486–93.
10. Reiter EO, Morris AH, MacGillivray MH, Weber D. Variable estimates of serum growth hormone concentrations by different radioassay systems. J Clin Endocrinol Metab 1988;66:68–71.
11. Hindmarsh P, Smith PJ, Brook CGD, Matthews DR. The relationship between height velocity and growth hormone secretion in short prepubertal children. Clin Endocrinol 1987;27:581–91.
12. Spadoni GL, Cianfarani S, Bernardini S, et al. Twelve-hour spontaneous nocturnal growth hormone secretion in growth retarded patients. Clin Pediatr 1988;27:473–8.
13. Rochiccioli P, Messina A, Tauber MT, Enjaume C. Correlation of the parameters of 24-hour growth hormone secretion with growth velocity in 93 children of varying height. Horm Res 1989;31:115–8.
14. Zadik Z, Chalew SA, Kowarski A. Assessment of growth hormone secretion in normal stature children using 24-hour integrated concentration of GH and pharmacological stimulation. J Clin Endocrinol Metab 1990;71:932–6.
15. Lin T-H, Kirkland RT, Sherman BM, Kirkland JL. Growth hormone testing in short children and their response to growth hormone therapy. J Pediatr 1989;115:57–63.
16. Albertsson-Wikland K, Rosberg S. Analyses of 24-hour growth hormone profiles in children: relation to growth. J Clin Endocrinol Metab 1988; 67:493–500.
17. Costin G, Kaufman FR, Brasel JA. Growth hormone secretory dynamics in subjects with normal stature. J Pediatr 1989;115:537–44.
18. Martha PM Jr, Rogol AD, Veldhuis JD, Kerrigan JR, Goodman DW, Blizzard RM. Alterations in the pulsatile properties of circulating growth hormone concentrations during puberty in boys. J Clin Endocrinol Metab 1989;69:563–70.
19. Rose SR, Municchi G, Barnes KM, et al. Spontaneous growth hormone secretion increases during puberty in normal girls and boys. J Clin Endocrinol Metab 1991;73:428–35.
20. Veldhuis JD, Iranmanesh A, Ho KKY, Waters MJ, Johnson ML, Lizarralde G. Dual defects in pulsatile growth hormone secretion and clearance subserve the hyposomatotropism of obesity in man. J Clin Endocrinol Metab 1991;72:51–9.
21. Blatt J, Bercu BB, Gillin JC, Mendelson WB, Poplack DG. Reduced pulsatile growth hormone secretion in children after therapy of acute lymphoblastic leukemia. J Pediatr 1984;104:182–6.
22. Romshe CA, Zipf WB, Miser A, Miser J, Sotos JF, Newton WA. Evaluation of growth hormone release and human growth hormone treatment in children with irradiation-associated short stature. J Pediatr 1984;104:177–81.
23. Costin G. Effects of low-dose cranial radiation on growth hormone secretory dynamics and hypothalamic-pituitary function. Am J Dis Child 1988;142: 847–52.
24. Ross JL, Long LM, Loriaux DL, Cutler GB. Growth hormone secretory dynamics in Turner syndrome. J Pediatr 1985;106:202–6.
25. Lin T-H, Kirkland JL, Kirkland RT. Growth hormone assessment and short-term treatment with growth hormone in Turner syndrome. J Pediatr 1988; 112:919–22.

26. Zadik Z, Landau H, Chen M, Altman Y, Lieberman E. Assessment of growth hormone (GH) axis in Turner's syndrome using 24-hour integrated concentrations of GH, insulin-like growth factor-1, plasma GH-binding activity, GH binding to IM9 cells, and GH response to pharmacological stimulation. J Clin Endocrinol Metab 1992;75:412–6.

Part VI

Poster Presentation Manuscripts

23

Expression of the Short Isoform of the Growth Hormone Receptor in Adipocytes

H. MAURICE GOODMAN, G. PETER FRICK, TOVA BICK, LIH-RUEY TAI, AND JACK L. LEONARD

Growth hormone (GH) acts on a variety of terminally differentiated cells, including adipocytes, myocytes, hepatocytes, and pancreatic beta cells, to regulate energy balance (1). In adipocytes GH produces an immediate insulin-like response that is most readily demonstrated by accelerated glucose metabolism in cells that have been deprived of GH for at least 3 h (2) and a delayed anti-insulin-like response typified by increased lipolysis (3). Because these responses are independently affected by chemical modifications of GH (4, 5) or the GH receptor (6), it was suggested that different responses may require different hormone receptor interactions (5, 6). However, consistent findings of linear Scatchard isotherms argue for a single class of binding sites (6–9). In the absence of any definitive mechanism for how GH might produce any of its biological actions, the question of whether GH interacts with a single receptor or a variety of receptor complexes cannot yet be answered.

The isolation and cloning of cDNA for the GH receptor from rabbit and human livers by Wood and his colleagues at Genentech (10) make it possible to study the nature of GH receptor interactions directly. The nucleotide sequence of the cDNA predicts a protein of about 70 kd comprised of 620 amino acid residues with a single putative transmembrane domain separating the ~250-residue aminoterminal extracellular domain from the ~350-residue carboxyterminal intracellular domain. Data from cross-linking studies suggest a molecular size of about 105–110 kd (11–13) and indicate that the extracellular domain is heavily glycosidated (14). In addition, the hepatic receptor appears to be ubiquinated (15). Of further interest is the finding that the circulating GH binding protein in human blood that had been described earlier by Baumann et al. (16) and by Herington et al. (17) is identical in amino acid sequence with the extracellular portion of the membrane-bound receptor and may arise by posttranslational processing or proteolytic cleavage of the receptor at the extracellular surface (10).

Studies by Smith et al. (18) and Baumbach et al. (19) revealed that the *GH binding protein* (GHBP) in rodent blood is encoded in hepatocytes by a ~1.2-kb mRNA transcript that is formed by alternate splicing of the transcript from the same gene as the transmembrane receptor protein, which in rats is encoded in a ~4.5-kb mRNA transcript. The soluble GHBP in the rat lacks the entire transmembrane and cytosolic domains found in the full-length receptor and contains instead at its carboxyl terminus a unique sequence of 17 hydrophilic amino acids.

We studied expression of the GH receptor in rat adipocytes using a cDNA fragment consisting of the partial sequence of the rabbit hepatic GH receptor kindly provided by Dr. Wood. In Northern blots of total RNA from adipocytes, the GH receptor cDNA probe hybridized with at least two mRNA transcripts, corresponding to 4.5 and 1.2 kb (20). These two transcripts were also present in mRNA extracted from muscle and other tissues (20–22). When Northern blots of adipocyte total or polyadenylated RNA are examined with a probe containing the nucleotide sequence of the extracellular portion of the rat GH receptor, the 4.5- and 1.2-kb species are present in nearly equal abundance (Fig. 23.1). The 4.5-kb transcript appeared to coincide with the mRNA that encodes the transmembrane receptor reported for liver (23). The apparent abundance

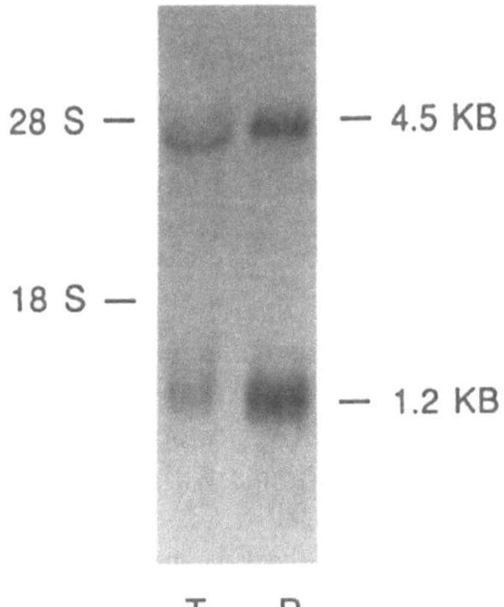

FIGURE 23.1. Northern blots of total (T) and polyadenylated (P) RNA extracted from rat epididymal adipocytes. RNA samples were analyzed by electrophoresis on agarose gels containing formaldehyde, immobilized on nitrocellulose membranes, and hybridized with a ^{32}P-labeled cDNA probe that contains the sequence for the short isoform of the rat GH receptor.

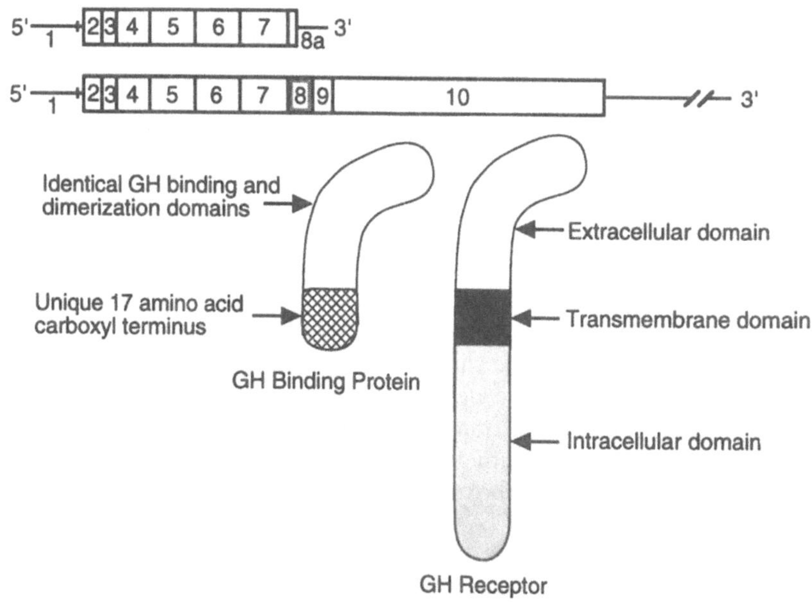

FIGURE 23.2. Proposed exon structures for RNA transcripts that encode the long and short forms of the GH receptor and the corresponding proteins. The extracellular domains of both the full-length receptor and the GH binding protein are encoded in exons 1–7. In the short isoform, exon 8a, corresponding to the unique 17-amino acid carboxyl terminus, replaces exon 8, which corresponds to the transmembrane domain, and exons 9 and 10, which correspond to the intracellular domain of the long form of the receptor.

of the 1.2-kb transcript raised the possibility that adipocytes might produce an additional receptor isoform that might account for some of the diverse responses to GH. We therefore isolated, cloned, and sequenced the 1.2-kb transcript and found that although there are some differences in the 5′ untranslated region, its coding region and most of the 3′ untranslated end are identical to the hepatic mRNA that encodes the circulating GHBP (24).

The structures of the two alternately spliced mRNA transcripts of the GH receptor gene and the exons from which they are thought to be derived (25) are shown in Figure 23.2. There are at least three variations at the 5′ end reflecting multiple start sites and/or alternate splicing. Exons 2–7 are identical in the transmembrane form of the GH receptor and the short isoform of the GH receptor, but the two diverge after exon 7. Exons 8, 9, and 10 encode a short stretch of the extracellular domain, the transmembrane domain, and the cytosolic domain, followed by a long untranslated tail at the 3′ end. In the short isoform of the GH receptor, exons 8, 9, and 10 are replaced with a short exon that codes for a unique 17-amino acid hydrophilic carboxyl terminus. Figure 23.2 also shows the predicted structures of the two proteins encoded by these alternately spliced mRNA transcripts. Because the short isoform of the GH receptor lacks a hydrophobic transmembrane domain and contains a leader sequence that directs its amino terminus across cisternal membranes, and is found in blood, this isoform of the GH receptor is generally regarded as a secreted protein. Thus, it is possible that adipocytes and various other rat tissues may contribute to the GH binding activity of normal rat plasma, or that fortuitously formed 1.2-kb message is not translated in adipocytes, or that the translated product is rapidly degraded.

To determine whether the short isoform of the GH receptor is actually produced in adipocytes, we developed an antibody that recognizes the short isoform, but not the transmembrane isoform of the GH receptor. Rabbits were immunized with a synthetic 17-amino acid peptide that corresponds to the unique carboxyl terminus of the short isoform. Because *antiserum from rabbit 1615* (As1615) had a high titer and affinity for the short isoform of the GH receptor, affinity-purified antibodies from this rabbit were used in subsequent studies.

One way that we verified that the short isoform of the GH receptor is expressed in rat adipocytes is shown in Figure 23.3. Adipocytes freshly isolated from normal epididymal fat were pipeted onto polylysine-coated coverslips and fixed in 4% paraformaldehyde for 15 min. After thorough washing, the cells were made permeable by treatment for 15 min with 0.1% Triton X-100. The coverslips were then incubated for 60 min with a 1:50 dilution of As1615. As a control, half the coverslips were incubated with As1615 that had been pretreated with 100 μg/ml of the peptide antigen. After thorough washing, the coverslips were then incubated with rhodamine-labeled secondary antibodies and again thoroughly washed. The coverslips were then examined by digital imaging fluorescence microscopy using a Biorad model MRC600 confocal microscope.

A control fat cell is shown in the upper left panel of Figure 23.3 with the gain setting on the microscope adjusted to just permit visualization of the cell. A thin rim of cytosol containing endogenous fluorescent material surrounds the large central fat droplet, giving the cell a typical signet ring appearance. The nuclear bulge is seen in the upper right of the cell. A fat cell stained with As1615 is shown at the identical gain setting in the upper right panel. Fluorescence intensity measured in the perinuclear region was 17-fold greater than in the control cell. The cytosol of an adjacent cell is visible at the upper right. The lower panels in the figure show the perinuclear regions of the same cells in higher magnification. These observations indicate that adipocytes express the short isoform of the GH receptor and that it appears to be present throughout the cytosol, but is excluded from the central fat droplet and probably from the nucleus. The resolution is too low to permit more precise localization.

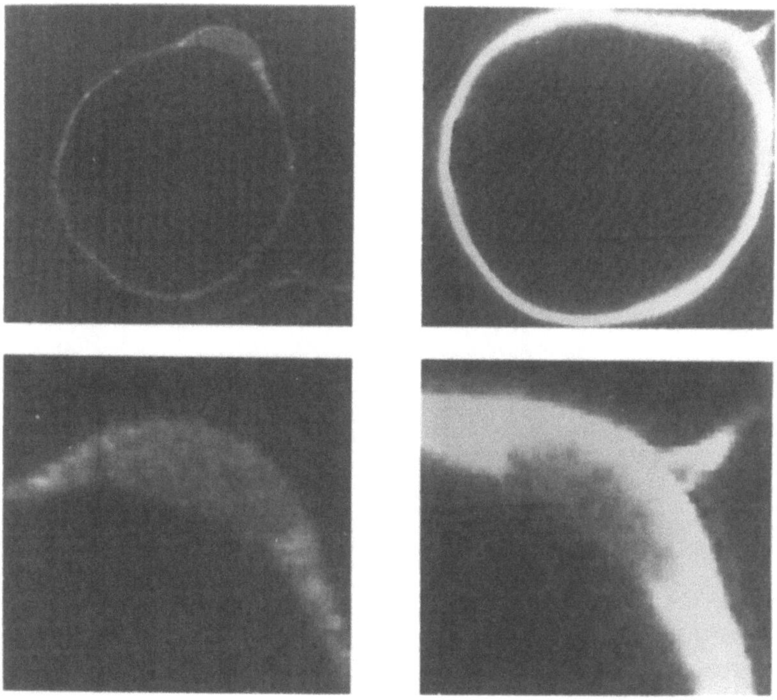

FIGURE 23.3. Distribution of the short isoform of the GH receptor in normal rat adipocytes. Adipocytes were fixed on coverslips, permeabilized, and reacted with affinity-purified As1615 (right panels) or As1615 that had been preadsorbed with excess peptide antigen (left panels). The cells were then treated with goat anti-rabbit antibodies that were labeled with rhodamine and examined with a Biorad Model MRC600 confocal microscope. The control cell in the upper left panel has a diameter of 37.9 micra, and the experimental cell on the upper right has a diameter of 52.0 micra. The gain control on the microscope display was set just to allow visualization of the control cell. At this gain setting, the image of the cell treated with As1615 saturated. The mean intensity of the cell treated with As1615 was >17 times greater than the control cell. The lower panels show the perinuclear regions of both cells at higher magnification. These images were obtained through the courtesy of the University of Massachusetts Medical School Imaging Facility with the generous assistance and cooperation of Mr. Douglas Bowman, Dr. Ronald Lynch, and Dr. Fredric Fay.

Using the more traditional biochemical approach of immunoprecipitation, we also demonstrated that the short isoform of the GH receptor is present in adipocytes. Rat epididymal adipocytes were incubated for 3 h with carrier-free ^{35}S-L-methionine to label endogenous proteins. The cells were homogenized in 0.3 M sucrose containing 1 mM EDTA, and particulate and soluble fractions were obtained by centrifugation at

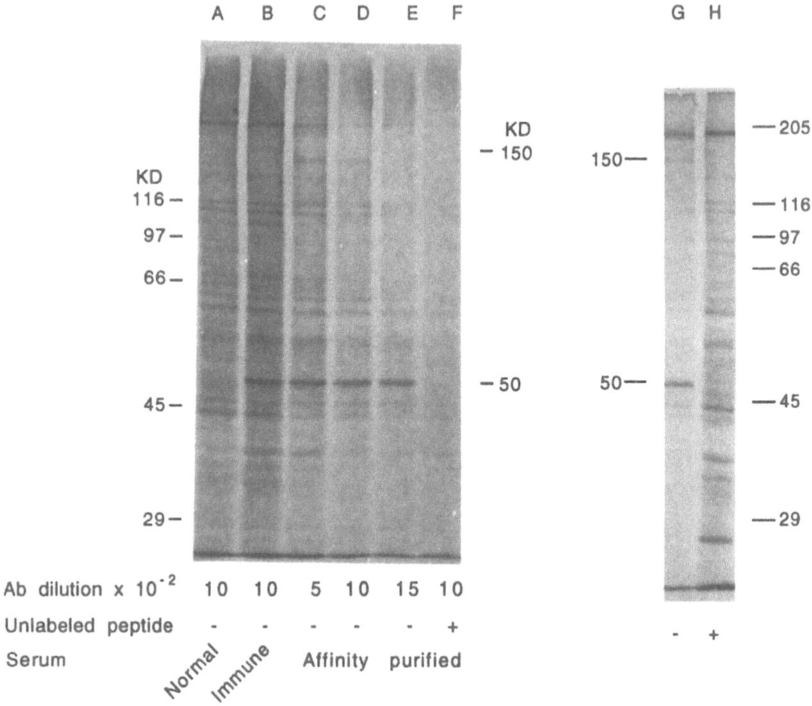

Figure 23.4. Imunoprecipitation of ^{35}S-labeled proteins from rat adipocytes. Proteins of adipocytes isolated from 20 g of epididymal fat tissue were metabolically labeled by incubation of the cells for 3 h with ^{35}S-methionine. The adipocytes were homogenized in 0.25 M sucrose containing 1 mM EDTA and centrifuged at 150,000 × g to obtain a high-speed supernatant and a high-speed particulate fracton. The high-speed supernatant fraction was concentrated by ultrafiltration, and 5 × 10^6 cpm of ^{35}S-labeled proteins were incubated overnight with normal (*A*), immune (*B*), or affinity-purified As1615 at several dilutions (*C–F*). Unlabeled antigen (100 μg/ml) was added to block immunoprecipitation (*F*). Proteins in the particulate fraction were solubilized with Triton X-100 and immunoprecipitated in the absence (*G*) or presence (*H*) of 100 μg/ml of unlabeled peptide antigen. The immunoprecipitates were collected using protein A cross-linked to agarose beads and analyzed by electrophoresis on 7.5% polyacrylamide under reducing conditions. Labeled proteins were visualized by autoradiography. Reprinted with permission from Frick and Goodman (24), © The Endocrine Society, 1992.

150,000 × g for 60 min. Proteins present in the soluble fraction were immunoprecipitated with normal, immune, or affinity-purified As1615. Immune complexes were recovered using protein A linked to Sepharose beads, analyzed by SDS polyacrylamide gel electrophoresis, and detected by autoradiography (Fig. 23.4). A prominent band of about 50 kd and a larger band at about 150 kd were specifically immunoprecipitated.

Similar results were obtained with the particulate fraction. After the pellets were extracted with 1% Triton X-100 in 50 mM Tris buffer, aliquots of the extract were immunoprecipitated with affinity-purified As1615 in the absence or presence of unlabeled antigen. Once again, As1615 specifically precipitated proteins with mobilities corresponding to 50 and 150 kd. Although the immunoprecipitate prepared in the presence of unlabeled antigen appears to have been washed less well and to contain a variety of contaminating proteins, it lacked bands at 50 and 150 kd. Although the 1.2-kb mRNA encodes a peptide with a predicted molecular size of about 32 kd, the circulating GHBP appears on Western blots as a doublet of 48 and 52 kd as a result of its carbohydrate content (26). We do not know if the 150-kd band associates with the short isoform of the GH receptor and therefore coprecipitates with it or if it is an artifact.

To determine if the immunoprecipitated proteins can bind GH, cells were treated as before, except that prior to immunoprecipitation the solubilized extract was incubated overnight with hGH that was cross-linked to Affigel beads or with Affigel beads without hGH. Proteins that adsorbed to the beads were eluted with 4.5 M urea, concentrated to remove the urea and immunoprecipitated with As1615. No band corresponding to 50 kd was seen, but two bands of 38 and 42 kd were found (Fig. 23.5), although immunoprecipitation of the same extract prepared prior to adsorption onto the GH beads produced bands of 50, 42, and 38 kd. Further studies are needed to determine why the 50-kd band failed to bind to the GH beads.

Using affinity-purified As1615, we developed an assay to quantitate the short isoform of the GH receptor present in adipocytes. Extracts of the soluble and particulate fractions of fat cell homogenates were incubated overnight in a mixture that also contained a trace of ^{125}IhGH, along with varying amounts of unlabeled hGH, As1615, and protein A cross-linked to agarose beads (Fig. 23.6). Immune complexes consisting of ^{125}IhGH, the short isoform of the GH receptor, and As1615 were collected on protein A-coated agarose beads and counted. Displaceability of ^{125}IhGH by unlabeled hGH served as a measure of specificity of binding to the short isoform of the GH receptor. Ninety percent of the short isoform present in adipocytes sedimented with the particulate fraction (Fig. 23.7).

To determine if this form of the GH receptor is secreted by adipocytes, we incubated a 1:3 suspension of fat cells for 1 h and measured GHBP in the incubation medium. Less than 10% of the short isoform of the GH receptor found in the cells was present in the incubation medium, and of this, more than 60% was sedimented by centrifugation at 150,000 × g and could therefore be attributed to the inevitable breakage of cells.

Scatchard analysis of binding activity detected by As1615 indicated that the short isoform that sedimented with the particulate fraction has an affinity for hGH that is similar to that found for the GH receptor (7–9) and is present in an amount sufficient to bind about 60 fmoles of GH/g of

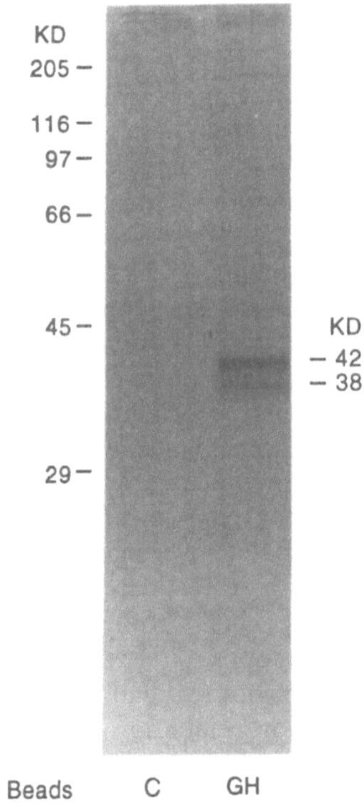

FIGURE 23.5. Immunoprecipitation of ^{35}S-labeled adipocyte proteins after elution from immobilized GH. Adipocyte proteins were labeled and extracted from the particulate fraction as described in Figure 23.4. Sedimented proteins were solubilized in Tris-buffered saline that contained 0.1% Triton X-100 and 0.1% hemoglobin. After clarification by centrifugation at 12,000 × g for 5 min, soluble proteins were incubated overnight with Affigel 15 beads cross-linked to ethanolamine (C = control) or hGH followed by ethanolamine. Proteins that were selectively eluted with 4.5 M urea were concentrated by ultrafiltration to remove the urea before immunoprecipitation and analysis as described in Figure 23.4.

tissue (Fig. 23.8). Assuming that a 2:1 BP:hormone complex is formed and that there are about 10^7 cells/g, we estimate that each cell contains about 0.3 fg or about 7,000 copies of the short isoform of the GH receptor. If accessible to GH present in the incubation medium, short isoform of the GH receptor could therefore account for 30% of the binding activity of intact fat cells (7–9).

Several lines of evidence suggest that at least some of the short isoform of the GH receptor might indeed be present on the fat cell surface. To examine this possibility, we added the bifunctional cross-linking agent disuccinimidyl suberate to a suspension of adipocytes that had incubated with ^{125}IhGH to covalently link closely associated proteins. Membrane

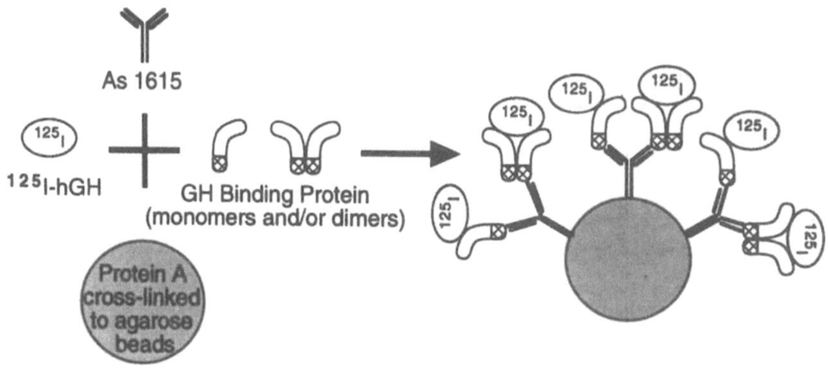

FIGURE 23.6. Assay for GH binding protein. Affinity-purified As1615 (1:500 dilution), ^{125}IhGH, and cell extracts containing both isoforms of solubilized GH receptor proteins were incubated overnight. Immune complexes were collected by incubation for 2 h with 10 μl of protein A-Sepharose beads, followed by centrifugation at 15,000 × g for 1 min. The beads were then washed twice with 1 ml of buffer and counted.

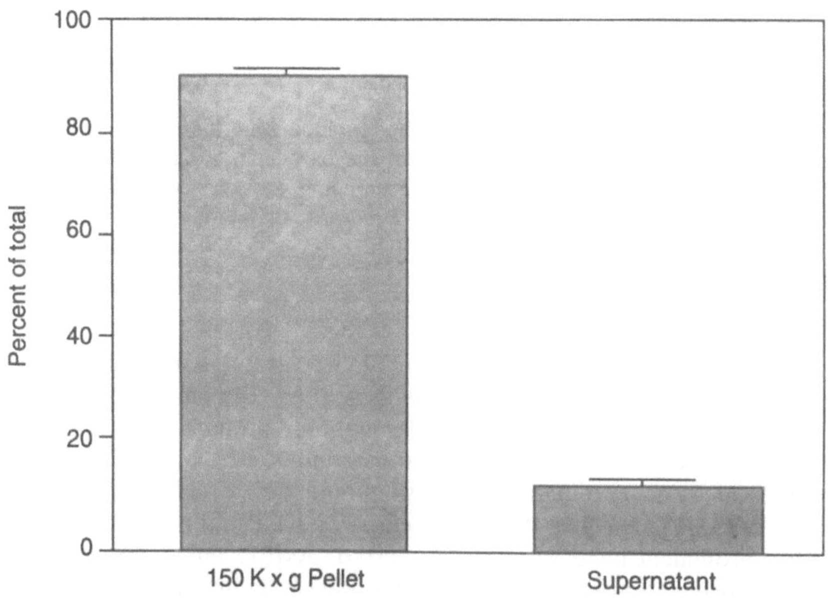

FIGURE 23.7. Recovery of the short isoform of the GH receptor in adipocyte extracts. The short isoform of the GH receptor was measured in the soluble and particulate fractions of adipocyte extracts in the presence of 10 ng/ml of ^{125}IhGH. Each bar represents the mean of seven independent experiments. The standard errors of the means are indicated on each bar, and the difference is highly significant ($P < 0.001$).

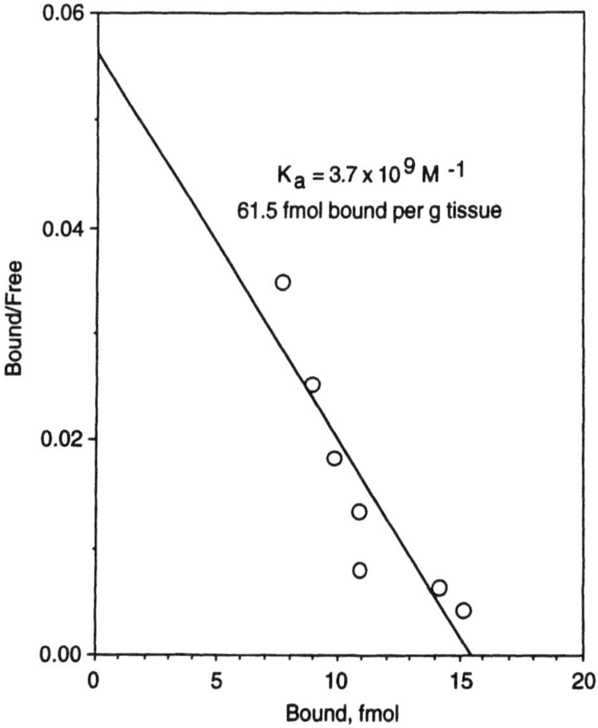

FIGURE 23.8. Scatchard analysis of ^{125}IhGH binding to the short isoform of the GH receptor in solubilized extracts of the particulate fraction of adipocyte homogenates. Proteins were extracted and binding was assayed as described in Figures 23.6 and 23.7, except the concentration of ^{125}IhGH ranged from 2.5 ng/ml to 80 ng/ml. Reprinted with permission from Frick and Goodman (24), © The Endocrine Society, 1992.

proteins were dissolved in a mixture of detergents (RIPA buffer) and then immunoprecipitated overnight with As1615. The immunoprecipitate and a sample of the total solubilized membrane proteins were separated by SDS polyacrylamide gel electrophoresis under reducing conditions and visualized by autoradiography. As we found before (11), a prominent band of radioactive cross-linked proteins was evident at about 130 kd and a less-prominent band was seen at ~60 kd. As1615 immunoprecipitated ~60-kd proteins cross-linked to ^{125}IhGH, indicating that the short isoform of the GH receptor was present at a cellular locale that was accessible to hGH.

In earlier experiments (27) we found that incubation of adipocytes with a low concentration of trypsin for 10 min removed all of the specific binding sites for GH from the adipocyte surface. These cells recovered their capacity to bind GH within 2 h after removal of trypsin. Following

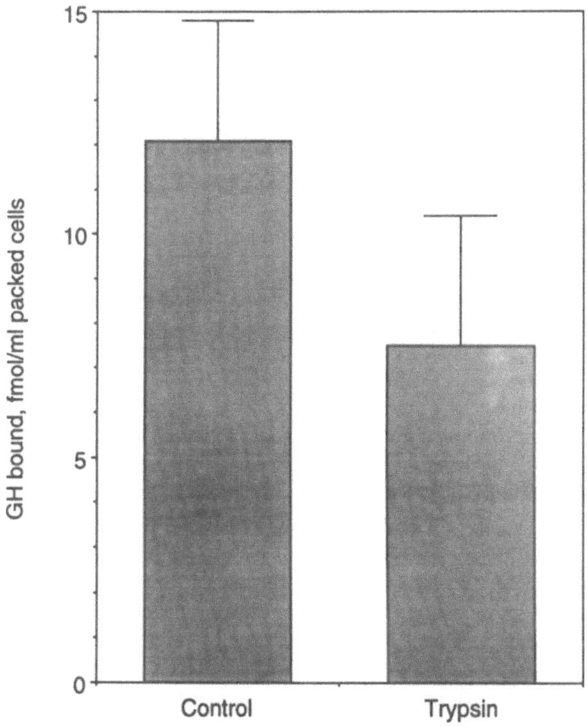

FIGURE 23.9. Loss of the short isoform of the GH receptor in extracts of adipocytes after exposure of intact cells to 0.1 mg/ml of trypsin for 10 min at 37 °C. The short isoform was quantitated as described in Figure 23.6. Each bar represents the mean value obtained in three independent experiments. The brackets indicate the SEMs. The effects of trypsin are statistically significant ($P < 0.05$) by paired analysis.

the same trypsinization protocol, we measured total binding of ^{125}IhGH and the short isoform of the GH receptor in cell extracts prepared immediately after trypsin treatment and again after 2 h. The abundance of the short isoform of the GH receptor was decreased by nearly 50% immediately after trypsin treatment (Fig. 23.9), while total binding fell virtually to zero. Recovery of full binding capacity 2 h later indicates that the surfaces of the adipocytes were not breached by trypsin treatment and, therefore, that at least some of the short isoform of the GH receptor must be at the cell surface.

The foregoing experiments strongly suggest that the short isoform of the GH receptor is available to interact with GH at the fat cell surface. To gain further insight into this issue, and in preparation for study of the possible physiological role of the short isoform of the GH receptor in adipocytes, we stably transfected 3T3-L1 cells with the cDNA that encodes

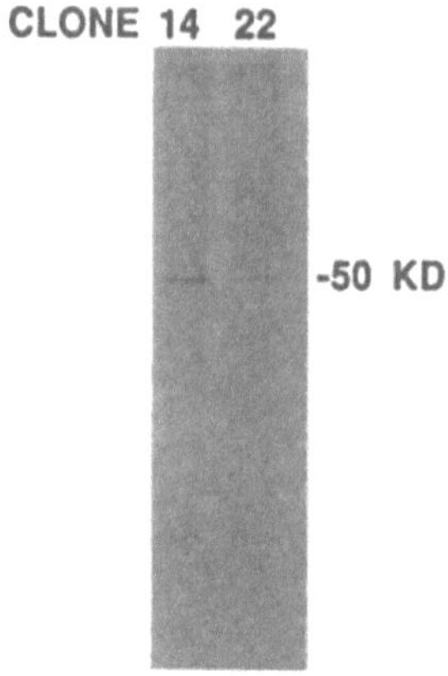

FIGURE 23.10. Immunoprecipitates of ^{35}S-labeled proteins extracted from 3T3-L1 cells that were stably transfected with the short isoform of the rat GH receptor. The cells were metabolically labeled by incubation for 4h with 100mM ^{35}S-L-methionine. The labeled cells were washed and extracted with RIPA buffer (150 mM NaCl, 1% NP-40, 0.5% deoxycholate, 0.1% SDS). The extract was clarified by centrifugation at 15,000 × g for 1min, and the short isoform of the GH receptor present in the supernatants was immunoprecipitated as in Figure 23.5, separated by electrophoresis, and visualized by autoradiography. Clone 14 is a stably transfected cell line that overexpresses the short isoform of the GH receptor. Clone 22 harbors the vector and expresses endogenous (mouse) short isoform, but does not express the rat isoform.

the short isoform of the GH receptor. These cells offer several advantages in this regard. First, when differentiated into adipose-like cells they can carry out many of the processes that are affected by GH in adipocytes, including both insulin-like and anti-insulin-like responses. In addition, 3T3-L1 cells express both the transmembrane form of the GH receptor and the short isoform. Therefore, they are likely to contain all of the postreceptor elements of the GH transduction pathway, as well as the machinery needed to express GH responses.

Stable transfections were achieved by the standard calcium phosphate method using a pcDNAI/neo/GHBP construct containing the 1.2kb transcript. Several neomycin-resistant clones were selected for further study based on the appearance of increased amounts of GH binding protein in the culture medium. Transfected cells were incubated with ^{35}S-L-methionine for 4h and extracted in RIPA buffer. The extracted proteins were immunoprecipitated with As1615, separated by SDS polyacrylamide gel electrophoresis and visualized by autoradiography (Fig. 23.10). The antibody precipitated a prominent 50-kd band in extracts of both transfected clones, but the intensity of the labeled band was at least twice as great for clone 14, whose production of the short isoform was markedly increased compared to clone 22, which exhibited no increase in expres-

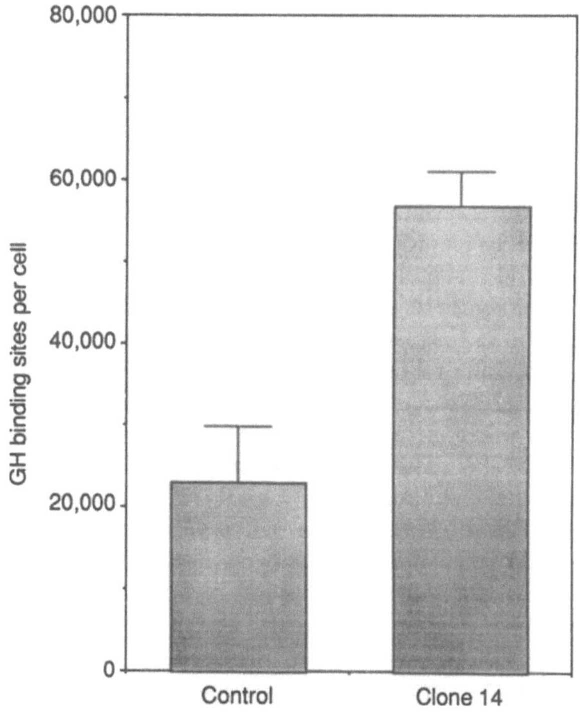

FIGURE 23.11. Binding of ^{125}IhGH by 3T3 cells that express the short isoform of the GH receptor. 3T3 cells stably transfected with vector that harbored the cDNA encoding the short isoform of the GH receptor (clone 14) and control cells transfected with vector alone were grown to confluence in 6-well culture dishes. The cells were incubated for 90 min with 1 ng/ml of ^{125}IhGH and various concentrations of hGH, washed three times in ice-cold buffer, and then scraped into counting vials. Abundance of specific binding sites was determined by Scatchard analysis. Each bar is the mean of three independent experiments. The vertical brackets indicate the SEMs. Cells of clone 14 had significantly greater numbers of binding sites than control cells ($P < 0.05$).

sion of the short isoform of the GH receptor. These findings are consistent with those of Figure 23.4 and support the idea that the initial translation product of the 1.2-kb mRNA is a protein of about 50 kd.

To determine whether increased expression of the short isoform increases binding to the surface of transfected cells, clones of 3T3-L1 cells bearing the 1.2-kb transcript were grown to confluence in 6-well plates. Trace amounts of ^{125}IhGH (~100,000 cpm) were added to each well, along with various amounts of unlabeled hGH, and the cells were incubated for 90 min at 37 °C. After washing three times, the cells were scraped into counting tubes and counted in a gamma counter (Fig. 23.11). Specific binding of ^{125}IhGH was increased threefold in mouse 3T3-L1 cells that expressed rat short isoform of the GH receptor. These findings argue strongly that at least some of the short isoform of the GH receptor is retained in or on cells in a manner that renders it accessible to hormone present in the extracellular medium.

These studies indicate that rat adipocytes produce at least two GH binding molecules. Compelling evidence from a variety of studies supports the idea that the translation product of the 4.5-kb mRNA transcript functions as a GH receptor. The present findings indicate that the short isoform of the GH receptor also contributes to specific GH binding at the cell surface and may therefore also function as a receptor for GH. How a soluble glycoprotein that lacks a transmembrane domain associates with the plasma membrane remains to be answered. It is likely that the short isoform of the GH receptor associates through some protein:protein interaction with other membrane proteins to form a multimeric receptor complex. It is for this reason, perhaps, that transfected 3T3-L1 cells that contain all of the components of the GH transduction apparatus can retain the transfected short isoform of the GH receptor, while similarly transfected CHO cells that lack endogenous GH receptors secrete virtually all of the short isoform of the GH receptor that they are induced to synthesize (28).

The GH receptor is a member of the receptor superfamily that also includes the prolactin receptor and receptors for many of the interleukins. Many of these receptors are expressed in more than one isoform, and some combine to form multimeric complexes. Experiments with the IL-6 receptor, for example, revealed that a truncated cytokine recognition unit that has no transmembrane or cytosolic domains complexes with a related transmembrane protein (gp 130) to form a functional receptor (29, 30). The short isoform of the GH receptor may behave in a similar fashion.

To date, formation of an alternately spliced short transcript of the GH receptor gene has been described only in rodents and may therefore represent only a species-specific anomaly. If, indeed, the short isoform is part of a functional GH receptor complex, we would expect to find it or its equivalent in other species, including humans. To date we know of no responses to GH that are seen only in rodents. The equivalent of the rat short isoform of the GH receptor in humans is a circulating proteolytic fragment of the GH receptor. It is possible that what the rat achieves by alternate splicing of mRNA, other species achieve by posttranslational modification of the translation product and that a truncated form of the GH receptor remains associated with the surface of target cells and participates in signal transduction. It is also possible that an alternatively spliced mRNA will be found in extrahepatic GH target cells in humans and other species.

Acknowledgments. This work was supported by Grant DK-19392 from the National Institutes of Health. The authors are pleased to acknowledge generous gifts of hGH and cDNA for the rabbit GH receptor from the Genentech Corp., South San Francisco, CA.

References

1. Goodman HM. Growth hormone and metabolism. In: Schreibman MP, Scanes CG, Pang PKT, eds. The endocrinology of growth, development and metabolism in vertebrates. San Diego: Academic Press, 1993:93–115.
2. Goodman HM, Coiro V, Frick GP, et al. Effects of growth hormone on the adipocyte: a model for studying direct actions of growth hormone. Endocrinol Jpn 1987;34(suppl 1):59–72.
3. Goodman HM, Gorin E, Honeyman TW. Biochemical basis for the lipolytic activity of growth hormone. In: Sherman B, Underwood L, eds. Perspectives in growth hormone research. New York: Marcel Dekker, 1988:75–111.
4. Goodman HM, Levy LK. Preparation and biological reactivity of polyiodinated human growth hormone. Endocrinology 1983;113:2017–23.
5. Goodman HM, Kostyo JL. Altered profiles of biological activity of growth hormone fragments on adipocyte metabolism. Endocrinology 1981;108:553–8.
6. Chipkin SR, Szecowka J, Tai L-R, Kostyo JL, Goodman HM. Different growth hormone-receptor interactions mediate insulin-like and lipolytic responses of rat adipose tissue. Endocrinology 1989;125:450–8.
7. Grichting G, Levy LK, Goodman HM. Relationship between binding and biological effects of human growth hormone in rat adipocytes. Endocrinology 1983;113:1111–20.
8. Fagin KD, Lackey SL, Reagan CR, DiGirolamo M. Specific binding of growth hormone by rat adipocytes. Endocrinology 1980;107:608–15.
9. Gavin JR III, Saltman RJ, Tollefsen SE. Growth hormone receptors in isolated adipocytes. Endocrinology 1982;110:637–43.
10. Leung DW, Spencer SA, Cachianes G, et al. Growth hormone receptor and serum binding protein: purification, cloning and expression. Nature 1987; 330:537–43.
11. Gorin E, Goodman HM. Covalent binding of growth hormone to surface receptors on rat adipocytes. Endocrinology 1984;114:1279–86.
12. Carter-Su C, Schwartz J, Kikuchi G. Identification of a high-affinity growth hormone receptor in rat adipocytes. J Biol Chem 1984;259:1099–104.
13. Donner DB. Covalent coupling of human growth hormone to its receptor on rat hepatocytes. J Biol Chem 1983:258:2736–43.
14. Szecowka J, Tai L-R, Goodman HM. Effects of tunicamycin on growth hormone-binding in rat adipocytes. Endocrinology 1990;126:1834–41.
15. Spencer SA, Hammonds RG, Henzel WJ, Rodriguez H, Waters MJ, Wood WI. Rabbit liver growth hormone receptor and serum binding protein: purification, characterization, and sequence. J Biol Chem 1988;263:7862–7.
16. Baumann G, Stolar MW, Amburn K, Barsano CP, De Vries BC. A specific growth hormone-binding protein in human plasma: initial characterization. J Clin Endocrinol Metab 1986;62:134–41.
17. Herington AC, Ymer S, Stevenson J. Identification and characterization of specific binding proteins for growth hormone in normal human sera. J Clin Invest 1986;77:1817–23.

18. Smith WC, Kuniyoshi J, Talamantes F. Mouse serum growth hormone (GH) binding protein has GH receptor extracellular and substituted transmembrane domains. Mol Endocrinol 1989;3:984–90.
19. Baumbach WR, Horner DL, Logan JS. The growth hormone-binding protein in rat serum is an alternatively spliced form of the rat growth hormone receptor. Genes Dev 1989;3:1199–205.
20. Frick GP, Leonard JL, Goodman HM. Effect of hypophysectomy on growth hormone receptor gene expression in rat tissues. Endocrinology 1990; 126:3076–82.
21. Carlsson B, Billig H, Rymo L, Isaaksson OGP. Expression of the growth hormone-binding protein messenger RNA in the liver and extra-hepatic tissues in the rat: coexpression with the growth hormone receptor. Mol Cell Endocrinol 1990;73:R1–6.
22. Walker JL, Moats-Staats BM, Selub SE, Lund PK, Stiles AD, Underwood LE. GH receptor (GHR) and binding protein (GHBP) mRNAs are widespread in rat fetal and neonatal tissues [Abstract]. 73rd annu meet Endocr Soc, Washington, DC, 1990:247.
23. Mathews LS, Engberg B, Norstedt G. Regulation of rat growth hormone gene expression. J Biol Chem 1989;264:9905–10.
24. Frick GP, Goodman HM. Characterization of the short form of the growth hormone receptor synthesized by rat adipocytes. Endocrinology 1992;131: 3083–90.
25. Godowski PJ, Leung DW, Meacham LR, et al. Characterization of the human growth hormone receptor gene and demonstration of a partial gene deletion in two patients with Laron-type dwarfism. Proc Natl Acad Sci USA 1989;86:8083–7.
26. Sadeghi H, Wang BS, Lumanglas AL, Logan JS, Baumbach WR. Identification of the origin of the growth hormone-binding protein in rat serum. Mol Endocrinol 1990;4:1799–805.
27. Gorin E, Goodman HM. Turnover of growth hormone receptors in rat adipocytes. Endocrinology 1985;116:1796–805.
28. Möller C, Hansson A, Enberg B, Lobie PE, Norstedt G. Growth hormone (GH) induction of tyrosine phosphorylation and activation of mitogen-activated protein kinases in cells transfected with rat GH receptor DNA. J Biol Chem 1992;267:23403–8.
29. Taga T, Hibi M, Hirata Y, et al. Interleukin-6 triggers the association of its receptor with a possible signal transducer, gp130. Cell 1989;58:573–81.
30. Yasukawa K, Saito T, Fukunaga T, et al. Purification and characterization of soluble human IL-6 receptor expressed in CHO cells. J Biochem 1990; 108:6473–6.

24

Effects of Porcine Somatotropin on Thymic Weight, Thymosin α1, and Thymosin β4 in Gilts and Barrows

T. WISE, J. KLINDT, F.C. BUONOMO, AND J.T. YEN

Extensive data document the effects of supplementary somatotropin (growth hormone) in altering the efficiency and productivity of farm and meat animals—that is, increased milk production (1, 2), enhanced growth efficiency and leanness in swine (3–5), and improved feed efficiency and growth in fish (6). As animals mature, growth rate decreases and eventually stops in conjunction with the declining secretion of somatotropin. The thymus gland reaches maximum size in early maturity and then regresses in size coordinate with the decline in function and responsiveness of the immune system. Along with age-associated atrophy of the thymus, thymic secretory peptides also decline in the circulation (7–9).

The immune system is sensitive to somatotropin stimulation (10–14), and the potential exists for positive health effects as an associated response to administration of exogenous somatotropin to food animals for the purpose of increasing their productivity. Initial reports on immune responses from *porcine somatotropin* (pST) treatment have been conflicting (15–18). The objectives of this study were to (i) define, determine, or quantitate the thymic weight changes and responses of thymic secretory peptides thymosin α1 and β4 to exogenous pST, (ii) relate the pattern of circulating pST to thymosin β4 changes throughout the trial, and (iii) determine influence of sex, genetic line, and method of pST delivery on changes in thymic weight and thymosin concentrations.

Materials and Methods

Genetically lean and obese gilts (females) and barrows (castrate males) from inter se matings of crossbreds produced by within-line matings of purebred Duroc and Yorkshire pigs selected solely for high and low backfat thickness over multiple generations (19) were used to define thymic responses to recombinant pST (Monsanto Agriculture Company). In experiment I, obese and lean gilts and barrows (n = 96; 60–70 kg body weight; 166 days of age) were injected in the fat over the extensor muscle of the neck with 0, 2, or 4 mg of pST in 1 ml of bicarbonate buffer daily for 42 days (n = 8/line/sex/dose). Blood was sampled (10 ml, jugular vein) at times 0, 6, and 24 h after the initial injection and subsequently at 14, 28, and 42 days. Blood samples were taken prior to the daily injection. In experiment II, lean and obese gilts and barrows were administered pST by sustained-release implants inserted subcutaneously at the base of the ear (5), designed to deliver 2 mg pST/day for 42 days. Blood was sampled at time 0 and 7, 14, 28, and 42 days postimplant. Doses of pST administered were 0 mg (trocar inserted, no implant), 2 mg (1 implant), and 4 mg (2 implants).

The study consisted of 48 animals (4/line/sex/dose). At the end of the 42-day treatment period, animals were slaughtered (90–110 kg body weight) and thymic weights recorded. Serum concentrations of pST and thymosin α1 and β4 were measured by radioimmunoassay (8, 20). The data were analyzed by GLM procedure (21), in which serum hormonal changes were examined by split-plot analysis of variance. Final body weight was used as a covariate when analyzing main effects (dose pST, line, sex) on thymic weights.

TABLE 24.1. Thymic weight and thymosin β4 changes due to dose of pST in injected (experiment I) and implanted (experiment II) gilts and barrows.

	pST dose		
	Control	2 mg/day	4 mg/day
		Thymic weight (g)	
Injected (n = 96)			
Barrows	127.5 ± 16.5	214.1 ± 16.4[b]	234.4 ± 16.5[b]
Gilts	124.9 ± 16.4	222.4 ± 16.5[b]	279.1 ± 17.1[b]
Implanted (n = 48)			
Barrows	159.3 ± 20.1	233.3 ± 20.1[b]	274.9 ± 20.3[b]
Gilts	173.6 ± 20.4	225.4 ± 20.1	298.9 ± 20.3[b]
		Thymosin β4 (ng/ml serum)	
Injected			
Barrows[a]	47.0 ± 15.3[b]	77.9 ± 15.2	93.1 ± 14.8[b]
Gilts	108.7 ± 15.0	113.9 ± 15.2	152.6 ± 15.5[b]
Implanted			
Barrows[a]	79.4 ± 29.0	126.9 ± 29.2	171.1 ± 29.9[b]
Gilts	169.9 ± 29.3	212.9 ± 29.9	225.2 ± 29.6

[a] Sex different ($P < 0.05$).
[b] Mean ± SEM, different from control ($P < 0.05$).

Results

In experiment I, there were no line or sex effects on the increases in thymic weight from pST. Thymic weight increased ($P < 0.01$) in a dose-responsive manner in both gilts and barrows (Table 24.1). In serum collected at slaughter, no treatment effects were noted in thymosin α1

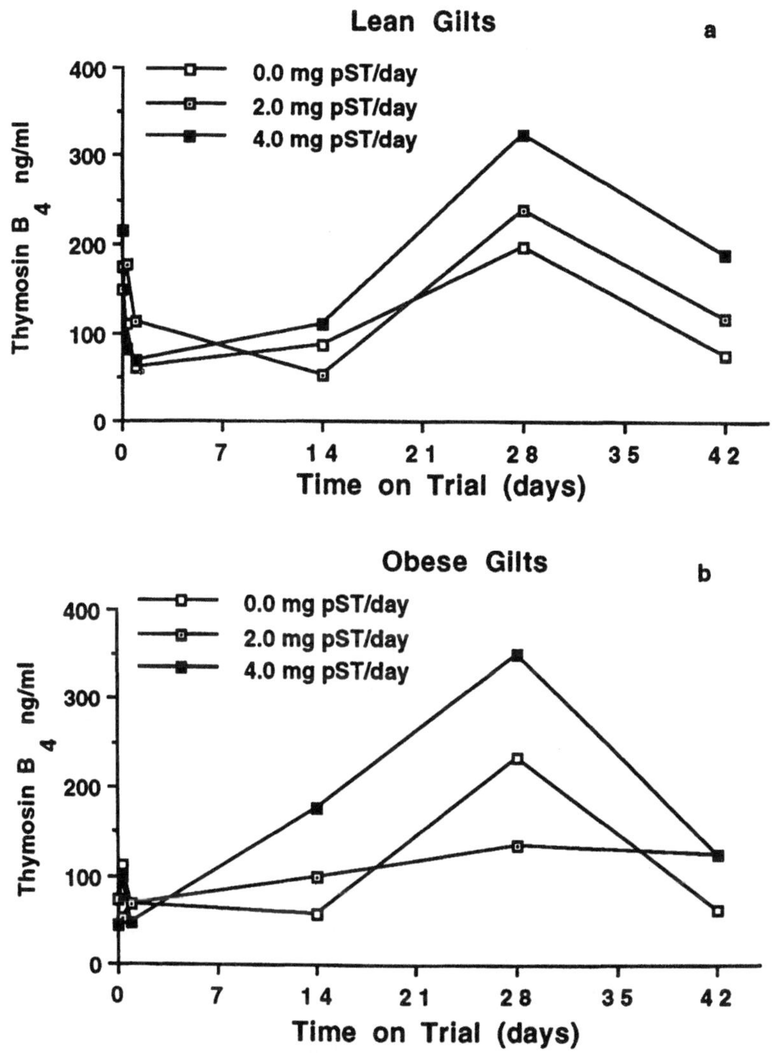

FIGURE 24.1. Changes in thymosin β4 in lean (*a*) and obese (*b*) gilts injected daily with 0, 2, or 4 mg of porcine somatotropin (pST) for 42 days. Thymosin β4 concentrations were dose related and increased up to day 28, then declined ($P < 0.05$).

concentrations. Thymosin α1 concentrations were greater ($P < 0.01$) in gilts (1.4 ± 0.06 vs. 0.8 ± 0.06 ng/ml). Thymosin β4 concentrations showed a dose-dependent increase ($P < 0.01$) in response to pST (Table 24.1). Analysis of blood samples throughout the times of injection of pST showed significant increases in thymosin β4 concentrations due to pST treatment, particularly after day 14 ($P < 0.05$; Figs. 24.1–24.2). No differences were detected between genetic lines, but sex and sex × time

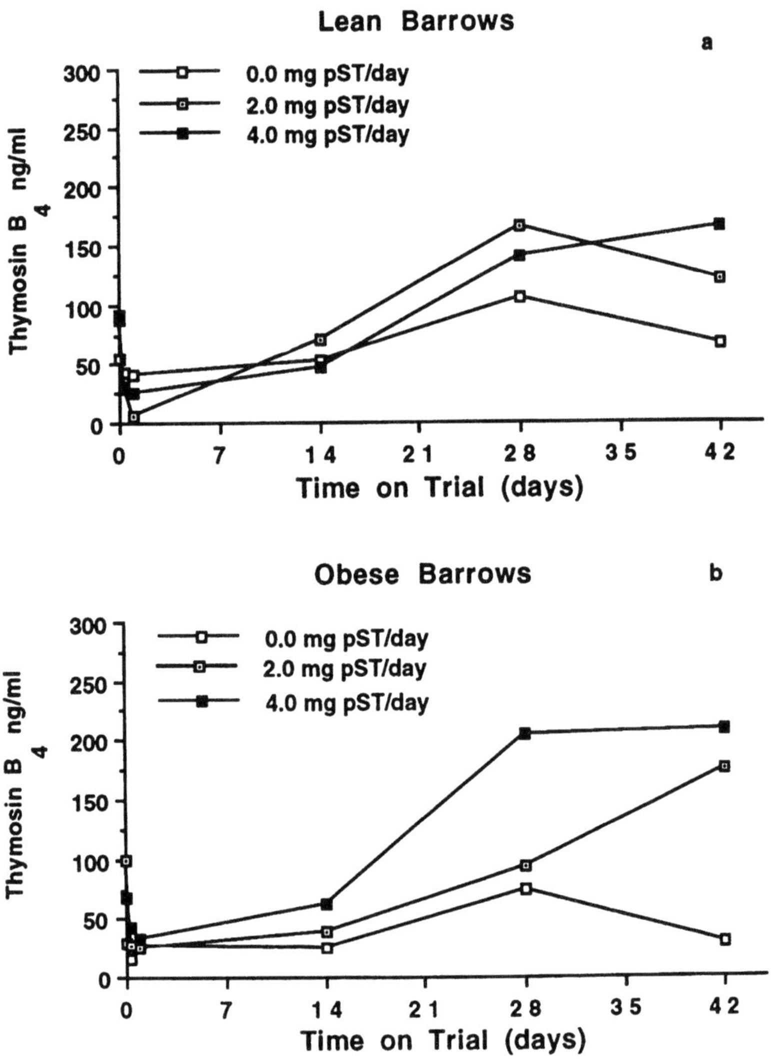

FIGURE 24.2. Changes in thymosin β4 in lean (*a*) and obese (*b*) barrows injected daily with 0, 2, or 4 mg of porcine somatotropin (pST) for 42 days. Thymosin β4 concentrations were dose related and increased throughout the trial ($P < 0.05$).

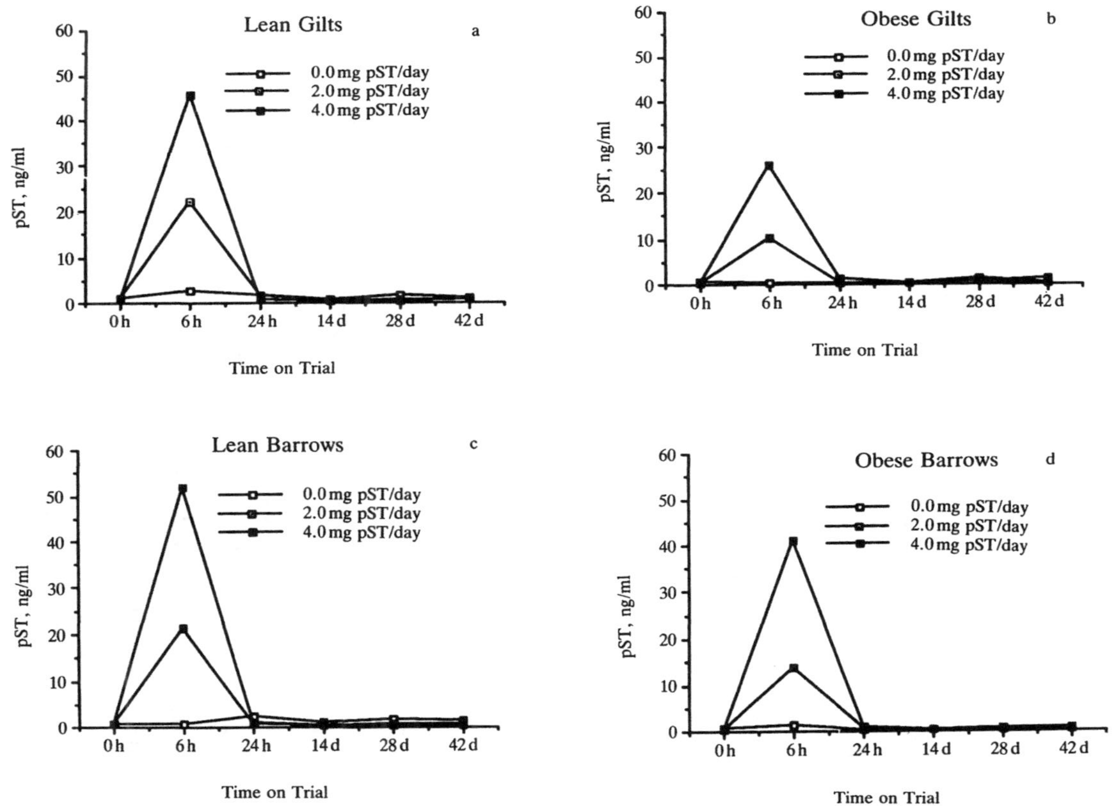

FIGURE 24.3. Changes in porcine somatotropin (pST) in lean and obese (*a*, *b*) gilts and barrows (*c*, *d*) injected daily with 0, 2, or 4 mg of porcine somatotropin (pST) for 42 days. Porcine somatotropin increased 6 h postinjection ($P < 0.05$), but by 24 h and throughout the rest of the study remained low and not different from controls.

differences were noted (Figs. 24.1 and 24.2; $P < 0.05$). While thymosin β4 concentrations significantly declined in gilts at the end of the study (day 42; Fig. 24.1), concentration in barrows continued to increase (Fig. 24.2). Concentrations of pST in experiment I are shown in Figure 24.3. Within 6h after initial injection, pST peaked, but by 24h postinjection, pST was comparable to controls (Fig. 24.3).

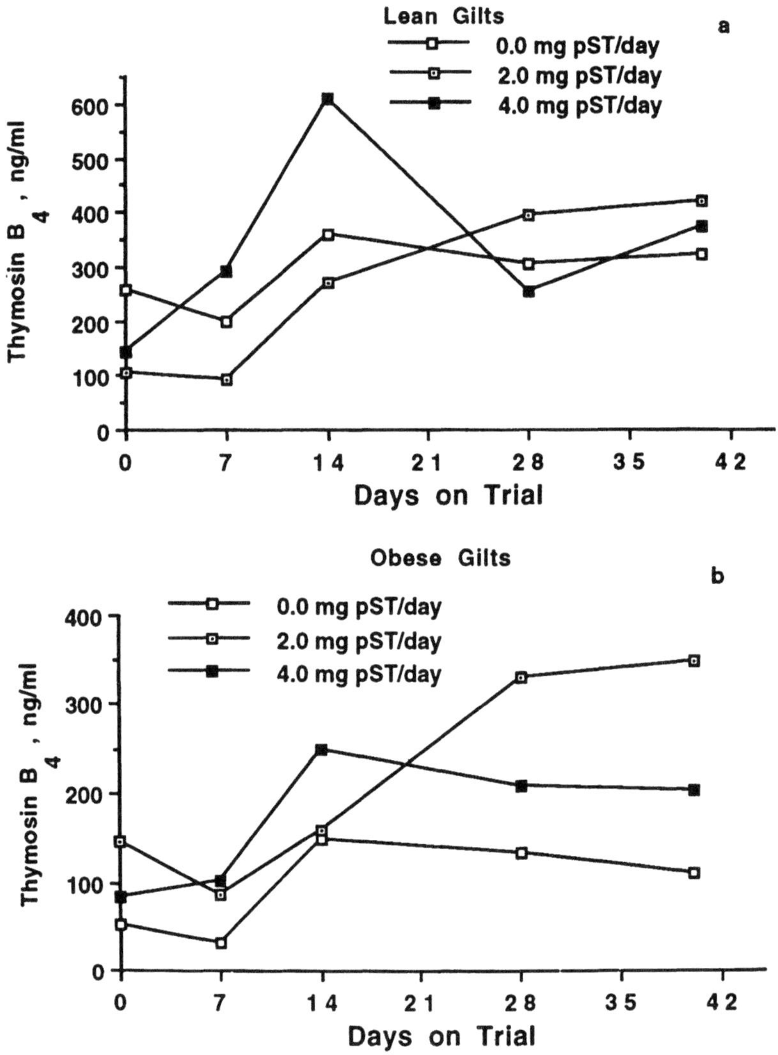

FIGURE 24.4. Changes in thymosin β4 in lean (*a*) and obese (*b*) gilts with implants that released 0, 2, or 4 mg/day of pST for 42 days (controls received no implant). Thymosin β4 concentrations significantly increased during the trial period ($P < 0.05$) and were related to dose of pST ($P < 0.05$).

In experiment II involving implants of pST, thymic weights increased ($P < 0.01$) in response to dose of pST in all lines (Table 24.1). Thymosin β4 concentrations increased ($P < 0.01$) in relation to dose of pST (Table 24.1). Changes in thymosin β4 concentrations throughout the duration of pST implants were similar to responses in experiment I including the high dose of pST increasing serum thymosin β4, but differences due to dose

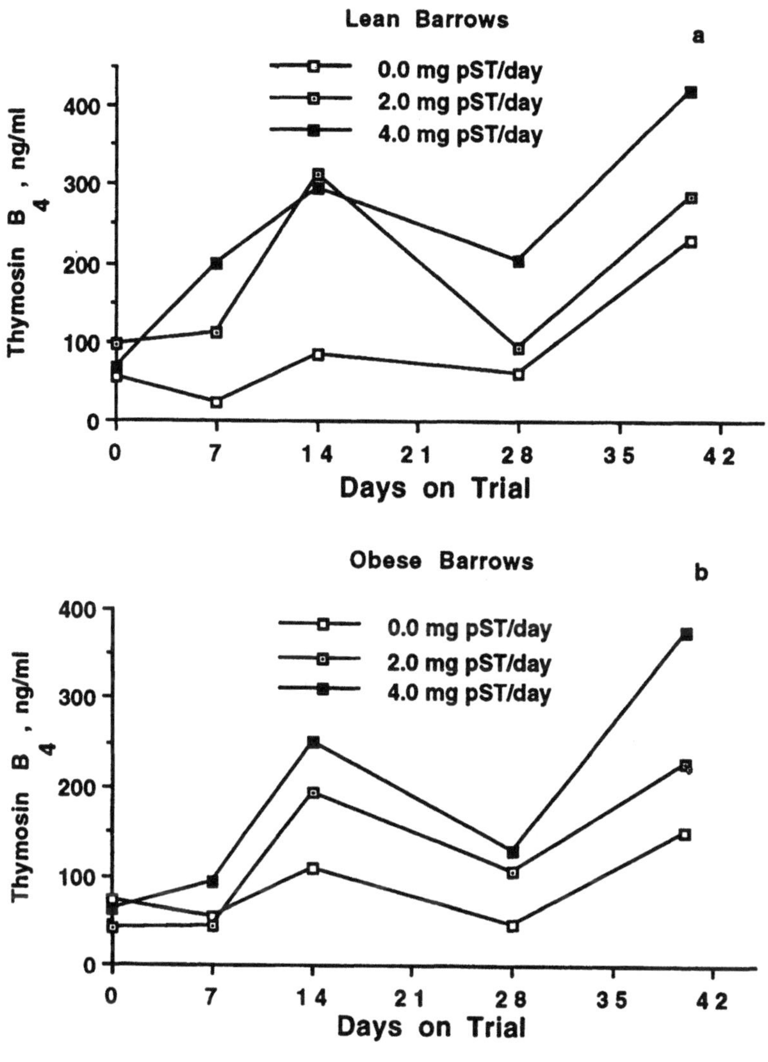

FIGURE 24.5. Changes in thymosin β4 in lean (*a*) and obese (*b*) barrows with implants that released 0, 2, or 4 mg/day of pST for 42 days (controls received no implant). Thymosin β4 concentrations significantly increased during the trial period ($P < 0.05$) and were related to dose of pST ($P < 0.05$).

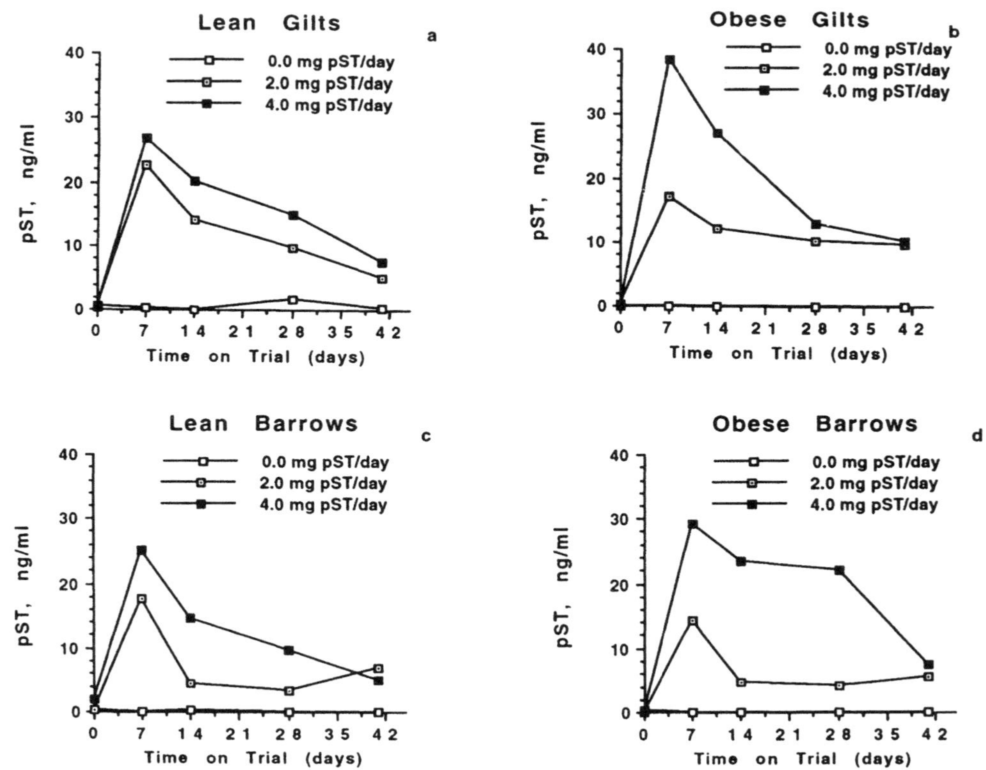

FIGURE 24.6. Changes in porcine somatotropin (pST) in lean and obese gilts (*a*, *b*) and barrows (*c*, *d*) with implants that released 0, 2, or 4 mg/day of pST for 42 days (controls received no implant). Detectable amounts of pST were dose related and sustained throughout the trial period.

were evident at 7 days (Figs. 24.4–24.5). As in experiment I, sex differences in thymosin β4 trends were evident. Concentrations of thymosin β4 in barrows continued to increase throughout the trial, while in gilts there was evidence of decreasing or leveling of concentrations from days 28 to 42 (Figs. 24.4–24.5). Concentrations of pST measured in implanted pigs were different from those of experiment I injected pigs (Fig. 24.6) and showed sustained release of pST over the duration of the trial. Although response in increased thymic weights were very comparable between the two methods of administering pST (daily injection or implant), overall concentrations of thymosin β4 in pST-implanted animals were considerably greater (Table 24.1).

Conclusion

The increases in thymic weights and thymosin β4 concentrations due to pST treatment (injection or implants) support the hypothesis that increased immune function may be related to the increased growth performance, rate and efficiency of weight gain, induced by pST supplementation in swine. The response is dependent on the sex, genotype, and method of pST administration. The pST implants that provided a chronic elevation of pST (Fig. 24.6) also induced earlier increases in thymosin β4 by dose of pST (day 14; Figs. 24.4–24.5) than daily injection of pST, which resulted in single daily pST peaks of extreme amplitude. Decreases in thymosin β4 noted at the end of the trials in gilts (Figs. 24.1 and 24.4) probably reflect negative feedback by gonadal steroids on thymic secretion (22, 23) as these females were approaching puberty and initiating follicular development by slaughter. Barrows that were castrated at three weeks of age showed no decline in thymosin β4 levels. Thus, the physiological status of the animal also influences responses to pST. The implications for improved health for both humans and animals from somatotropin treatment make this an area in which further evaluation may yield information to maximize the responses (whether growth, milk production, or altered health) to somatotropin.

References

1. Peel CJ, Bauman DE. Somatotrophin and lactation. J Dairy Sci 1987; 70:474–86.
2. Johnsson ID, Hart IC. Manipulation of milk yield with growth hormone. In: Haresign W, Cole DJA, eds. Recent advances in animal nutrition. Proc 20th Nutrition Conference for Feed Manufacturers. London: Butterworths, 1986: 105–23.
3. Noblet J, Herpin P, Dubois S. Effect of recombinant porcine somatotropin on energy and protein utilization by growing pigs: interaction with capacity for lean tissue growth. J Anim Sci 1992;70:2471–84.

4. Dunshea FR, Bauman DE, Boyd RD, Bell AW. Temporal response of circulating metabolites and hormones during somatotropin treatment of growing pigs. J Anim Sci 1992;70:123–31.
5. Klindt J, Buonomo FC, Yen JT. Administration of porcine somatotropin by sustained release implant: growth and endocrine responses in genetically lean and obese barrows and gilts. J Anim Sci 1992;70:3721–33.
6. Cloud JG. Strategies for introducing foreign DNA into the germ line of fish. J Reprod Fertil [Suppl] 1990;41:107–16.
7. Lewis M, Twomey JJ, Bealmer P, Goldstein G, Good RA. Age, thymic involution and circulating thymic hormone activity. J Clin Endocrinol Metab 1978;47:145–50.
8. Wise T, Klindt J, Ford JJ, Christenson RK. Changes in thymosin β4 and gonadal function during puberty in boars and gilts immunized against estrone. Anim Reprod Sci 1991;24:139–52.
9. Voetsch W, Anderer A. The ontogeny of bovine thymostimulin production in fetal and postnatal age. Thymus 1992;20:127–33.
10. Mocchegiani E, Paoluici P, Balsamo A, Cacciari E, Fabris N. Influence of growth hormone on thymic endocrine activity in humans. Horm Res 1990;33:248–55.
11. Fu YK, Arkins S, Wang BS, Kelley KW. A novel role of growth hormone and insulin-like growth factor I. J Immunol 1991;146:1602–8.
12. Weigent DA, Blalock JE, LeBoeuf RD. An antisense oligo deoxynucleotide to growth hormone messenger ribonucleic acid inhibits lymphocyte proliferation. Endocrinology 1991;128:2053–7.
13. Wise T, Klindt J, Macdonald GJ, Ford JJ. Effects of neonatal sexual differentiation, growth hormone and testosterone on thymic weights and thymosin β4 in hypophysectomized rats. J Reprod Immunol 1991;19:43–54.
14. Timsit J, Savino W, Safieh B, et al. Growth hormone and insulin-like growth factor-I stimulate hormonal function and proliferation of thymic cells. J Clin Endocrinol Metab 1992;75:183–8.
15. Reddy PG, Shirley JE, Frey RA, Blecha F. Effect of recombinant bovine somatotropin (rBST) on the immune system of dairy cows [Abstract]. J Dairy Sci 1990;73:150.
16. Goff BL, Flaming KP, Frank DE, Roth JA. Recombinant porcine somatotropin: an immunological study. J Anim Sci 1991;69:4523–37.
17. Burton JL, McBride BW, Kennedy BW, Burton JH, Elasser TH, Woodward B. Hematological profiles in dairy cows treated with recombinant bovine somatotropin. J Anim Sci 1992;70:1488–95.
18. Burton JL, McBride BW, Kennedy BW, Burton JH, Elasser TH, Woodward B. Contact sensitivity and systemic antibody responses in dairy cows treated with recombinant bovine somatotropin. J Dairy Sci 1992;75:747–55.
19. Hetzer HO, Harvey WR. Selection for high and low fatness in swine. J Anim Sci 1967;26:1244–51.
20. Klindt J, Ford JJ, Berardinelli JG, Anderson LL. Growth hormone secretion after hypophysial stalk section in pigs. Proc Soc Exp Biol Med 1983;172: 508–13.
21. SAS. User's Guide, vol 2. Cary, NC: SAS Institute Inc., 1989.

22. Wise T, Day ML, Vakharia DD. Endocrine relationships of thymosin α1, thymosin β4, and luteinizing hormone throughout the prepubertal period of development in heifers, ovariectomized heifers and ovariectomized heifers with estradiol implants. J Reprod Immunol 1990;18:225–35.
23. Wise T, Maurer RR. Characterization of thymosin α1 and β4 during the bovine estrual period: effects of elevated estradiol and progestin. Biol Reprod 1991;45:57–63.

25

Effects of an Acetylcholine Precursor on GH Secretion in Elderly Subjects

Gian Paolo Ceda, Gian Piero Marzani, Valeria Tontodonati, Emanuela Piovani, Alberto Banchini, Maria Teresa Baffoni, Giorgio Valenti, and Andrew R. Hoffman

GH secretion is under complex neurotransmitter and hormonal control. *GH releasing hormone* (GHRH) is released from the hypothalamus into the hypothalamic-hypophyseal circulation, stimulating pulsatile GH secretion. Hypothalamic somatostatin, on the other hand, inhibits GH release. GH stimulates the synthesis of insulin-like growth factor I in several peripheral tissues, which mediates some, but not all, of GH hormone action. Completing the feedback-loop paradigm, IGF-I directly interacts with specific IGF receptors on somatotropes, inhibiting further GH secretion. Moreover, IGF receptors are present in brain, and there is evidence that locally synthesized IGF-I can inhibit GHRH release and stimulate somatostatin secretion, thereby suppressing GH secretion (1). Finally, GH inhibits its own secretion at the level of the hypothalamus.

The cause for the decline in the activity of the GHRH-GH-IGF network with aging is complex (2–4) (Fig. 25.1). We have shown that the pituitary response to GHRH is decreased in aged rats (5) and humans (6). This desensitization may be accounted for in part by a decreased pituitary cAMP response to GHRH administration (5, 7). Moreover, there are decreased concentrations of pituitary GHRH receptors (8), hypothalamic GHRH mRNA (9), and pituitary GH mRNA (3). In elderly people, there is a decreased number of GH secretory pulses, as well as a decline in GHRH-induced GH (10, 11) and IGF-I (12) secretion. The release of GHRH and GH after L-dopa stimulation is diminished in the elderly (13). Serum GH binding protein levels have been shown to fall (14), and the half-life of endogenous GH is diminished in older subjects (15). Serum IGF-I levels are decreased in the elderly, but may increase in response to GHRH (16) or GH. In women, *IGF binding protein* (IGFBP)

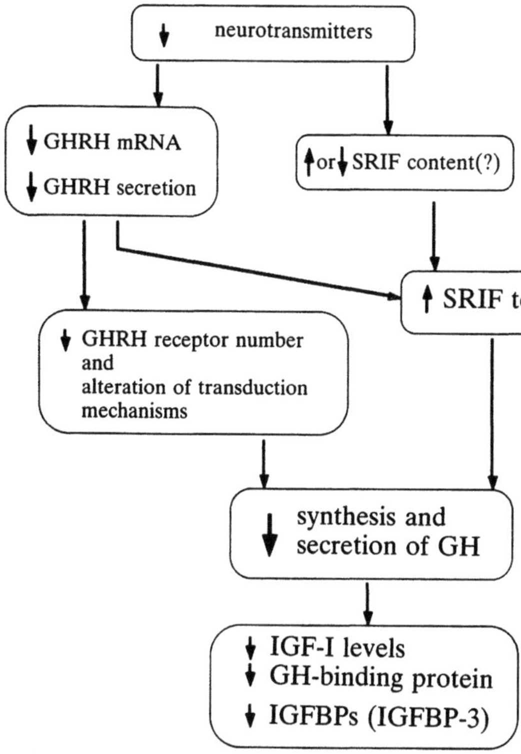

FIGURE 25.1. Schematic representation of the mechanisms leading to reduced GH secretion during aging.

levels also change with aging, with a relative decrease in IGFBP-3 (17). With aging, the imbalance between stimulatory (GHRH) and inhibitory (SRIF) activities on GH secretion, together with a depression of the hypothalamic cholinergic tone (18), leads to an enhanced somatostatin action and decreased GH release.

Cholinergic neurotransmission has been shown to play an important role in the control of GH secretion from the pituitary. Anticholinergic drugs diminish the GH response to a variety of stimuli, while drugs able to increase cholinergic transmission, like cholinesterase inhibitors and acetylcholine precursors, potentiate the stimulatory effect of GHRH (19, 20) through a decrease of somatostatin release. Since brain acetylcholine synthesis declines with aging, it is theoretically possible to increase GH levels in the elderly by giving drugs that increase *acetylcholine* (Ach) synthesis or that enhance cholinergic tone. When pyridostigmine was administered to old people alone or in combination with GHRH, there was an increase and a potentiation of GHRH-induced GH secretion, respectively (21). Nonetheless, GH secretion was not restored to the level seen in normal young subjects. Similar results were obtained with CDP-choline, a putative acetylcholine presursor (20).

More recently we have investigated the effects on GH secretion of another acetylcholine precursor, *alpha-glycerylphosphorylcholine* (alpha-GFC) (22). This drug, currently used for the treatment of amnestic and cognitive disorders of aging, increases Ach release from specific brain areas after intraperitoneal administration to rats (23). In order to learn what effect alpha-GFC had on GH secretion, GHRH was given to eight young and seven old human volunteers, with or without the addition of 1 g of alpha-GFC. Alpha-GFC potentiated the stimulatory effect of GHRH on GH secretion in both young and old subjects (22). However, in the elderly subjects the low basal response of GH to GHRH was only partially restored by alpha-GFC administration.

The purpose of this study was to investigate further the role of cholinergic transmission in the regulation of GH secretion during aging using the acetylcholine precursor alpha-GFC. The specific aims were: (i) to investigate whether there was any further potentiation of the GH secretion in response to GHRH using a dose of alpha-GFC higher than that previously shown to potentiate GH secretion; (ii) to assess the role of alpha-GFC during intermittent administration of GHRH (that is, every 2 h), a procedure that has been shown to induce in normal adult subjects somatotrope refractoriness to GHRH (24); and (iii) to learn whether the potentiating effect of alpha-GFC on GHRH-induced GH secretion was maintained for a prolonged period of time with the repetitive administration of the drug.

Protocols

Three different protocols were used. In protocol 1, ten normal elderly subjects, aged 79.4 ± 1.7 years, with a BMI of 23.1 ± 1.4 were studied twice. Each subject received a 15-min infusion of 2 g alpha-GFC (Brezal, Sandoz, Italy) dissolved in 100 ml normal saline, followed immediately by a bolus injection of GHRH (1 μg/kg). Blood samples for GH determinations were taken at −30, 0, 15, 30, 45, 60, and 90 min after GHRH injection. The results were compared with those obtained in eight young and seven old subjects of comparable age who received both GHRH and alpha-GFC (at a dose of 1 g) plus GHRH.

In protocol 2, 13 normal subjects, aged 78 ± 1.5 years, received two consecutive (every 120 min) bolus IV injections of GHRH (1 μg/kg). In six subjects a second set of bolus GHRH injections was performed, but the second GHRH bolus was preceded (by 15 min) by the infusion of 2 g alpha-GFC in 100 ml normal saline. Blood samples for GH determination were drawn every 15 min before the first GHRH bolus and for 180 min thereafter.

In protocol 3, eight elderly subjects, aged 80.6 ± 1.5 years, were treated with daily alpha-GFC (2 g) plus GHRH (1 μg/kg) for 15 days. The GH responses to the combined stimulation after the fifteenth day of treatment were compared with those obtained before treatment.

All tests were started between 0800 and 0900 h after an overnight fast and 30 min after cannulation of a cubital vein. GH was measured with an IRMA method using reagents provided by Sorin (Saluggia, Italy). The intra- and interassay coefficients of variation were 4% and 8%, respectively. The GH secretory responses were expressed as area under the curve (AUC; ng/ml/h) calculated by trapezoidal integration. Statistical evaluation was performed using analysis of variance for repeated measures and Student's t-test where applicable.

Results

Protocol 1

GHRH injection caused a blunted GH response in the older subjects in comparison with that observed in the young subjects (Table 25.1). When 2 g of alpha-GFC was administered with GHRH, the GH responses were significantly higher than those found after GHRH alone, but not significantly different from those seen in the elderly subjects who received 1 g of alpha-GFC (804 ± 182 vs. 977 ± 203 ng/ml/h), suggesting that a maximal potentiation of GHRH-induced GH responses could be achieved by the lower dose of the drug.

Protocol 2

Administration of GHRH in the elderly subjects at a 2-h interval resulted in blunted GH responses to both stimuli. GH AUC after the first GHRH bolus was 230.1 ± 65.9 ng/ml/h, a value that was not significantly different from that elicited by the second GHRH bolus of 108.3 ± 14.9 ng/ml/h. This GH secretory pattern was clearly different from that seen in seven

TABLE 25.1. Area under the curve (AUC = ng/ml/h) of GH after GHRH and GHRH + alpha-GFC in groups of young subjects treated with 1 g of alpha-GFC (n = 8), elderly people treated with 1 g alpha-GFC (n = 7) and elderly subjects treated with 2 g alpha-GFC (n = 10).

GH AUC (ng/ml/h)	Young (1 g)	Old (1 g)	Old (2 g)
GHRH	1375 ± 263	425 ± 119	191 ± 57
GHRH + alpha-GFC	1849 ± 215	977 ± 273	804 ± 182

Note: Values are expressed as mean ± SEM.

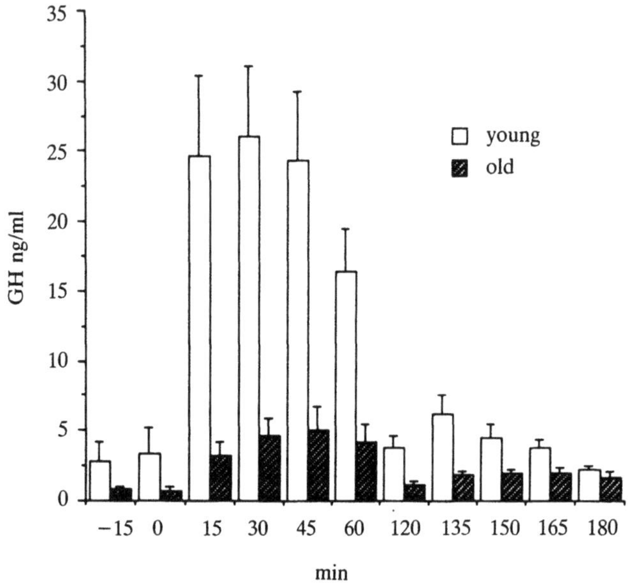

FIGURE 25.2. GH responses to two consecutive GHRH injections (at time 0 and 120 min) in 7 young (open columns) and 13 elderly subjects (hatched columns). Values are expressed as mean ± SEM.

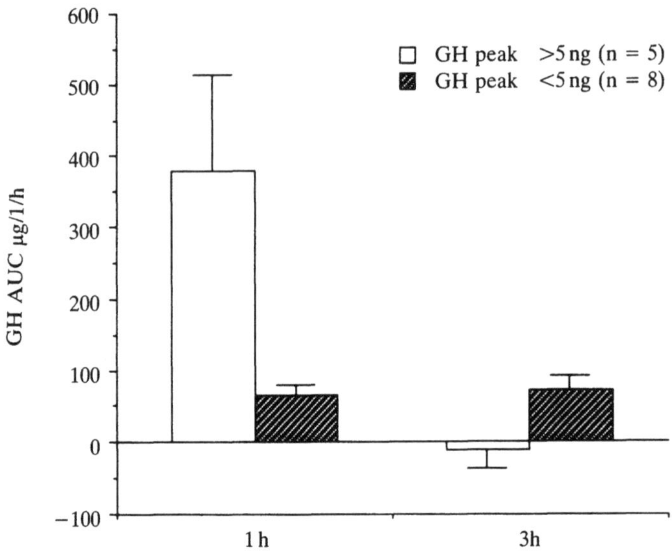

FIGURE 25.3. Area under the curve of GH over 60 min after the first GHRH bolus (1 h) and the second GHRH bolus (3 h) in the elderly subjects, divided according to a GH peak response after the first GHRH stimulation higher than 5 ng/ml (open columns; n = 5 subjects) or lower than 5 ng/ml (hatched columns; n = 8 subjects). Values are expressed as mean ± SEM.

young normal adult subjects, in whom the second GHRH stimulation led to a nearly complete suppression of GH secretion (GH AUC after the first dose of GHRH of 1155 ± 339 ng/ml/h vs. GH AUC after the second GHRH bolus of 41 ± 20 ng/ml/h) (Fig. 25.2). However, the analysis of the individual responses obtained in the old subjects showed two different patterns of GH secretion. The subjects, who showed a consistent increase of GH values (i.e., >5 ng/ml) to the first GHRH stimulation, had a suppressed GH response to the second GHRH bolus, while the others had similar GH responses to the two consecutive GHRH injections (Fig. 25.3). In the six subjects studied with alpha-GFC, the administration of the drug significantly potentiated the GH responsiveness to the second GHRH stimulation (304 ± 98 vs. 110 ± 74 ng/ml/h; $P < 0.05$).

Protocol 3

In order to learn if the potentiating effect of alpha-GFC on GH secretion was maintained during prolonged treatment, eight elderly subjects were treated each day for 15 days with GHRH (1 μg/kg) plus 2 g alpha-GFC. Pituitary responses of GH to the combined administration after treatment were similar to those seen before treatment (814 ± 230 vs. 968 ± 442 ng/ml/h), but with a high degree of variability between subjects (Fig. 25.4).

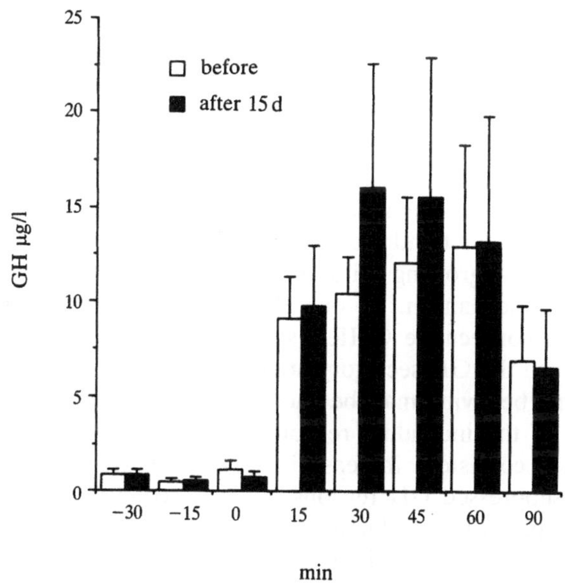

FIGURE 25.4. GH responses to the combined administration of alpha-GFC (2 g) plus GHRH (1 μg/kg) before (open columns) and after (hatched columns) a treatment with daily administration of the two drugs for 15 days.

Discussion

These results confirm that GH responses both to a single dose of GHRH and to two consecutive GHRH injections are reduced in elderly subjects. The mechanism of this blunted GH secretion is probably multifactorial. In the aged rats there is evidence for an enhancement of somatostatin tone (25). Since hypothalamic somatostatin release is inhibited by cholinergic agonists, it is possible that the decreased cholinergic activity during aging is responsible for the increased GH inhibitory tone. In addition, GHRH action also declines with age. There are lower levels of hypothalamic GHRH mRNA (9), fewer pituitary GHRH receptors (8), and decreased pituitary responsiveness to GHRH in vitro (5, 7). Indirect evidence for a reduced GHRH action with aging comes from the observation that in old human subjects the administration of cholinergic agents, known to increase GH secretion through an inhibition of SRIF release, can only partially restore the reduced GH responsiveness to GHRH. In this study, we have demonstrated that increasing the dose of alpha-GFC does not further potentiate the GH response to GHRH stimulation, suggesting that the choline acetyltransferase activity is probably fully activated by the lower dose of the drug.

The effects of consecutive doses of GHRH on GH secretion have recently been explored in detail in normal adults. Repeated administration of GHRH (at doses between 50 and 200 μg/bolus) at intervals of 2 h consistently leads to blunted GH response to the second GHRH stimulation (26). This reduced somatotrope responsiveness to repeated GHRH doses is likely due to a GH autofeedback mechanism mediated by somatostatin release, as suggested by the finding that the enhancement of cholinergic transmission restores the GH response to the second GHRH bolus (27). A different pattern of GH responses to two consecutive GHRH injections has been observed in children with similar GH responses to both the first and the second GHRH bolus, thus suggesting the existence in normal growing children of a neural refractoriness to the GH autoregulatory mechanism. In our elderly subjects the pattern of GH secretion after consecutive GHRH stimulations is blunted, with no difference between the GH secretory area after the first and second GHRH injection. This behavior may be due to diminished GH autofeedback. The analysis of the individual responses demonstrates that the subjects who showed a consistent increase in GH levels after the first GHRH bolus had a suppressed GH response to the second GHRH dose with a GH secretory pattern resembling that of normal adults.

With aging the decline of the GH-IGF-I axis leads to a syndrome that we have termed the "somatopause," characterized by decreased lean body mass, increased adiposity, osteopenia, declining renal function, and sleep disturbances. Recent data on GH and/or IGF-I replacement therapy suggest that such treatment may reverse or prevent some of these sequelae of aging. Fewer data are available on the use of GHRH as a therapeutic agent. Alternate-day IV injections of GHRH for 12 days have been shown to restore the GHRH responsivity of older men to that of young men (28), while daily IV injections of GHRH for 14 days increased GH responses, IGF-I, and osteocalcin levels in postmenopausal women (29). However, we have been unable, using daily injections of GHRH (50–100 μg IV for 14 days) in a group of older subjects, to document any potentiating effect on GH secretion or an increase of IGF-I levels (Ghigo et al., unpublished observations).

More recently Corpas et al. (16), using high-dose subcutaneous injections of GHRH (2 mg per day) for 14 days, showed increased spontaneous diurnal GH secretion and increased IGF-I and phosphate levels in older men. Since the acute administration of the cholinergic precursor alpha-GFC potentiates the GH response to a standard GHRH test, we explored the possibility of maintaining this effect for prolonged periods of time by administering 2 g alpha-GFC followed by GHRH bolus injection on a daily basis. Our results clearly demonstrate that after 14 days of treatment the pituitary responsiveness to the combined administration of these agents was maintained at the level seen before therapy, suggesting that the use of an Ach precursor together with GHRH could represent, in some patients, a useful tool in the treatment of the catabolic changes of aging.

References

1. Rosenfeld RG, Ocrant I, Valentino KL, Hoffman AR. Interaction of IGF with the hypothalamus and pituitary. In: LeRoith D, Raizada MK, eds. Molecular and cellular biology of insulin-like growth factors and their receptors. New York: Plenum Press, 1990:39–56.
2. Rudman D, Kutner MH, Rogers CM, Lubin MF, Fleming GA, Bain RP. Impaired GH secretion in the adult population: relation to age and adiposity. J Clin Invest 1981;67:1361–9.
3. Hoffman AR, Griffin C, Kalinyak JE, Perkins SN, Ceda GP. The hypothalamic-somatotroph-somatomedin axis and aging. In: Valenti G, ed. Psychoneuroendocrinology of aging: basic and clinical aspects. Fidia Research Series, vol 16. Berlin: Springer-Verlag, 1988:43–60.
4. Ceda G, Denti L, Hoffman AR, Ceresini G, Valenti G. Aging and pituitary responses to hypothalamic peptides. In: Segal HL, ed. Protein metabolism in aging. New York: Wiley-Liss, 1990:335–44.
5. Ceda GP, Valenti G, Butturini U, Hoffman AR. Diminished pituitary responsiveness to growth hormone releasing factor in aging male rats. Endocrinology 1986;118:2109–14.
6. Ceda GP, Ceresini G, Denti L, Cortellini P, Hoffman AR, Valenti G. Androgens do not regulate the growth hormone response to GHRH in elderly men. Horm Metab Res 1989;21:695–6.
7. Robberecht P, Gillard M, Waelbroeck M, Camus JC, De Neef P, Christophe J. Decreased stimulation of adenylate cyclase by growth hormone releasing

factor in the anterior pituitary of old rats. Neuroendocrinology 1986;44: 429–32.
8. Abribat T, Deslauriers N, Brazeau P, Gaudreau P. Alterations of pituitary GHRH binding sites in aging rats. Endocrinology 1991;128:633–5.
9. De Gennaro Colonna V, Zoli M, Cocchi D, et al. Reduced growth hormone-releasing factor (GHRF)-like immunoreactivity and GHRF gene expression in hypothalamus of aged rats. Peptides 1989;10:705–9.
10. Ho KY, Evans WS, Blizzard RM, et al. Effects of sex and age on the 24-hour profile of growth hormone secretion in man: importance of endogenous estradiol concentrations. J Clin Endocrinol Metab 1987;64:51–8.
11. Shibasaki T, Shizume K, Nakahara M, et al. Age-related changes in plasma growth hormone response to growth hormone releasing factor in man. J Clin Endocrinol Metab 1984;58:212–4.
12. Pavlov EP, Harman SM, Merriam GR, Gelato MC, Blackman MR. Reponses of growth hormone (GH) and somatomedin-C to GH-releasing hormone in healthy aging men. J Clin Endocrinol Metab 1986;62:595–600.
13. Bando H, Zhang C, Takada Y, Yamasaki R, Saito S. Impaired secretion of growth hormone-releasing hormone, growth hormone and IGF-I in elderly men. Acta Endocrinol (Copenh) 1991;124:31–6.
14. Daughaday WH, Trivedi B, Andrews BA. The ontogeny of serum GH binding protein in man: a possible indicator of hepatic GH receptor development. J Clin Endocrinol Metab 1987;65:1072–4.
15. Iranmanesh A, Lizarralde G, Veldhuis JD. Age and relative adiposity are specific negative determinants of the frequency and amplitude of growth hormone (GH) secretory bursts and the half-life of endogenous GH in elderly men. J Clin Endocrinol Metab 1991;73:1081–8.
16. Corpas E, Harman SM, Pineyro MA, Roberson R, Blackman MR. Growth hormone (GH)-releasing hormone-(1–29) twice daily reverses the decreased GH and insulin-like growth factor-I levels in old men. J Clin Endocrinol Metab 1992;75:530–5.
17. Donahue LR, Hunter SJ, Sherblom AP, Rosen C. Age-related changes in serum insulin-like growth factor-binding proteins in women. J Clin Endocrinol Metab 1990;71:575–9.
18. Gibson GE, Peterson C, Jenden DJ. Brain acetylcholine synthesis declines with senescence. Science 1981;213:674–6.
19. Massara F, Ghigo E, Demislis K, et al. Cholinergic involvement in the growth hormone releasing factor-induced growth hormone release: studies in normal and acromegalic subjects. Neuroendocrinology 1986;43:670–5.
20. Ceda GP, Ceresini G, Denti L, et al. Effects of cytidine 5′-diphosphocholine administration on basal and growth hormone-releasing hormone-induced growth hormone secretion in elderly subjects. Acta Endocrinol (Copenh) 1991;124:516–20.
21. Ghigo E, Goffi S, Arvat E, et al. Pyridostigmine partially restores the GH responsiveness to GHRH in normal aging. Acta Endocrinol (Copenh) 1990;123:169–72.
22. Ceda GP, Ceresini G, Denti L, et al. Alpha-glycerylphosphorylcholine administration increases the GH responses to GHRH to young and elderly subjects. Horm Metab Res 1992;24:119–21.

23. Imperato A, De Mei C, Scrocco MG, Angelucci L. Attività colinergica di alfa-GFC a livello ippocampale e striatale. Studio "in vivo" mediante microdialisi cerebrale. Le Basi Razionali della Terapia 1990;20(suppl 1):17–22.
24. Ghigo E, Goffi S, Mazza E, et al. Repeated GHRH administration unravels different GH secretory patterns in normal adults and children. Acta Endocrinol (Copenh) 1989;120:598–601.
25. Sonntag WE, Forman LJ, Miki N, et al. Effects of CNS active drugs and somatostatin antiserum on GH release in young and old male rats. Neuroendocrinology 1981;33:73–8.
26. Sartorio A, Spada A, Arosio M, Conti A, Faglia G. Effects of consecutive doses of GHRH on GH secretion. Clin Endocrinol (Oxf) 1991;35:187–8.
27. Massara F, Ghigo E, Molinatti P, et al. Potentiation of cholinergic tone by pyridostigmine bromide re-instates and potentiates the GH responsiveness to intermittent administration of GHRH in man. Acta Endocrinol (Copenh) 1986;113:12–6.
28. Iovino M, Monteleone P, Steardo L. Repetitive GHRH administration restores the attenuated GH responses to GHRH testing in normal aging. J Clin Endocrinol Metab 1989;69:910–2.
29. Franchimont P, Urban-Choffray D, Lambelin P, Fontaine MA, Frangin G, Reginster JY. Effects of repetitive administration of GHRH on GH secretion, IGF-I and bone metabolism in postmenopausal women. Acta Endocrinol (Copenh) 1989;120:121–4.

26

Even Short-Term hGH Treatment Induces Endocrine Changes in the Elderly

F. Strollo, A. Semprini, G. Strollo, M. Morè, L. Bollanti, A. Ciarmatori, D. Pizzabiocca, N. Mangrossa, and G. Riondino

Growth hormone (GH) and *insulin-like growth factor* (IGF-I) circulating levels decrease with age (1–7), and some evidence has accumulated regarding the role of hypothalamic bioamine dysregulation (8–15), both in experimental animals and in humans. One might then suppose that IGF-I levels induce, rather than merely reflect, the onset of aging processes: A recent report on progeria lends support to such a hypothesis (16). Increased *body mass index* (BMI) and osteopenia are typical features of aging; also, a possible role of low IGF-I levels in the genesis of such phenomena has been hypothesized (16–21). During the last few years some evidence has accumulated that elderly subjects can be successfully treated with *recombinant human growth hormone* (rhGH). The first studies were in GH-deficient adults (22, 23). Some side effects were recognized, such as a degree of insulin resistance, but on the whole the results were clinically good: Well-being self-rating and muscle strength improved, while lean body mass increased and fat distribution changed from prevalently abdominal (high *waist to hip* [W/H] ratio) to peripheral (lower W/H ratio) and circulating cholesterol levels decreased. These observations suggested that rhGH therapy might lower the risk for developing cardiovascular diseases. When trying to transfer the information learned from GH-deficient adult treatment protocols to normal aging subjects, we first had to assess the dosage and second evaluate possible side effects (24–27). It is well known that in acromegaly, a spontaneously occurring GH-excess condition, hypertension, diabetes mellitus, impotence and osteoporosis develop after an early phase characterized by a general feeling of well-being and increase of strength. Trials were also done in chronically ill elderly subjects, leading to conflicting results (28).

The aim of our study was to monitor the effects of rhGH treatment in a small group of healthy hyposomatotropinemic elderly subjects and compare them to control untreated subjects in order to better design longer-term studies in our Mediterranean country, whose lifestyle and dietary habits (including vitamin intake) significantly differ from those in Scandinavia or the USA, where the first reported trials occurred.

Materials and Methods

Ten subjects aged 61 to 85 years entered the study. All had resting and fasting GH levels less than 0.5 ng/ml (normal levels for our laboratory were 0.5–4 ng/ml in men, 0.6–8 ng/ml in women). They were kept on a standard diet (25 kcal/kg per day; CHO 55%, fat 30%, protein 15%) for at least one week before the beginning of the study. They were then randomly allotted to two equally distributed treatment groups: Five were treated with rhGH 0.06 U/kg b.w. (Norditropin, kindly provided by Novo Nordisk; 4 U/ampoule, where 2.6 U = 1 mg) on alternate days for two weeks (group A), and the others were given placebo (group B). Six hours after exogenous administration, GH reached 8.4 ± 1.8 ng/ml (ranging from 4.4 to 13.1 ng/ml), never exeeding the spontaneuous pulse levels for young adult subjects.

Blood was drawn on days 0, 7, and 14 during the treatment period and serum analyzed for GH, glucose, *insulin* (IRI), IGF-I, *calcium* (Ca), *intact parathormone* (PTH), *osteocalcin* (OC), *sex hormone binding globulin* (SHBG), *total and free T3 and T4* (TT3, FT3, TT4, and FT4), and TSH. Glucose and Ca measurements were performed by a Kodak Autoanalyzer system. Serum samples for hormone determinations were kept frozen until they were processed in duplicate during a single session by means of highly sensitive commercially available kits distributed by Nichols Institute (California, USA) (GH and PTH), Diagnostic Systems Laboratories, Inc. (Texas, USA) (IGF-I), Diagnostic Products Corporation (California, USA) (SHBG, FT3, and FT4), Henning (OC) and Serono (T3, T4, and TSH) with intra-assay CVs all less than 5%. A standard questionnaire was submitted to all subjects in order to get subjective ratings for back pain and muscle strength before and during treatment.

Statistical evaluation was performed following the minimum squares method for paired t-test (single-tailed when there was an expected change) and ANOVA. Results were then expressed as mean ± SEM for percent variation compared to baseline. Wilcoxon signed rank and Mann-Whitney U tests were applied to the analysis of self-rating symptom scales within and between groups.

Results

Group A

No adverse effects were reported by the treated subjects. Self-rating of back pain improved slightly immediately after starting rhGH treatment ($P < 0.05$), while muscle strength increased only on day 14 ($P < 0.05$). Such changes were also significant when compared to those in group B ($P < 0.03$ on day 7 and $P < 0.01$ on day 14).

SHBG initially decreased ($-22.2 \pm 5.5\%$, $P = 0.05$) but quickly returned to baseline concentrations while treatment was still ongoing (Fig. 26.1). IGF-I slowly increased and became significantly higher than the baseline levels on day 14 ($33.9 \pm 10.2\%$, $P < 0.01$) (Fig. 26.1).

Glucose levels also increased slightly, as expected, by day 7 ($9.5 \pm 3.1\%$, $P < 0.05$) and stayed higher than baseline concentrations on day 14 ($12.9 \pm 2.6\%$, $P < 0.02$), when IRI also increased significantly ($37.6 \pm 8.8\%$, $P < 0.005$) (Fig. 26.2). Baseline glucose/insulin ratio (G/IRI) decreased significantly from 5.46 ± 0.54 on day 0 to 4.50 ± 0.53 on day 14 ($P < 0.02$) (Fig. 26.4). Nevertheless, fasting hyperglycemia or hyperinsulinemia were never observed throughout the study.

Calcium levels followed a similar pattern within a narrower range of variation ($4.0 \pm 0.3\%$, $P < 0.01$ on day 7 and $3.2 \pm 0.6\%$, $P < 0.03$ on day 14). PTH levels did not change significantly, although a decremental trend was noticed, while OC increased markedly and significantly on day 14 ($96.7 \pm 29.9\%$, $P < 0.01$) (Fig. 26.3). PTH/OC mean ratio was lower than baseline on day 7 (6.2 ± 1.5 vs. 9.7 ± 2.6, $P < 0.05$), as well as on day 14 (4.0 ± 1.1, $P < 0.03$) (Fig. 26.4). No significant changes in thyroid function occurred during treatment.

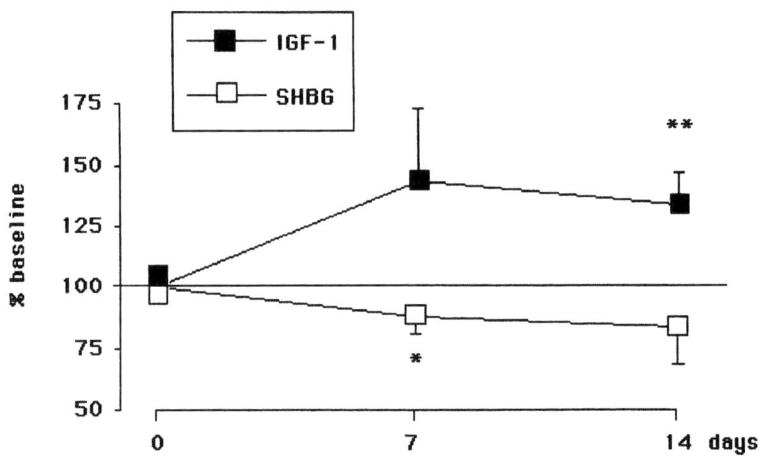

FIGURE 26.1. IGF-I and SHBG percent changes during rhGH treatment. (Mean and SEM are represented; $^*P < 0.05$; $^{**}P < 0.01$.)

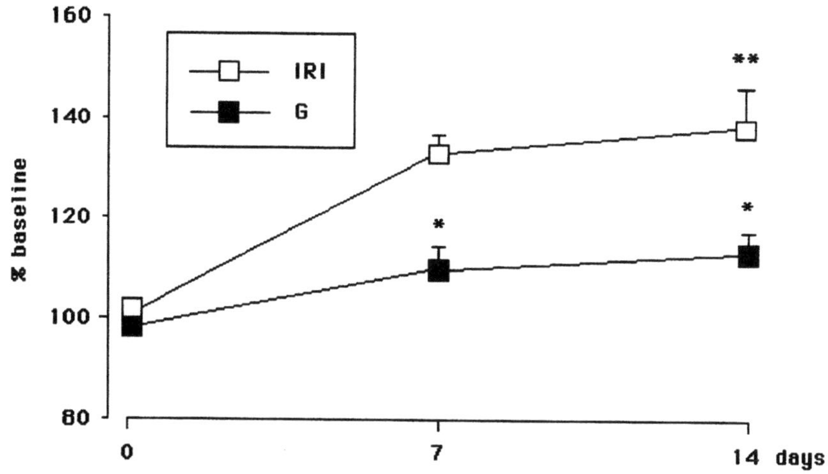

FIGURE 26.2. Glucose and IRI percent changes during rhGH treatment. (Mean and SEM are represented; $*P < 0.05$; $**P < 0.01$.)

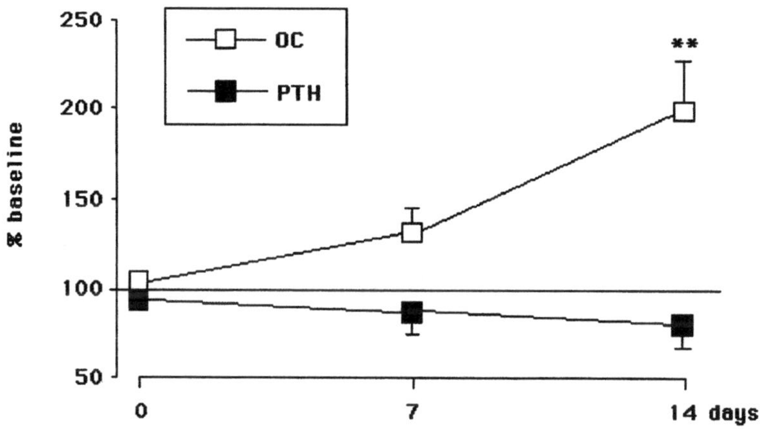

FIGURE 26.3. OC and PTH percent changes during rhGH treatment. (Mean and SEM are represented; $**P < 0.01$.)

Group B

No changes in any parameters studied were found.

Discussion

As already described by others, there was a slight increase in glucose levels relative to insulin concentrations, with a significant decrease of baseline glucose/insulin ratio on day 14, suggesting that treated subjects

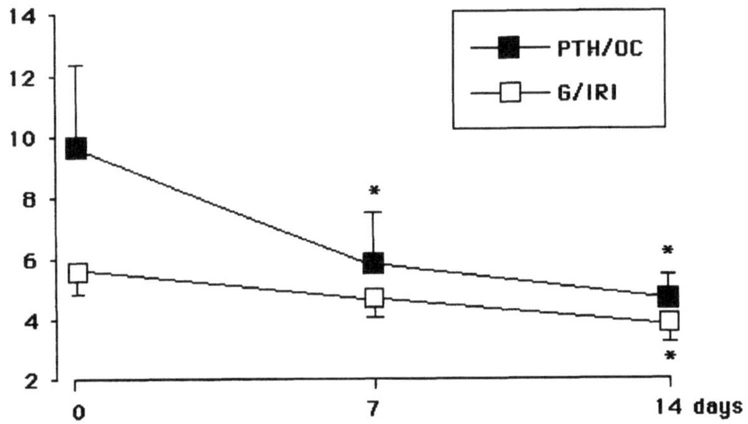

FIGURE 26.4. Changes in some derived parameters (ratios) during rhGH treatment. (Mean ± SEM are represented; $*P < 0.05$.)

quickly developed some insulin resistance. On the other hand, there was a 33.9% increase in IGF-I concentrations to a level that was similar to those commonly found in young adult subjects. This seems to be a very important result because at least some of the well-known signs of the aging process (fat distribution, fat body mass increase, muscle atrophy, and osteoporosis) are thought to depend, to a large extent, on decreased production of IGF-I during senescence.

With respect to senile osteoporosis (type II, as opposed to postmenopausal or type I), we found no significant changes in PTH levels associated with the significant increase in both Ca and OC concentrations (21, 29). This suggested that there was quick activation of *bone remodeling units* (BRU) with an overall prevailing anticatabolic effect, as witnessed by the steady decrease in PTH/OC ratio during treatment when compared to baseline. Indirect support for this hypothesis is also suggested by the rapid relief of back pain in treated subjects, even if a psychoneurogenic mechanism involving pain threshold level increase by rhGH cannot be ruled out. The possibility that the SHBG decrease observed during the first week of treatment might contribute to BRU activation through an increase in sex steroid availability cannot be ruled out as well, but such changes do not seem to play a major role in our study, due to minimal significance on bone metabolism in a study of short duration.

Muscle strength increased gradually, using a subjective rating scale. This was also an expected result. We unfortunately did not use any methods to assess muscle strength objectively, but we plan to introduce some objective methodology in future protocols.

Finally, no significant changes in thyroid function were observed during treatment. This may not occur during long-term rhGH therapy, but it was reassuring that, at least in our very short duration study, no overt changes occurred in peripheral availability of thyroid hormones as expressed by total and free T3 and T4, or in pituitary TSH secretion.

Summary

GH and IGF-I circulating levels decrease with age: This has been hypothesized to be the cause of pathophysiological changes characterizing senescence, rather than a secondary effect. During the last few years evidence has accumulated regarding successful treatment of elderly subjects with rhGH. The aim of our study was to monitor the effects of rhGH treatment in a small group of healthy hyposomatotropinemic elderly subjects in order to better design longer-term studies in our Mediterranean country, whose lifestyle and dietary habits (including vitamin intake) significantly differ from those in Scandinavia or the USA, where the first reported trials occured. Five 61- to 85-year-old healthy subjects were given rhGH 0.06 U/kg body weight on alternate days for two weeks and tested for subjective rating of well-being and muscle strength, as well as for IGF-I, glucose, IRI, SHBG, PTH, OC, calcium, T3, T4, FT3, FT4, and TSH circulating levels on days 0, 7, and 14; five age-matched subjects were used as controls.

No adverse effects were noticed in the treated group; they demonstrated a significant increase in IGF-I (33.9 ± 10.2%; $P < 0.01$) and in OC (96.7 ± 29.9%, $P < 0.01$) on day 14. Glucose and IRI also increased, while the glucose/IRI ratio decreased from 5.46 ± 0.54 to 4.50 ± 0.53, $P < 0.02$. Other minor changes were noticed that are referred to in the text. The control group showed no significant changes in the parameters studied during the observation period.

These observations suggest that even short-term rhGH treatment induces significant hormonal changes in the elderly. These changes can be beneficial since an increase in IGF-I and OC levels might help counteract some of the typical effects of senescence, but a decrease in insulin sensitivity seems to be the price for that at the customary treatment dose. New protocols, involving lower rhGH doses are now under evaluation in our institute in order to optimize future long-term studies on a wider scale.

Conclusions

The results of this pilot study gave us confidence to utilize rhGH in healthy elderly Mediterranean subjects on a wider scale. On the other hand, we were able to confirm the reports of other groups on the change in glucose metabolism during rhGH therapy. Insulin resistance is strongly associated with obesity and with hypertension and is considered a risk factor for increased cardiovascular morbidity and mortality. A possible solution to prevent such a major side effect might be the reduction of the rhGH dosage to the therapeutic lowest level, allowing beneficial effects for a longer period of time and thus reducing the development of overt insulin resistance. This is theoretically possible when IGF-I is slowly and substantially increased over the long run without inducing a sharp increase in GH, but comparative studies on a larger scale are needed in order to substantiate such a hypothesis. Another approach to the reversal of the decrease in IGF-I levels in the elderly is, of course, the administration of IGF-I itself, but this possibility is less easily accomplished at this time.

Our goal is to evaluate possible prevention of deterioration of some functional systems during aging by monitoring the endocrine effects of 0.015 and 0.03 U/kg b.w. administered on alternate days for longer periods of time to the elderly. We hope to be able to provide data about optimum rhGH treatment in the future.

References

1. Rudman D. Growth hormone, body composition, and aging. J Am Geriatr Soc 1985;33:800–7.
2. Lang I, Schernthaner G, Pietschmann P, Kurz R, Stephenson JM, Templ H. Effects of sex and age on growth hormone response to growth hormone-releasing hormone in healthy individuals. J Clin Endocrinol Metab 1987; 65:735–40.
3. Mayerhofer A, Weis J, Bartke A, Yun JS, Wagner TE. Effects of transgenes for human and bovine growth hormones on age-related changes in ovarian morphology in mice. Anat Rec 1990;227:175–86.
4. Abribat T, Deslauriers N, Brazeau P, Gaudreau P. Alterations of pituitary growth hormone-releasing factor binding sites in aging rats. Endocrinology 1991;128:633–5.
5. Wheeler MD, Schutzengel RE, Barry S, Styne DM. Changes in basal and stimulated growth hormone secretion in the aging rhesus monkey: a comparison of chair restraint and tether and vest sampling. J Clin Endocrinol Metab 1990;71:1501–7.
6. Van Coevorden A, Mockel J, Laurent E, et al. Neuroendocrine rhythms and sleep in aging men. Am J Physiol 1991;260:E651–61.
7. Ceda GP, Ceresini G, Denti L, et al. Effects of cytidine 5′-diphosphocholine administration on basal and growth hormone-releasing hormone-induced secretion in elderly subjects. Acta Endocrinol (Copenh) 1991;124:516–20.
8. Coiro V, Volpi R, Cavazzini U, et al. Restoration of normal growth hormone responsiveness to GHRH in normal aged men by infusion of low amounts of theophylline. J Gerontol 1991;46:M155–8.
9. Meites J, Goya R, Tekahashi S. Why the neuroendocrine system is important in aging processes. Exp Gerontol 1987;22:1–15.

10. Ge F, Tsagarakis S, Rees LH, Besser GM, Grossman A. Relationship between growth hormone-releasing hormone and somatostatin in the rat: effects of age and sex on content and in-vitro release from hypothalamic explants. J Endocrinol 1989;123:53–8.
11. Panzeri G, Torsello A, Cella SG, Muller EE, Locatelli V. Age-related modulatory activity by a cholinergic agonist on the growth hormone responses to GH-releasing hormone in the rat. Proc Soc Exp Biol Med 1990;193:301–5.
12. Meites J. Role of hypothalamic catecholamines in aging processes. Acta Endocrinol (Copenh) 1991;125(suppl 1):98–103.
13. Raskind MA, Peskind ER, Veith RC, Wilkinson CW, Federighi D, Dorsa DM. Differential effects of aging on neuroendocrine responses to physostigmine in normal men. J Clin Endocrinol Metab 1990;70:1420–5.
14. Bando H, Zang C, Takada Y, Yamasaki R, Saito S. Impaired secretion of growth hormone-releasing hormone and IGF-I in elderly men. Acta Endocrinol (Copenh) 1991;124:31–6.
15. Tanaka K, Inoue S, Shiraki J, et al. Age-related decrease in plasma growth hormone: response to growth hormone-releasing hormone, arginine, and l-Dopa in obesity. Metabolism 1991;40:1257–62.
16. Brown WT. Progeria: a human disease model of accelerated aging. Am J Clin Nutr 1992;55:1222S–4S.
17. Riggs BL, Wahner HW, Dunn WL, Mazess RB, Offord KP, Melton LJ. Differential changes in bone mineral density of the appendicular and axial skeleton with aging. J Clin Invest 1981;67:328–35.
18. Rudman D. Impaired growth hormone secretion in the adult population: relationship to age and adiposity. J Clin Invest 1981;67:1361–9.
19. Copeland KC, Colletti RB, Devlin JT, Auliffe TL. The relationship between insulin-like growth factor-I, adiposity, and aging. Metabolism 1990;39:584–7.
20. Iranmanesh A, Lizarralde G, Veldhuis JD. Age and relative adiposity are specific negative determinants of the frequency and amplitude of growth hormone (GH) secretory bursts and the half-life of endogenous GH in healthy men. J Clin Endocrinol Metab 1991;73:1081–8.
21. Kaufman J-M, Taelman P, Vermeulen A, Vandeweghe M. Bone mineral status in growth hormone-deficient males with isolated and multiple pituitary deficiencies of childhood onset. J Clin Endocrinol Metab 1992;74:118–23.
22. Christiansen JS, Jorgensen JO, Pedersen SA, et al. GH-replacement therapy in adults. Horm Res 1991;36(suppl 1):66–73.
23. Weissberger AJ, Cuneo RC, Salomon F, McGauley G, Lowy C, Sonksen PH. Growth hormone treatment in adults with growth hormone deficiency. J Endocrinol Invest 1992;15(suppl 4):36.
24. Rudman D, Feller AG, Nagraj HS, et al. Effects of human growth hormone in men over 60 years old. N Engl J Med 1990;323:1–6.
25. Rudman D, Feller AG, Cohn L, Shetty KR, Rudman IW, Draper MW. Effects of human growth hormone on body composition in elderly men. Horm Res 1991;36(suppl 1):73–81.
26. Marcus R, Butterfield G, Holloway L, et al. Effects of short term administration of recombinant human growth hormone to elderly people. J Clin Endocrinol Metab 1990;70:519–27.
27. Marcus R. Effects of growth hormone on bone and calcium metabolism in elderly people. J Endocrinol Invest 1992;15(suppl 4):36.

28. Kaiser FE, Silver AJ, Morley JE. The effect of recombinant human growth hormone on malnourished older individuals. J Am Geriatr Soc 1991;39: 235–40.
29. Donahue LR, Hunter SJ, Sherblom AP, Rosen C. Age-related changes in serum insulin-like growth factor-binding proteins in women. J Clin Endocrinol Metab 1990;71:575–9.
30. Belgorosky A, Ferraris JR, Ramirez JA, Jasper H, Rivarola MA. Serum sex hormone-binding globulin and serum non sex hormone-binding globulin-bound testosterone fractions in prepubertal boys with chronic renal failure. J Clin Endocrinol Metab 1991;73:107–10.

27

Dichotomic Role of Glucocorticoids in the Regulation of Growth Hormone Secretion In Vivo

ANDREA GIUSTINA, ANNA ROSA BUSSI, AND
WILLIAM B. WEHRENBERG

Growth hormone (GH) synthesis and secretion are regulated by the hypothalamic peptides *GH releasing hormone* (GHRH), which has an excitatory role, and somatostatin, which has an inhibitory role (1). Glucocorticoids interfere with normal somatic growth in laboratory animals (2), children receiving high-dose glucocorticoid treatment (3), and children with Cushing's disease (4). Numerous studies have demonstrated that normal subjects, after administration of pharmacological doses of glucocorticoids, and patients with Cushing's disease have reduced GH responses to various stimuli (5–7).

In vitro, GH secretion by somatotrophs obtained from human pituitaries (8) and from normal monkey and rat pituitaries (9) is markedly increased in the presence of glucocorticoids. The enhanced secretion occurs after several days of co-incubation and is amplified if the cells are first cultured for long periods without glucocorticoids. Glucocorticoids also increase the GH response to GHRH in vitro. Pretreatment of rat pituitary cells for 24 h with the long-acting glucocorticoid dexamethasone increases the GH response to GHRH more than fivefold (10). Glucocorticoids are hypothesized to increase the GH response to GHRH by significantly increasing GHRH receptors on the somatotrophs (11). The glucocorticoid-induced increase in pituitary secretion is abolished by cycloheximide or actinomycin D (12) and has been hypothesized to be mediated by the binding of the activated glucocorticoid receptor protein to the first intron of the GH gene, which enhances transcription rates and finally results in intracellular increase in GH mRNA levels (13) (Fig. 27.1).

The aim of our study was to investigate the mechanisms underlying the effects of glucocorticoids in vivo. Therefore, we examined the effects of

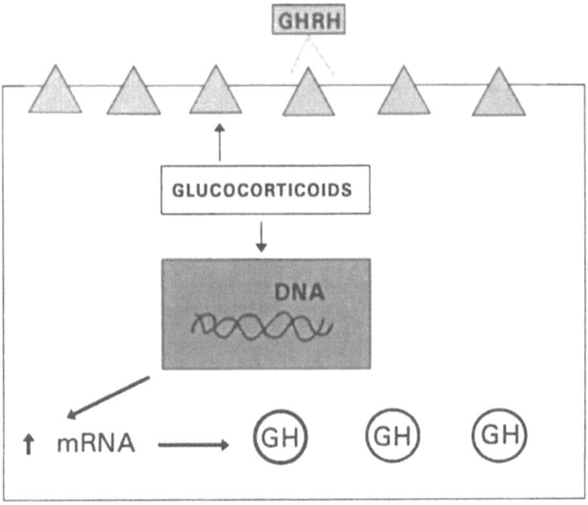

FIGURE 27.1. Hypothetical mechanisms of the direct action of glucocorticoids on somatotrophs.

hypoadrenalism and hyperadrenalism on GH secretion in both rats and humans.

Rat Studies

Hypoadrenal Rats

Adrenalectomy, without steroid replacement therapy, significantly decreases the pituitary GH response to a submaximal dose of GHRH in vivo (14). This decreased response can be reversed by dexamethasone replacement therapy. When a supramaximal dose of GHRH is used, there is no difference between normal and adrenalectomized rats. This suggests that adrenalectomy results in a decreased pituitary sensitivity to GHRH and not a change in the releasable pool of GH.

Previous studies (15) have reported that chronic dexamethasone treatment suppresses GH concentrations, increases hypothalamic somatostatin concentrations, and decreases hypothalamic GHRH concentrations in conscious female rats. In male rats, trough, or baseline, GH values in conscious glucocorticoid-treated rats are lower for a longer time interval than in normal animals (16, 17).

Hyperadrenal Rats

Wehrenberg et al. (18) conducted a series of experiments in normal, conscious, freely moving rats receiving chronic saline or dexamethasone treatment. The GH response to a physiological dose of GHRH was significantly lower ($P < 0.01$) in glucocorticoid-treated rats than in saline-treated rats. When these animals were passively immunized with somatostatin antiserum, however, the GH response was completely reversed; that is, glucocorticoid-treated rats had a significantly higher GH response to GHRH than saline-treated animals. Further corroborating the fact that glucocorticoids increase somatostatin tone is the observation that chronic somatostatin antiserum treatment can partially block glucocorticoid inhibition of growth in rats (19).

Human Studies

Hypoadrenal Humans

The role played by normal levels of circulating cortisol in the physiology of GH secretion in humans is still unclear. In order to find a clinical model suitable for clarifying this key point, we examined GH secretion during glucocorticoid replacement in patients with *adrenocorticotropin hormone* (ACTH) deficiency who showed impaired GH response to stimuli before beginning replacement therapy.

Three women (age 21, 42, and 55 years) with associated ACTH and GH deficiency were studied after obtaining their informed consent. All patients presented with mild symptoms of hypoadrenalism (weakness, anorexia, loss of weight, humor disturbances) of duration ranging from two to more than five years. None of the patients had been previously treated for adrenal insufficiency. On admission, each patient underwent evaluation of anterior pituitary function and had a *computed tomography* (CT) scan of the brain performed. In no patients were detectable lesions found on CT scan; thus, our diagnosis was "idiopathic ACTH and GH deficiency." After the diagnosis of secondary hypoadrenalism, each patient was given glucocorticoid replacement with cortisone acetate (12.5–25 mg/day); dosages were divided in twice-daily administration at 0800 and 1600 h. Patients underwent (i) *insulin-tolerance test* (ITT), with intravenous (IV) injection of 0.01 U/kg short-acting insulin with blood samples obtained 0, 15, 30, 45, 60, 90, and 120 min after the injection and (ii) *arginine-infusion test* (AIT), with blood samples obtained 0, 15, 30, 45, 60, 90, and 120 min after beginning the infusion. At the time of the diagnosis all patients had basal cortisol levels below the detection limit of our assay (55 nmol/L) and no ACTH and cortisol response to ITT; moreover, their GH response to ITT and AIT was absent or severely impaired.

Pituitary function evaluation was repeated using the same methodology after a period of replacement that was variable for any single patient (3–24 months). During this period none of the patients had other medical therapy. Basal cortisol levels were measured at the beginning of the tests, during which glucocorticoid replacement was started 1 h after the administration of the morning cortisone acetate dose, and ranged from 195 to 331 nmol/L (mean 248 ± 55 nmol/L). The GH response to the stimuli returned to normal (peak levels greater than 7 μg/L) in all three patients after replacement therapy.

Confirming these findings is the observation that when the glucocorticoid replacement is suspended, patients show a progressive loss of their GH response to GHRH (20). On the basis of these studies, it has been hypothesized that the impaired GH response is a functional consequence of the low circulating cortisol levels. Therefore, physiological levels of glucocorticoids may be hypothesized to play an important role in regulating somatotroph function in humans, as well as in rats.

Hyperadrenal Humans

The mechanisms underlying the inhibitory effect of hyperadrenalism on GH secretion in humans have been studied by utilizing the following pharmacological tools (Fig. 27.2): (i) *pyridostigmine* (PD), an acetylcholinesterase inhibitor that enhances central cholinergic tone; (ii) the amino acid arginine: both PD and arginine are thought to decrease hypothalamic somatostatin secretion (21, 22); (iii) galanin, a 29-amino acid neuropeptide that acts as a neurotransmitter in the brain; and (iv) clonidine, an alpha-2 adrenergic receptor agonist: galanin and clonidine are hypothesized to increase hypothalamic GHRH secretion (23, 24).

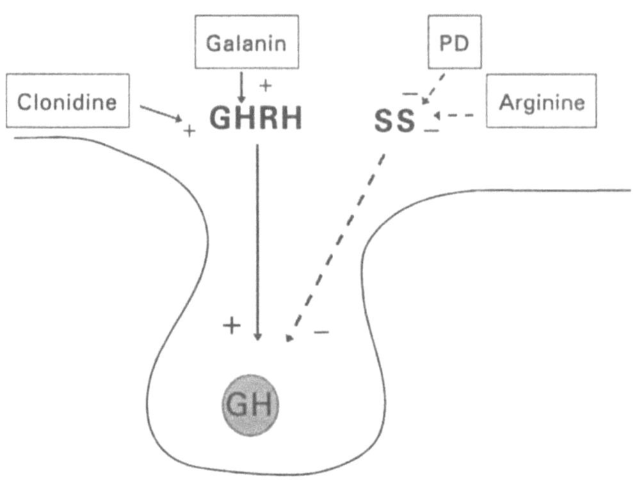

FIGURE 27.2. Hypothetical mechanisms of the GH stimulatory action of clonidine, galanin, pyridostigmine (PD), and arginine. (GHRH = growth hormone releasing hormone; SS = somatostatin.)

Acute Effects

The acute effect of exogenous glucocorticoids in normal humans is the suppression of GH secretion. Cortisone inhibits the GH response to GHRH during the first hour following administration (25), and the GH response is still inhibited 8h after treatment (26). To clarify the mechanisms underlying this acute inhibition, we studied five male and one female adult volunteers with no history of endocrine or metabolic disease. Their ages ranged from 24 to 29 years, and their body mass index ranged from 19 to 24 kg/m^2. Six different experimental trials were conducted in each of the subjects at no less than 7-day intervals: (i) human GHRH(1-29)NH_2, 100 μg in 1 ml of saline injected as an IV bolus; (ii) cortisone acetate 50 mg administered p.o. 60 min before a hGHRH IV bolus injection; (iii) PD, 120 mg administered p.o. 60 min before a hGHRH IV bolus injection; (iv) PD and cortisone acetate administered p.o. 60 min before a hGHRH IV bolus injection; (v) PD, administered p.o. 60 min before a saline IV bolus injection; and (vi) PD and cortisone acetate administered p.o. 60 min before a saline IV bolus injection.

Mean absolute GH levels at times 15, 30, 45, 60, and 90 min were significantly lower after GHRH + cortisone than after GHRH alone. Peak values occurred between 15 and 90 min and averaged 7.3 ± 0.8 μg/L after cortisone + GHRH and were significantly lower than the peak values observed after GHRH alone (25.1 ± 1.8 μg/L, $P < 0.001$). However, PD completely removed the inhibitory effect of cortisone on GH secretion. In fact, mean absolute GH levels at 15, 30, 45, 60, 90, and 120 min and GH peaks were not significantly different after PD + cortisone + GHRH and after PD + GHRH. Moreover, mean absolute GH values at 15, 30, 45, 60, 90, and 120 min and mean GH peaks were similar after PD + cortisone and after PD alone. Overall, it can be hypothesized that acute glucocorticoid administration to normal humans inhibits GH release through an increase in hypothalamic somatostatin tone.

Chronic Effects in Children and Adults

An interesting human model for studying the interactions between chronic hypercortisolemia and GH secretion is provided by children and adults undergoing chronic treatment with glucocorticoids for nonendocrine diseases. We studied six pediatric patients who underwent orthotopic liver transplantation, receiving long-term immunosuppressive glucocorticoid treatment (5.92 ± 0.63 mg/day, mean dose ± SEM). The group included three males and three females; age at the start of the study was 7.6 ± 1.5 years (mean ± SEM). All patients were studied within one year of liver transplantation (6.6 ± 1.6 months, range 4–11.5 months). At the time of the studies all patients were medically stable with normal liver function. Five normal children, matched for age with the liver-transplanted subjects and not undergoing any therapy, served as controls. Nocturnal spontaneous GH secretion was evaluated after the administration of PD, 2 mg/kg p.o., or placebo, 1 tablet p.o. After placebo administration in liver-transplanted children, spontaneous GH secretory activity was slightly, even if not significantly, lower compared to normal children (5.4 ± 1.0 mU/L vs. 6.4 ± 0.6 μg/L). After PD administration in the transplanted children the mean nocturnal GH concentration and maximal GH peak were all significantly increased compared to placebo treatment ($P < 0.05$). In normal children, PD had no effect on nocturnal GH secretion. Interestingly, the maximal GH secretory capacity during the night was significantly higher after PD administration in the steroid-treated compared to normal children.

Patients and controls also received human GHRH(1-29)NH_2, 1 μg/kg in 1 ml of saline as an IV bolus 60 min after PD, 2 mg/kg p.o., or placebo, 1 tablet p.o. Liver-transplanted patients had blunted GH responses to GHRH (GH peak 13.6 ± 5.4 μg/L) after placebo administration compared to normal subjects (GH peak 40.8 ± 5.0 μg/L) ($P < 0.05$). In contrast, after PD pretreatment no significant difference in the GH responses to GHRH was found between controls (GH peak 127.8 ± 11.0 μg/L) and steroid-treated patients (94.0 ± 15.2 μg/L).

GH secretion in adults under chronic immunosuppressive treatment has also been investigated. The results obtained are consistent with observations made in rats.

Arginine Study

We studied eight nonobese (body mass index <28 kg/m^2) adult patients (three males and five females; 39.1 ± 3.7 years, mean age ± SEM) undergoing chronic daily immunosuppressive steroid treatment for autoimmune diseases (one mixed connective tissue disease, one Churg-Strauss disease, two systemic lupus erythematosus, one chronic pericarditis, one scleroderma, two polymyalgia rheumatica). Six normal adult nonobese volunteers (four males and two females) with no history of endocrine or metabolic diseases, matched for age (35 ± 3.5 years, mean ± SEM) with the steroid-treated subjects, and not undergoing any therapy served as controls. After a 30-min stabilization period, the steroid-treated patients received their morning oral dose of prednisone, and all subjects initiated the following two tests in random order: (i) infusion of arginine hydrochloride 30 g IV in 100 ml saline, or (ii) infusion of saline 100 ml IV from −30 to 0 min together with human GHRH(1-29)NH_2 100 μg in 1 ml saline, injected as an IV bolus at 0 min.

Glucocorticoid-treated patients had blunted GH responses to GHRH (GH peak 8.7 ± 2.4 μg/L) after saline administration compared to normal controls (GH peak 23.8 ± 3.9 μg/L) ($P < 0.05$). After arginine pretreatment, a significant increase in the GH responses to GHRH compared to

TABLE 27.1. GH peak (μg/L) after GHRH and GHRH + arginine in six normal subjects and eight steroid-treated patients.

	GH peak (μg/L)	
Test	Normals	Steroid treated
GHRH	23.8 ± 3.9*	8.7 ± 2.4
GHRH + arginine	36.6 ± 4.0†	28.4 ± 5.5†

* $P < 0.05$ vs. steroid treated.
† $P < 0.05$ vs. GHRH + saline.

saline pretreatment was found in both control subjects (GH peak 36.6 ± 4.0 μg/L) ($P < 0.05$) and glucocorticoid-treated patients (GH peak 28.4 ± 5.5 μg/L) ($P < 0.05$). Mean absolute GH levels at 15, 30, 45, and 60 min and GH *area under curve* (AUC) measurements were significantly lower after placebo in glucocorticoid-treated patients than those in normal controls; the same parameters after arginine pretreatment were not significantly different in glucocorticoid-treated patients compared to normal subjects (Table 27.1). The effect of arginine compared to saline on the GH response to GHRH was significantly greater in the steroid-treated subjects than in the normal subjects when either the percent peak ratio GHRH + arginine/GHRH + saline (443.1 ± 104.7% vs. 165.3 ± 19%) ($P < 0.05$) or the AUC ratio GHRH + arginine/GHRH + saline (319.3 ± 45.8% vs. 189.2 ± 27%) ($P < 0.05$) was considered.

Galanin Study

We studied seven adult patients undergoing long-term immunosuppressive glucocorticoid treatment for nonendocrine diseases. The group was composed of two men and five women. Mean age was 38.7 ± 4.5 years (mean ± SEM), their body mass index was less than 28 kg/m^2, and blood glucose levels were in the normal range. Prednisone (twice daily, p.o.) was the only drug the patients received. Six normal adult nonobese volunteers (four men and two women), matched for age (35 ± 3.5 years, mean ± SEM) with the steroid-treated subjects, served as controls. All the subjects underwent in random order from −10 to 30 min (i) infusion of synthetic porcine galanin IV, 500 μg in 100 ml of saline, or (ii) infusion of saline IV, 100 ml. A bolus of human GHRH(1-29)NH_2, 100 μg in 1 ml of saline was injected IV at 0 min. Glucocorticoid-treated patients demonstrated blunted GH response to GHRH (GH peak, 8.1 ± 2.8 μg/L) during saline infusion compared with normal controls (GH peak, 23.8 ± 3.9 μg/L) ($P < 0.05$). During galanin infusion, a significantly enhanced GH response to GHRH compared with saline infusion was observed in control subjects (GH peak, 46.6 ± 9.4 μg/L) ($P < 0.05$), but not in glucocorticoid-treated patients (GH peak, 16.6 ± 6.5 μg/L) (Table

TABLE 27.2. GH peak (μg/L; mean ± SEM) after GHRH and GHRH + galanin in six normal subjects and seven steroid-treated patients.

	GH peak (μg/L)	
Test	Normals	Steroid treated
GHRH	23.8 ± 3.9†	8.1 ± 2.8
GHRH + galanin	46.6 ± 9.4*†	16.6 ± 6.5

* $P < 0.05$ vs. GHRH + saline.
† $P < 0.05$ vs. steroid treated.

27.2). Mean absolute GH levels at 15, 30, 45, and 60 min and GH AUCs were significantly lower after both saline and galanin in glucocorticoid-treated patients compared with normal controls.

Clonidine Study

We studied seven adult glucocorticoid-treated nonobese patients (four males and three females, age 39 ± 4 years, body mass index 25 ± 1 kg/m^2). Eleven normal adults (seven males and four females) matched for age (mean 32 ± 1 years) served as controls. All the subjects underwent (i) IV infusion of clonidine 150 μg in 10 ml of saline, from time 0 to 10 min, or (ii) bolus IV injection of human GHRH(1-29)NH_2, 100 μg in 1 ml of saline at time 0. Steroid-treated patients initiated both treatments 30 min after receiving their morning oral dose of prednisone. Steroid-treated patients had blunted GH peaks after clonidine (5.8 ± 2.8 μg/L) compared to normal controls (17.6 ± 2.3 μg/L) ($P < 0.01$). GH peaks after GHRH were also significantly lower in steroid-treated (8.3 ± 3 μg/L) with respect to normal subjects (19.3 ± 2.4 μg/L) ($P < 0.01$) (Table 27.3). Mean absolute GH levels after 15, 30, 45, 60, and 90 min and GH AUCs were also significantly lower in glucocorticoid-treated patients both after clonidine and after GHRH with respect to normal controls. No significant differences between the GH responses to clonidine and GHRH were observed within the two groups. Taken together, the evidence obtained by studying GH responses to the stimuli in patients under long-term

TABLE 27.3. GH peak (μg/L; mean ± SEM) after GHRH or clonidine in eleven normal subjects and seven steroid-treated patients.

	GH peak (μg/L)	
Test	Normals	Steroid treated
GHRH	19.3 ± 2.4*	8.3 ± 3
Clonidine	17.6 ± 2.3*	5.8 ± 2.8

* $P < 0.01$ vs. steroid treated.

treatment with glucocorticoids suggests that excess levels of these steroids in the circulation may cause impairment of pituitary GH secretion through an increase in hypothalamic somatostatin tone.

Discussion

Long-term treatment with glucocorticoids is known to cause growth failure in children (27) and protein wasting, together with an increase in lipid deposits, in adult subjects (28). The suppression of linear growth in children is almost certainly multifactorial. Potential factors include direct action on peripheral tissue and inhibition of IGF-I synthesis and release (27). Although some studies have reported that administration of human GH is not able to prevent the steroid-induced growth failure, alterations in GH secretion may play an important role. For example, passive immunization with antibodies against somatostatin is able to partially abolish growth deficiency in rats receiving dexamethasone (19). Following this line of research and the availability of recombinant human GH, it has been shown that exogenous GH treatment may be useful in improving growth in glucocorticoid-treated children after kidney transplantation (29).

Protein catabolism induced by the chronic administration of glucocorticoids is implicated in many, if not most, of the complications associated with use of these drugs. Identification of the factors leading to glucocorticoid-induced protein losses and of therapeutic modalities that might ameliorate or prevent these losses may have major implications for the large number of patients receiving long-term glucocorticoid therapy. GH is a potent anabolic agent that stimulates protein synthesis in peripheral tissues. Interestingly, a recent study using nitrogen balance and isotope dilution techniques demonstrated in normal volunteers that short-term GH administration may be useful in the treatment of glucocorticoid-enhanced protein catabolism (28). Despite this clinically interesting perspective, there are several potential drawbacks to the combined use of GH and glucocorticoids. They include: (i) occurrence or deterioration of carbohydrate intolerance (worsening of insulin resistance); (ii) potential immunostimulation of GH, which may antagonize some of the therapeutic immunosuppressive effects of glucocorticoids; and (iii) development of acromegaly.

Due to these potential side effects with administration of exogenous GH, an intriguing perspective in patients under chronic steroid therapy may be treatment with neuroactive drugs able to reduce somatostatinergic tone at the hypothalamic level. Results of several studies show that both PD and arginine acutely administered are able to remove the glucocorticoid-mediated inhibition of GH secretion. The advantage of this kind of treatment should be to reinstate a physiological pattern of GH secretion in patients under chronic glucocorticoid treatment. It remains to be established whether the somatostatin-lowering effect of PD and arginine could be maintained on a long-term basis.

Conclusions

The effects of glucocorticoids on GH secretion and the mechanisms underlying their action are very similar in humans and laboratory rats. We feel that it is possible to organize the available data into a unique pathophysiologic model explaining these complex interactions (Fig. 27.3). Physiologic levels of circulating glucocorticoids are necessary to maintain normal pituitary GH secretion due to their stimulatory actions on both GH synthesis and GHRH receptors. When serum concentrations of glucocorticoids increase above the normal range, there is an increase in hypothalamic somatostatin tone, which in turn suppresses pituitary GH secretion and overrides any stimulatory effects. This GH inhibitory effect has significant clinical implications. The aim of future clinical studies should be to assess the efficacy and safety of the administration of exogenous GH and/or somatostatin-lowering drugs to reverse the inhibitory effects of glucocorticoid therapy on growth and GH secretion.

Acknowledgment. We wish to thank the Centro Studi e Ricerche di Neuroendocrinologia, Brescia, Italy, for their helpful support.

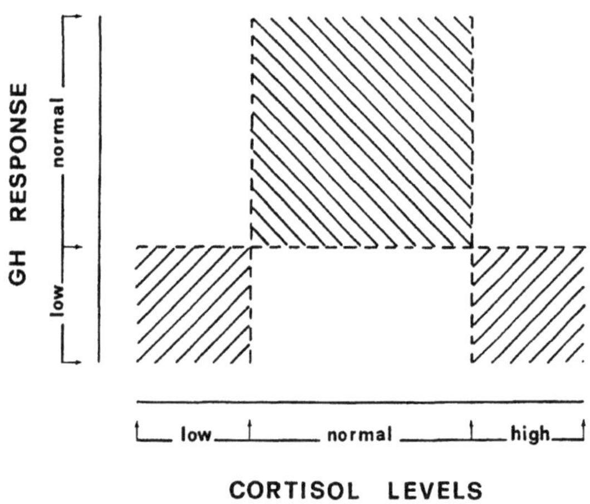

FIGURE 27.3. Proposed model for the relationship between glucocorticoid levels and GH secretory capacity.

References

1. Dieguez C, Page MD, Scanlon MF. Growth hormone neuroregulation and its alterations in disease states. Clin Endocrinol (Oxf) 1988;28:109–43.
2. Hodges JR, Vernikos J. A comparision of the pituitary inhibitory effects of prednisone, prednisolone and hydrocortisone. Br J Pharmacol Chemother 1958;13:98–102.
3. Blodgett FM, Burgin L, Iezzoni D, Gribetz D, Talbot NB. Effects of prolonged cortisone therapy on the statural growth, skeletal maturation and metabolic status of children. N Engl J Med 1956;254:636–41.
4. Solomon IL, Schoen EJ. Juvenile Cushing syndrome manifested primarily by growth failure. Am J Dis Child 1976;123:207–13.
5. Hartog M, Gaafar MA, Fraser R. Effect of corticosteroids on serum growth hormone. Lancet 1964;ii:376–8.
6. Nakagawa K, Horiuchi Y, Mashimo K. Responses of plasma growth hormone and corticosteroids to insulin and arginine with or without prior administration of dexamethasone. J Clin Endocrinol Metab 1969;29:35–40.
7. Giustina A, Bossoni S, Bodini C, et al. Pyridostigmine enhances even if it does not normalize the growth hormone responses to growth hormone-releasing hormone in patients with Cushing's disease. Horm Res 1991;35: 99–103.
8. Bridson WE, Kohler PO. Cortisol stimulation of growth hormone production by human pituitary tissue in culture. J Clin Endocrinol Metab 1970;30: 538–40.
9. Kohler PO, Bridson WE, Rayford PL. Cortisol stimulation of growth hormone production by monkey adenohypophysis in tissue culture. Biochem Biophys Res Commun 1968;33:834–40.
10. Vale W, Vaughan J, Yamamoto G, Spiess J, Rivier J. Effects of synthetic human pancreatic (tumor) GH releasing factor and somatostatin, triiodothyronine and dexamethasone on GH secretion in vitro. Endocrinology 1983; 112:1553–5.
11. Seifert H, Perrin M, Rivier J, Vale W. Growth hormone-releasing factor binding sites in rat anterior pituitary membrane homogenates: modulation by glucocorticoids. Endocrinology 1985;117:424–6.
12. Oosterom R, Verleun J, Zuiderwijk J, Lamberts SWJ. Growth hormone secretion by cultured rat anterior pituitary cells: effects of culture conditions and dexamethasone. Endocrinology 1983;113:735–41.
13. Evans RM, Birnberg NC, Rosenfeld MG. Glucocorticoid and thyroid hormones transcriptionally regulate growth hormone gene expression. Proc Natl Acad Sci USA 1982;79:7659–63.
14. Wehrenberg WB, Brazeau P, Luben R, Bohlen P, Guillemin R. Inhibition of the pulsatile secretion of growth hormone by monoclonal antibodies to hypothalamic growth hormone releasing factor (GRF). Endocrinology 1982; 111:2146–8.
15. Nakagawa K, Ishizuka T, Obara T, Matsubara M, Akikawa K. Dichotomic action of glucocorticoids on growth hormone secretion. Acta Endocrinol (Copenh) 1987;116:165–71.
16. Mosier HD, Jansons RA. Increase in pulsatile secretion of growth hormone during failure of catch-up growth following glucocorticoid-induced growth inhibition. Proc Soc Exp Biol Med 1985;178:457–61.

17. Wehrenberg WB, Baird A, Klepper R, Mormede P, Ling N. Interactions between growth hormone-releasing hormone and glucocorticoids in male rats. Regul Pept 1989;25:147–55.
18. Werhenberg WB, Janowski B, Piering AW, Culler F, Jones KL. Glucocorticoids: potent inhibitors and stimulators of growth hormone secretion. Endocrinology 1990;126:3200–3.
19. Wehrenberg WB, Bergman PJ, Stagg L, Ndon J, Giustina A. Glucocorticoid inhibition of growth in rats: partial reversal with somatostatin antibodies. Endocrinology 1990;127:2705–8.
20. Giustina A, Romanelli G, Girelli A, Pizzocolo G, Valentini U, Zuccato F. Effects of short-term glucocorticoid deprivation on growth hormone response to GH-releasing hormone in man. Horm Metab Res 1989;21:693–4.
21. Giustina A, Bodini C, Bossoni S, et al. Effects of calcitonin on GH response to pyridostigmine in combination with hGHRH(1–29)NH_2 in normal adult subjects. Clin Endocrinol 1990;33:375–80.
22. Alba-Roth J, Muller OA, Schopohl J, von Werder K. Arginine stimulates growth hormone secretion by suppressing endogenous somatostatin secretion. J Clin Endocrinol Metab 1988;67:1186–9.
23. Murakami Y, Kato Y, Shimatsu A, et al. Possible mechanisms involved in growth hormone secretion induced by galanin in the rat. Endocrinology 1989;124:1224–9.
24. Miki N, Ono M, Shizume K. Evidence that opiatergic and alpha-adrenergic mechanisms stimulate rat growth hormone release via growth hormone-releasing factor. Endocrinology 1984;114:1950–2.
25. Giustina A, Doga M, Bodini C, et al. Acute effects of cortisone acetate on growth hormone response to growth hormone-releasing hormone in normal adult subjects. Acta Endocrinol (Copenh) 1990;122:206–10.
26. Del Balzo P, Salvatori R, Cappa M, Gertner JM. Pyridostigmine does not reverse dexamethasone-induced growth hormone inhibition. Clin Endocrinol (Oxf) 1990;33:605–12.
27. Loeb JN. Corticosteroids and growth. N Engl J Med 1976;295:547–52.
28. Horber FF, Haymond MW. Human growth hormone prevents the protein catabolic side effects of prednisone in humans. J Clin Invest 1990;86:265–72.
29. Tejani A, Butt KMH, Rajpoot D, et al. Strategies for optimizing growth in children with kidney transplants. Transplantation 1989;47:229–33.

Author Index

A

Abe, H., 223
Abribat, T., 155
Arkins, S., 87

B

Baffoni, M.T., 328
Banchini, A., 328
Bell, G.I., 33
Bercu, B.B., 167, 286
Bick, T., 301
Blalock, J.E., 104
Bollanti, L., 338
Bowers, C.Y., 203
Brazeau, P., 155
Buonomo, F.C., 317
Bussi, A.R., 347
Butterfield, G.E., 150

C

Ceda, G.P., 328
Chihara, K., 223
Ciarmatori, A., 338
Clark, R., 242
Cohen, P., 279
Cohn, L., 133
Coy, D.H., 3
Cunningham, B., 231

D

de Vos, B., 231
Deslauriers, N., 155
Diamond, F.B., Jr., 279, 286

F

Fabris, N., 116
Feller, A.G., 133
Fielder, P.J., 279
Frick, G.P., 301
Fuh, G., 231

G

Gargosky, S.E., 279
Garrel, D.R., 155
Giustina, A., 347
Goodman, H.M., 301
Guevara-Aguirre, J., 279

H

Hayashi, S., 223
Hines, J., 33
Hocart, S.J., 3
Hoffman, A.R., 328
Holloway, L., 150
Hu, C.-S., 167

J

Jorgensen, E.V., 286

K

Kaji, H., 223
Kelley, K.W., 87
Klindt, J., 317
Kong, H., 33
Kraicer, J., 17

L

Law, S.F., 33
Leonard, J.L., 301
Li, Y.M., 87
Lowman, H., 231

M

Mangrossa, N., 338
Marcus, R., 150
Marzani, G.P., 328
Masuda, R., 167
Melmed, S., 73
Mitani, M., 223
Mocchegiani, E., 116
Morè, M., 338
Murphy, W.A., 3

O

Ohashi, S.-i., 223

P

Piovani, E., 328
Pizzabiocca, D., 338
Prager, D., 73
Preece, M.A., 269

R

Ramirez, G., 193
Raynor, K., 33
Reeves, I., 155
Reisine, T., 33
Rens-Domiano, S., 33
Riondino, G., 338
Robinson, I.C.A.F., 47
Root, A.W., 286
Rosenbloom, A.L., 279
Rosenfeld, R.G., 279
Rudman, D., 133
Rudman, I.W., 133

S

Semprini, A., 338
Sethumadhavan, K., 203
Shulman, D.I., 286
Sims, S.M., 17
Strollo, F., 338
Strollo, G., 338

T

Tai, L.-R., 301
Takeshima, Y., 223
Tallent, M., 33
Taylor, J., 3
Theveniau, M., 33
Tontodonati, V., 328

U

Ultsch, M., 231
Underwood, L.E., 259

V

Vaccarello, M.A., 279
Valenti, G., 328
Vance, M.L., 66
Veeraragavan, K., 203

W

Walker, R.F., 167
Wehrenberg, W.B., 347
Weigent, D.A., 104
Wells, J., 231
Wilson, K., 279
Wise, T., 317

Y

Yagi, H., 223
Yang, S.-W., 167
Yasuda, K., 33
Yen, J.T., 317

Subject Index

A2B5 antibody, 117
Ab 1-2 antibody, 80–81
Absorptiometry, 151
Acetylcholine (Ach), 69, 329–331, 333–335
Acetylcholinesterase, 169, 350
Acromegaly, 54, 69, 94, 121, 123, 125, 144–145, 147, 223, 227–228, 239, 259, 263–264, 338, 355
β-actin, 96
Actinomycin D, 347
Action potential, 20–22, 25, 27
Acute lymphoblastic leukemia (ALL) cell line, 261, 265
Acute myelocytic leukemia (AML) cell line, 261, 265
Adenocarcinoma, 262–263
Adenohypophysis, 3–4, 87
Adenoma, 56, 124, 181, 183, 186, 188, 228, 264
Adenylyl cyclase, 17–18, 20, 22, 24, 33, 38, 40, 55–56, 157, 159
Adipocytes, 88, 249, 301–302, 304–311, 314
Adiposity, 133–134, 136, 139–143, 145, 150–152, 159, 334, 338, 342
Adolescent growth spurt, 271–273
Adrenal, 36, 87, 118
  dysfunction, 120, 188
  weight, 181–182, 187–188
Adrenalectomy, 348
Adrenergic agonists, 69, 117–118
β-adrenergic response, 40–41, 120
Adrenocorticotropin (ACTH), 87, 104, 119, 126, 193–195
  deficiency, 349–350
Affigel beads, 307–308
Agarose beads, 306–308
Agarose gel, 302
Aging
  acetylcholine and, 329
  acromegaly in, 123
  alpha-GFC and, 330–331, 333–335
  cognitive disorders in, 330
  disease resistance in, 155–156, 162–163, 187
  GH and, 68, 70, 94–95, 117, 124, 126–127, 133, 150–151, 159, 162, 167–179, 181–188, 203–204, 328–330, 338, 343
  GHRH and, 156–158, 168–172, 174–188, 328–335
  GHRP-6 and, 168–188
  GHRPs and, 203, 223
  IGF-I and, 95, 124, 156, 159–162, 338, 342–343
  normal, 150, 153, 155
  somatostatin and, 156, 158
  thymic atrophy in, 94–95, 118, 124, 126–127, 167–168, 317
  tissue regeneration and, 160–163
  zinc deficiency, 118
Alanine-scanning mutagenesis, 232, 237–239
Alcohol abuse, 134
Aldosterone, 194–195
All Children's Hospital, 170, 286–287
Allopurinol, 135
Alpha-glycerylphosphorylcholine (αGFC), 330–331, 333–335
Altitude, effects of, 195–196
Alzheimer's disease, 43
American Association of Laboratory Animal Sciences, 170

American Diabetes Association, 137
Amino acid sequence
  GH receptor, 244-245, 301
  GHBPs, 244-245
  GHRH, 9-10
  neuro-GI peptides, 9-10
  SRIF, 5
  SRIF analogs, 5-8
  SRIF receptors, 34-35, 39, 41
Amino acids
  deletion of, 74-75, 78
  GH stimulation of, 69
  insertion of, 74-75
7-aminoheptanoic acid (Ahep), 5
Amnesia, 330
Amygdala, 54
Analogs
  cAMP, 17, 23-24
  GHRH, 11-13
  SRIF, 3-7
Androgen, 145-146. *See also* Testosterone
Anemia, 193-194, 265
Angina, 134
Anorexia, 69, 349
Anti-diuretic hormone (ADH), 195
Antibodies
  A2B5, 117
  Ab 1-2, 80-81
  As1615, 304-307, 309-310, 312
  FITC conjugated, 113-114
  to GH, 90-91, 105-107, 110, 113-114, 176, 228, 247, 251, 264, 279
  to GH receptor, 93, 304
  to GHBP, 234
  to GHRH, 48-50, 53-54, 169, 172, 177, 212, 225-226
  to IGF-I, 105, 113-114, 150, 261
  to IGF-I receptor, 79-81, 92-93, 110
  to interferon γ, 92
  monoclonal, 76-77, 79-81, 92-93, 117, 244, 251
  NIDDK-anti-rGH-s-5, 105-106
  to phosphotyrosine, 76-77
  polyclonal, 105, 247, 288
  to prolactin, 176
  to prolactin receptor, 235
  to SRIF, 158, 169, 211, 216, 225-226, 349, 355
  to SRIF receptors, 38-39, 41-42
  synthesis of, 109
  to thymic epithelial cells (A2B5), 117
Anticholinergic drugs, 329
Antihypertensives, 135
Antiserum. *See* Antibodies
Apoptosis, 95, 97, 99
AR42J cells, 39
Arachidonic acid, 23
Arginine, 69, 156, 158
Arginine treatment, 350, 352-353, 355-356
Arginine-infusion test (AIT), 287, 349
As1615, 304-307, 309-310, 312
Asphyxia at birth, 225
ATP, 18, 75
ATPase, 18-20
Atrial natriuretic factor (ANF), 195
AtT-20 cells, 27, 37-40, 42
Autocrine action, 95, 97, 210, 213, 217, 261
Autofeedback, 334
Autoimmunity, 88, 126, 352
Autophosphorylation, 76-78
Autoradiography, 106, 125, 306, 310, 312

B lymphocytes, 88, 97, 110, 114, 121
Back pain, 339-340, 342
Bacteria, 87, 91-92
Barrow
  pGH treatment, 318-325
  thymic weight, 318-319, 324-325
  thymosin in, 318, 320, 322-323, 325
Benzomorphan opiate 2549, 212, 215-216
Beta-adrenergic receptor kinase (BARK), 40-41
Binding constant, 238
Binding protein
  GH (GHBP). *See* Growth hormone binding proteins
  prolactin receptor, 232-233, 235
Binding sites
  GH, 232-235, 237-238, 249, 251, 301, 303, 310, 313
  GH receptor, 232-238
Bladder cancer, 261-262

Blood brain barrier, 43
Blood
  count (CBC), 134, 141
  GH in, 47-51, 53, 57-59, 67-68, 150, 167, 171-178, 180-182, 185, 223-227, 242, 288, 330-333, 338-339, 344
  GHBP in, 242-245, 302-303
  glucose in, 68, 176-177, 182-184, 339-341, 343, 353
  IGF-I in, 133, 135, 137-141, 143-146, 150, 156, 158-162, 224, 251, 287, 328-329, 335, 339-340, 342-343
  insulin in, 339-341, 343
  pressure, 134, 152
  prolactin in, 145-146, 176, 182-184, 186, 188, 194, 288, 291
  sampling, 170-171, 175-176, 287-288, 318, 320, 330, 339, 349
  thymosin in, 318-320, 322-325
  thymulin in, 121-123
  urea nitrogen (BUN), 283
  zinc in, 123, 125
Bloom syndrome, 265
Blue sclera, 280-281
Bmax, 211
Body composition index (BCI), 175, 179, 181
Body mass index (BMI), 204, 287, 290, 292, 295, 338, 351, 353
Body weight, 177, 179-180, 182, 187, 287, 291-292
Bombesin, 3-4, 8
Bone
  age, 280, 286, 290-291, 294
  demineralization, 159
  density, 134, 141, 150-154, 167
  growth, 251
  marrow, 90, 95-98, 104, 110, 119, 261
  remodeling, 153-154, 342
  turnover, 150-152
Brain, 37, 39-40, 88, 118, 226, 328, 350
  CT scan, 349
  SRIF in, 33-34, 36, 38, 40-42
  SSTR receptors in, 9
  tumor, 266, 287
Breast cancer, 147, 260-261
Breast development, 145, 272, 287
Breech delivery, 225
Bronchiectasis, 134
Bronchodilators, 135
Buffalo rat, 121, 124
Burkitt's lymphoma, 261
Burn patients, 156, 161-162
Butyl-N-(4-hydroxybutyl) nitrosamine [N-] (BBN), 262

c-*myc* gene, 261
C-terminal, 6, 11, 36, 80, 108, 236-237, 245, 301-302, 304
Cachexia, 262
Calcitonin, 153
Calcium, 17-21, 23-24, 27-28, 33, 38, 151-153, 339-340, 342-343
  channels, 218
  excretion, 283
  ionophores, 23-24, 28
Calcium phosphate transfection method, 312
cAMP, 17-20, 22-23, 38, 42, 55-56, 66, 74, 156, 169, 175, 178-179, 185, 208-209, 213-214, 328
  analogs, 17, 23-24
  -dependent protein kinase, 18-19
  response-element binding protein, 56
Carbohydrate intolerance, 355
Carbohydrate metabolism, 70, 175
Carcinogens, 262, 264
Carcinoma. *See also* Leukemia; Tumors
  acromegaly and, 263-264
  bladder, 261-262
  bone, 150
  breast, 147, 260-261
  colon, 260, 263-264
  fibrosarcoma, 262
  lung, 7, 262
  lymphoma, 40, 107, 261
  pancreatic, 7, 9
  prostate, 143, 262-263
  pulmonary, 262
  rectal, 264
Cardiovascular disease, 263
Cardiovascular risk, 154, 338, 345
Carpal tunnel syndrome, 134, 136-140, 143-147, 152
Cartilage, 88, 145
Catecholamines, 186

cDNA
  generation of, 106–107
  GH, 107–108
  GH receptor, 301–302, 311–313
  IGF-I, 260
  IGF-I receptor, 74–75, 78
  rat library, 34, 42
Cell culture, 210, 249, 260–261, 264–265, 283, 347
  hamster ovary, 36–42
  pituitary, 9, 12, 21–22, 347
  rat, 104–105, 108, 110–111, 157, 175, 177, 207–209, 226, 228
Central nervous system, 73, 88, 118
Cerebellum, 36
cGMP, 23
CGP 23996, 34, 37, 40
Channel blockers, 24, 27
Chemotactic response, 110
Chemotherapy, 265–266
Chicken, GH receptor, 282
Children
  cranially irradiated, 287–288, 292–293, 295–296
  GH deficient, 55, 70, 90, 121, 150, 204, 206, 223, 225, 227, 247, 259, 265, 294
  glucocorticoid treated, 347, 351–352, 355
  with leukemia, 259, 261, 264–265
  with liver transplants, 351–352
  obese, 292–293
  short, 224–225, 227, 247, 259, 264, 269–276, 280–281, 283, 286, 292–296
Chinese hamster ovary (CHO) cells, 36–42, 314
Cholera toxin, 55
Cholesterol, 176, 182–183, 187–188, 338
Cholestyramine, 135
Choline acetyltransferase, 334
Cholinergic neurotransmission, 329–330, 334
Cholinergic tone, 350
Cholinesterase, 329
Chromatography, 106–107, 243, 246
Chronic obstructive pulmonary disease (COPD), 195
Churg-Strauss disease, 352
Clones
  GH receptor, 231, 244–245, 301–303, 312–313
  PRL receptor, 231
  SRIF receptor, 33–42
Clonidine, 69, 287, 350, 354
CLUSTER, 66
Codons, 75, 110, 282
Collagen, 150
Colon cancer, 260, 263–264
Colonic polyps, 263–264
Colony-stimulating factor, 97, 110, 282
Complement, 228
Computed tomography (CT) scan, 349
Computer programs, GH pulses, 66
Conductance, 18–20, 25, 28, 33, 38
Connective tissue disease, 134, 145, 352
Consanguinuity, 282
Corticosteroids, 117–118, 176, 182–183, 188, 349–351
Corticotropin releasing hormone (CRH), 193–194
Cortisol, 194, 291, 349
COS cells, 38, 40
Cosecretagogue, 168–169, 185
Cow, 22, 239, 282
Cranial irradiation, 287–288, 292–293, 295–296
Cranial lesion, 287
Craniopharyngioma, 271
Creatinine, 134
Creatinine kinase MB (CKMB), 160–161
Crosslinking studies, 301–303
Crossreactivity, 105–106, 136, 226
Cunningham chamber, 228
Current-voltage relationship, 26–27
Cushing's disease, 347
Cycloheximide, 92, 347
Cytidine 5′-diphosphocholine (CDP), 330
Cytokine, 87, 90, 95, 99, 109, 116, 314
Cytokine receptors, 88, 246
Cytolysis, 109–110
Cytosolic domain. *See* Intracellular domain

Daughaday acid ethanol extraction method, 136, 141, 144, 146

Deconvolution analysis, 67
Dentate gyrus, 35
Dermorphin (DM) analog, 212-215
DETECT, 66
Dexamethasone, 108, 347-349
Dextran, 40
DG44 cells, 37
Diabetes, 69, 88, 95, 134, 137, 143, 147, 183, 231, 251, 338
Diagnostic Products Corporation, 339
Diagnostic Systems Laboratories Inc., 339
Dialysis, 193-194, 196
Diethylstilbestrol, 69
Dimerization, 232, 234-235, 237-238, 250
DISA 2000, 136
Dissociation constant, 211, 235
Disuccinimidyl suberate, 308
Disulfide bridges, 5-6
Diuresis, 195
Diurnal variation
  GH secretion, 289, 291, 295-296
  IGF-I level, 291, 296
  prolactin pool, 291, 296
Diverticulosis, 135
DNA synthesis, 111-112, 125, 261
Dog, GH deficient, 121-122
Dopamine, 33, 42, 69, 186, 224, 287
Dosage, GH therapy, 269-270, 272, 275-276
Down's syndrome, 118, 291
Drug abuse, 134
Dwarf rats, 175, 177, 179-180, 248, 251
Dwarfism, 56, 122, 244, 279-283

Ecuador, Laron syndrome, 279-284
Elbow limitation, 280-281
Electromyography (EMG), 136-137, 145
Electrophoresis, 302, 306, 310, 312
E_m, 18-21, 24-25
Endorphins, 69, 104
Energy balance, 301
Enkephalin, 69-70, 177
Enkephalin receptor, 70
Epididymus, 302, 304-306
Epilepsy, 43
Epiphyseal fusion, 295
Epithelial cells, 52, 94, 98-99, 117, 124-125
Epitope, of GH, 232-233, 235, 237-238
Erythroid cells, 92, 95, 97-98, 110, 196, 228
Erythroleukemia cells, 260, 264
Erythropoiesis, 109-110, 193, 196, 282
Escherichia coli, 232, 252
Estradiol, 67-68, 150
Estrogen, 145-146
Ethanolamine, 308
Eunuchoid body form, 271
Exercise, 66, 69, 153, 193, 195
Exocytosis, 19, 22, 28
Exons, 77, 96, 106, 111, 281-282, 303-304
Extracellular domain, GH receptor, 58, 231-232, 237, 244-245, 282, 301-303
Extracellular fluid (ECF), 59-60

FAb fragments, 234-235
Facteur thymique serique (FTS). *See* Thymulin
Fanconi anemia, 265
Farm animals, 317
Fasting, GH and, 68-69, 83, 206, 340
Fat deposition. *See* Adiposity
Fatty acids, 68-69, 176, 183
FDC-P1 cells, 232, 235, 238-239
Feedback regulation, 3, 47, 54, 83, 88, 121, 126, 210, 328, 334
Fibrinogen, 152
Fibroblast growth factor, 105
Fibroblasts, 283
Fibrosarcoma, 262
Fischer rats, 170, 182
Fish, GH treatment, 317
Fluid retention, 146, 151-152
Fluorescence microscopy, 304-305
Follicle stimulating hormone (FSH), 105, 194
Formic acid-acetone procedure, 158
Forskolin, 38, 74
Friend leukemia virus, 264
Furlanetto method, 135-136, 141, 144, 146

G-proteins, 17–20, 22, 28, 33–34, 37–38, 41–42, 157–158, 162
Galanin, 3–4, 350, 353–354
Gastrin peptides, 3, 8, 10
Gel electrophoresis, 282, 310, 312
Gel filtration, 107, 235
Gene for
  c-*myc*, 261
  dwarfism, 56, 119, 175
  GH, 56, 74, 78, 82, 106–107, 109, 111, 347
  GH receptor, 281–282, 284
  GHRH, 54, 68, 156–157
  GHRH receptor, 57
  IGF-I, 112
  interleukins, 87
  little, 56, 175
  metallothionein GH, 124–125
  nude, 120–121
  SRIF receptor, 22, 34
Gene expression, 54, 68, 82, 112, 156–157, 227
Genentech, 93, 244, 251, 301
Genetic counseling, Laron syndrome, 284
Geriatric medicine, 184, 188
GH3 cells, 94–95, 124
Gigantism, 54
Gilt
  pGH treated, 318–325
  thymic weight, 318–319, 324–325
  thymosin in, 318–320, 322, 325
GLM procedure, 318
Glucagon, 9–10, 33, 69
Glucocorticoid hormones, 87, 108, 134
  GH and, 347–356
  treatment with, 348–356
Glucose
  blood, 68, 176–177, 182–184, 339–341, 343, 353
  /insulin ratio, 340–343
  intolerance, 68, 137, 144–145, 183–184, 187
  metabolism, 90, 301
  tolerance test, 137, 175–176
α-glycerylphosphorylcholine (αGFC), 330–331, 333–335
Glycosylation, 34, 39–40, 301
Gonadotropin insufficiency, 155, 271–272
Gonadotropin releasing hormone (GnRH), 193–194
Gout, 135
Granulocyte colony-stimulating factor (G-CSF) receptor, 232, 238
Granulocytes, 110
GRF (Growth hormone releasing factor). *See* Growth hormone releasing hormone (GHRH)
Growth
  disorders, 69, 269–276, 286, 347, 349, 355–356
  efficiency, 231, 317, 325
  glucocorticoids and, 347, 355–356
  increase, 49, 54–55, 67, 69, 73, 88, 95
  velocity, 269–270, 273–276, 279, 283, 287–290, 294–296
Growth Disorder Clinic, 270
Growth factors, 105. *See also* Insulin-like growth factor-I
Growth hormone
  aging and, 117, 124, 133, 147, 150–151, 159, 162, 167–179, 181–188
  agonists, 231, 234–235, 238
  alpha-GFC and, 330–331, 333–335
  antagonists, 231, 234, 238–239
  antibodies to, 90–91, 105–107, 110, 113–114, 228, 247, 251, 264, 279
  assay, 66–67, 105, 288, 294
  binding sites, 232–235, 237–238, 249, 251, 301, 303, 310, 313
  circadian rhythm, 68
  clearance, 66–68, 250–252
  concentration, 48–49, 53, 57–59, 67–68, 83, 150, 167, 171–182, 185, 194–195, 204–206, 209, 211–217, 223–227, 235, 242, 250–251, 270, 279–283, 288–290, 292–294, 321–322, 330–333, 338–339, 344, 347–348, 351–354
  Cushing's disease and, 347
  decrease of, 155–157, 159, 162, 167–168, 317
  deficiency, 119–120, 122, 146–147, 150, 153, 170, 248, 259, 265, 269, 271, 279, 338, 349

deficient children, 55, 70, 90, 121, 150, 204, 206, 223, 225, 247, 259, 265, 294
disease recovery and, 156, 160, 162–163
diurnal variation, 289, 291, 295–296
function, 88, 90–92, 95, 110–114, 117–118, 150, 153, 162–163, 272, 301
galanin and, 3–4, 350, 353–354
gene, 55, 74, 78, 82, 106–107, 109, 227, 347–348
GH receptor and, 251, 301, 307–311, 313–314
GHBP and, 246–252, 308
GHRH and, 168–169, 171–188, 196, 203–219, 223–228, 293, 295, 328–335, 347–354
GHRP-6 and, 168–169, 171–188, 203–208, 210–212, 215–216
GHRPs and, 203–219, 223–228
glucocorticoids and, 347–356
human, 93, 133, 135–139, 142–146, 150–154, 159, 231–239, 242–244, 247, 249–250, 269–276, 282, 307–310, 313, 338–344, 355
hypersecretion, 121, 196
IGF-I and, 95, 133, 135, 137–139, 145–147, 150–153, 158–159, 162, 167, 184–185, 260–263, 290, 295–296
immune system and. *See* Immune system
implants in pigs, 318, 322–325
inhibition of, 3–5, 7, 9, 17, 19–20, 23–24, 27–28, 33, 41, 73–74, 78–83, 119, 249
injection, 88–90, 95, 110, 259, 261–263, 265, 269–276, 279, 282, 318–322, 325
insensitivity, 279, 281–282
insulin-like response, 301, 312
leukemia and, 259, 261, 264–266
lymphocyte-derived, 104–108, 110, 112–114
mRNA, 66, 106, 109, 226–227, 328, 347–348
mutants, 93–94, 232–233, 235, 238–239
nutrition and, 66–68, 83
peak/pulse, 48–51, 54–55, 57–59, 66–69, 73, 124, 169, 206, 225, 247, 251, 259, 264, 272, 286, 288–289, 292–296, 325, 328, 339, 351–354
pig research, 317–325
porcine, 91, 93, 318–321
prolactin receptor binding, 93, 231–233, 235, 238
recombinant, 90–94, 150–154, 159, 167, 261, 269–276, 318, 338–344, 355
secretion, 3–5, 8–9, 12, 17–20, 23–24, 27–28, 47–61, 66–70, 73–74, 78, 82–83, 94, 108–109, 111, 114, 126, 150, 153, 156–159, 162–163, 167–180, 183–188, 193–196, 203–219, 223–228, 248, 259, 264, 286, 288–296, 328–335, 347–356
spontaneous release studies, 286–296
SRIF and, 156, 158, 169, 171, 207–212, 216–219, 224–225, 228, 328–329, 334, 347, 349
stored, 22, 49, 56, 69–70, 156–157, 168, 179, 182, 185–186, 188, 195–196, 235
structure, 106–108, 231, 233, 236–238
sustained-release implants, 318, 322–325
synthesis, 55–56, 66, 97, 104–109, 114, 156, 210, 226–227, 329, 347, 356
therapy, 119–122, 125, 133–147, 150–154, 159, 163, 167–168, 252, 259, 264–266, 269–276, 279, 282–283, 286, 334, 338–344, 355–356
thymosin β4 and, 317–319, 324
tumorigenicity, 259, 261–265
variants, 93, 231, 238–239, 242, 301
zinc complex, 233, 235
zinc deficiency and, 118–119
Growth hormone binding protein (GHBP), 58–60, 93, 159, 280–281, 284, 232–238, 242–252, 301, 303, 307, 328–329
assay, 246–247, 309
cloning of, 244–246
concentration, 244, 246–251, 312

Growth hormone binding protein (GHBP) (*continued*)
 dimerization, 232, 234-235, 237-238, 250
 GH and, 232, 246-252, 308
 GH receptor and, 235, 244-249, 251
 history of, 243-246
 mRNA, 245, 302-303
 recombinant, 246, 249-252
 sequence, 244-245
 therapy with, 252
 transfectants, 312
Growth hormone binding protein complex, 232, 235-238, 242, 249-252
 clearance, 250-251
 species, 250
Growth hormone insufficiency (GHI), 269-271, 275-276
Growth hormone receptor, 58, 60-61, 88, 93, 107, 114, 125, 231-235, 237-238, 305
 antibodies to, 304-307, 309-310, 312
 cDNA, 301, 311-313
 cloning of, 231, 244-245, 301-303, 312-313
 gene for, 281-282, 284
 gene therapy for, 284
 GH and, 301, 307-311, 313-314
 GHBP and, 232-238, 244-249, 251
 mRNA, 245, 248, 302-304, 307, 314
 prolactin receptor and, 245
 proteolytic fragment, 245, 302, 314
 sequence, 244-245
 short isoform, 303-305, 307-314
Growth hormone receptor deficiency (GHRD), 279-284
Growth hormone releasing factor (GRF). *See* Growth hormone releasing hormone (GHRH)
Growth hormone releasing hormone (GHRH)
 aging and, 168-172, 174-188
 analogs, 11-13, 157, 226
 antibodies to, 169, 172-173, 212, 225-226
 arginine and, 156
 function, 3-4, 17-24, 47-48, 51, 53-56, 61, 66-67, 73, 108-109, 114, 328-329, 334
 GH and, 156-159, 168-188, 193, 196, 203-219, 223-228, 293, 295, 328-335, 347-354
 GHRP-6 and, 168-188, 204-205, 207-208, 212, 215-216
 GHRPs and, 203-205, 207-210, 212-219, 224-229
 glucocorticoids and, 347-348, 350-351
 IGF-I and, 74, 328, 335
 infusion, 49-51, 54-57, 68, 225, 227
 injection, 194, 196, 287-288, 293, 330-335, 351-354
 mRNA, 168, 185
 pulses, 51-52, 54, 67-68
 secretion, 47, 49-50, 52, 68, 104, 156, 158-159, 168, 295, 328-329, 350
 sequence, 9-10
 as signal, 47-61
 SRIF and, 187, 207-210, 212, 216-219, 225, 228
 synthesis, 156-158
 therapy, 55, 156, 168-188, 335, 348
 tone, 54
Growth hormone releasing hormone receptor (GHRH-R), 17-18, 56-57, 70, 73, 114, 175, 178-179, 185-186, 203, 207, 328-329, 334, 347, 355
 alteration of, 157-159, 162
 antagonist, 172-174
 GHRP-6 and, 169, 172-174, 177
Growth hormone releasing peptide-1. *See also* Growth hormone releasing peptides
 GH release, 203-205, 208-209
 GHRH and, 209-210, 217
Growth hormone releasing peptide-2, 204
Growth hormone releasing peptide-6 (GHRP-6)
 aging and, 168-188
 binding site, 169
 GH and, 156, 168-169, 171-188, 203-208, 210-212, 215-216
 GHRH and, 156, 168-188, 204-205, 207-208, 212, 215-216
 GHRH receptor and, 169, 172-174, 177
 intraperitoneal injection, 171, 175-176

rat studies, 156, 168–188
SRIF and, 187
Growth hormone releasing peptide-like ligand, 206–208, 210, 212–213, 216–219
Growth hormone releasing peptides (GHRPs)
binding sites, 210–211
GH and, 69–70, 203–219, 223–228
GHRH and, 53–54, 224–229
infusion, 53, 225
injection, 223–224, 226–227
intranasal administration, 206, 223-224
models of, 207–219
natural ligand, 206–207
potency, 203–205
SRIF and, 207–213, 216–217, 225-226, 228
Growth hormone releasing peptides receptors, 206–207, 227
Guanine nucleotide binding proteins. *See* G-proteins
Guanylyl nucleotides, 19, 37, 157
Guanylyl cyclase, 23
Gut, 33–34
Gynecomastia, 136–139, 144–147

Hassal's corpuscles, 125
Healing impairment, 160–163
Heart failure, 134
Height, 287, 290, 294
adult, 270–276, 280, 283
HEK-293 cells, 39
Helix structure, 237
Helodermin, 10
Hematocrit, 134
Hemodialysis, 193–194, 196
Hemolysis, 228
Hemopoiesis, 88, 92, 95, 97, 196
Hepatocytes, 301–302
Hexapeptides, 6, 8–9, 37, 54, 69. *See also* Growth hormone releasing peptide-6
High altitude, 195–196
HIK-293, 37
Hip, maintenance of, 152–153
Hippocampus, 34
Histamine, 69
Histomorphometry, 153
HL60 cell line, 261
Hodgkin-derived cell line, 265
Homolog-scanning mutagenesis, 232
Homology, GH receptor, 282
Hormones. *See also* Adrenocorticotropin; Gonadotropin
glucocorticoid, 74, 87, 108, 134, 347-356
gonadal steroids, 60, 67–68, 73, 117-118
mimics, 231
neuroendocrine, 87–88, 104
Hospital
All Children's, 170, 286–287
Hotel Dieu, 161–162
Notre Dame, 160–161
for Sick Children, London, 270
Hs-578T cell line, 260
Human. *See also* Children; Men; Women
adrenal dysfunction, 348–352
aging, 155–158, 167, 169, 328–335, 338–339, 343–344
alpha-GFC treatment, 330–331, 333-335
autopsy, 168
GH, 231–239, 242–244, 247, 249–250, 348
GH receptor, 282
GHRPs in, 203–207, 223–224
glucocorticoid therapy, 349–351
pituitary, 74, 347
/rat cell transfection, 74–82
sex differences, 68
short, 196, 223–225, 227, 247
somatostatin receptor genes, 34
thymus, 122
Hydrogen peroxide, 91–92
Hyperadrenalism, 348–352
Hyperglycemia, 69, 136–138, 145–147, 193, 340
Hyperimmunity, 121
Hyperinsulinemia, 69, 340
Hyperlipemia, 135
Hyperpolarization, 20, 23, 25, 27–28
Hyperprolactinemia, 186, 193, 239
Hypertension, 135, 146, 338

Hypoadrenalism, 348-350
Hypobaria, 195
Hypoglycemia, 193, 224, 281, 283
Hypokalemia, 283
Hypomuscularity, 280
Hypophyseal portal system, 51-52, 68, 73, 193-195, 207, 210, 328
Hypophysectomy, 88-92, 95, 110, 121, 125, 163-164, 248, 251, 261
Hypophysis, 3, 47, 51-52, 87, 119
Hypopituitarism, 119, 265
Hyposomatomedinemia, 133, 146-147
Hyposomatotropinemia, 68, 339, 343
Hypothalamic-hypophyseal axis, 193-195, 218
Hypothalamus, 9, 47, 52, 61, 66-69, 73, 108, 110, 114, 156-158, 168-169, 177, 185, 193-194, 204, 206-208, 210-213, 216, 218-219, 224, 226, 328-329, 334, 338, 348, 350, 355-356
Hypoxemia, 194-195
Hypoxia, 194-195

Idiopathic short stature (ISS), 269, 272-273, 275-276
IM-9 lymphocytes, 109, 235, 238, 245
Immune system
- balance, 116-117
- complexes, 306-309
- function, 167-168, 264, 325
- GH and, 88-95, 97, 99, 104, 106-114, 117-127, 155-156, 160-163, 167-168, 187
- IGF-I and, 88, 92, 95-99, 108-109, 112-114, 121, 125, 155-156, 160-162
- response, 87-88, 90, 155

Immunoblotting, 39, 42
Immunodeficiency, 119, 126, 167
Immunofluorescence, 106-107, 113
Immunoglobulin. *See* Antibodies
Immunohistochemistry, 22
Immunoneutralization, 212
Immunoprecipitation, 37-39, 76-77, 79-81, 145, 305-308, 310, 312
Immunoregulation, 87-88, 97-98
Immunosuppression, 351-353, 355
Impotence, 193, 338
Informed consent, 134, 287, 349
Initiation codon, 110
Inositol diacylglycerol, 23
Inositol polyphosphate, 23
Insulin, 9, 33, 41, 50, 68-69, 88, 106, 152, 176, 182-183, 187-188, 193, 224, 231, 261, 283, 287, 295, 301, 312, 338-342, 349
Insulin receptor, 77
Insulin resistance, 338, 342-344, 355
Insulin-like growth factor I (IGF-I)
- aging and, 95, 124, 156, 159-162, 342
- antibodies to, 113-114, 261
- assay, 105, 136
- binding of, 75-80
- binding proteins and, 145, 158-159, 194, 196, 244, 249, 252, 328-329
- binding protein 2, 281, 283
- binding protein 3, 280-281, 283-284
- carpal tunnel syndrome and, 137-140, 145-146
- disease and, 160-162
- displacement, 80
- diurnal variation, 291, 296
- function, 88, 99, 110, 150, 155, 162-163, 210
- gene, 112
- GH and, 3-4, 47, 66-67, 73, 82, 92, 108, 112-114, 125, 133, 135, 137-139, 145-147, 150-153, 158-159, 162, 167-168, 184-185, 260-263, 290, 295-296, 328, 335
- GHRH and, 178, 279-281, 283-284
- GHRP-6 and, 178
- gynecomastia and, 138-139, 147
- human, 283
- hypoglycemia and, 283
- immune system and. *See* Immune system
- injections, 283
- insulin and, 283, 295
- internalization, 77-78
- leukocytes and, 95-97, 112-114
- level, 133, 135, 137-146, 150, 156, 158-162, 176, 178, 182-183, 185, 188, 194, 196, 204, 224, 251, 279-281, 283, 286-288, 290-296, 328-329, 335, 338-340, 342-344

mRNA, 95-97, 113-114, 260-261
as osteotrophic agent, 150
pituitary action, 73-74
recombinant, 92, 106, 153, 283
secretion, 92, 183, 355
SRIF and, 328
suppression, 83
synthesis, 68, 95-99, 150, 355
therapy, 77, 79, 252, 283-284, 334, 344
thymulin and, 121
tumors and, 259-265
Insulin-like growth factor I receptor
antibodies to, 110
in bone marrow, 97-98
cDNA, 74-75, 78
function of, 74, 82-83, 328
hybrid, 78-82
immune system and, 114
internalization of, 77-78
mRNA for, 96
mutants of, 74-82
nonfunctional, 78-79
overexpression of, 74, 78, 80-81
phosphorylation of, 76
subunits, 75, 77-78, 80
transfection of, 74-76, 78
tumor cells, 147, 260-261
type I, 260
Insulin-like growth factor II, 73-74, 105-106, 136, 150, 159, 280-281, 283-284
Insulin-tolerance test (ITT), 349
Interferon γ (IFN-γ), 90-92, 95, 109, 168
Interleukins, 87-88, 95, 97, 282
Interleukin receptors, 246, 252, 314
Interleukin-1 (IL-1), 87, 90, 98-99, 109
Interleukin-2, 90, 109, 124, 168, 282
Internalization of IGF-I, 77-78
Intracellular domain, GH receptor, 74-75, 232, 237, 245-246, 282, 301-303
Intrauterine growth retardation (IUGR), 269, 273-274, 276
Introns, 34, 282, 347
Iodination, 37, 39-40, 75, 77-79, 107, 247, 307-310
Ionic mechanisms, 17-24, 38, 40-41
IR3 (alpha), 79-81
IRMA method, 331
IRS-1 protein, 77-78
Israel, Laron syndrome, 279

Jackson Laboratories, 175
Japan, leukemia and GH, 265
Joint disease, 135

Kd, 74-76, 211, 238, 260
Ketamine, 171, 176
Kidney, 34, 37, 41, 261, 283
atrophy, 167, 187
failure, 193-194
function, 167, 187, 250, 334
-pituitary implant, 124
transplant, 355
weight, 181-182, 187-188
Killer cells, 88, 109-110, 114

L-428KSA cell line, 261
L-dopa, 33, 42, 69, 186, 224, 287, 328
Lactation, 231
Lactotroph, 183, 186
Lanreotide, 6, 8
Laron syndrome, 244, 279-283
Lean body mass, 133-137, 139-143, 147, 151-154, 167, 175, 187, 334, 338
Leukemia, 143, 259-261, 264-266, 287
cells, 232, 235
GH and, 259, 261, 264-266
IGF-I and, 260-261, 264-265
relapse, 265-266
Leukocytes
GH production, 93, 97, 104, 107-109
IGF-I production, 95-97, 112-114
interleukin production, 87-88
prolactin receptor, 94, 97
superoxide production, 92
Leukotrienes, 23
Ligand-mediated immuno-functional assay (LIFA), 247, 251
Lipids, 70, 159, 187. *See also* Adipocytes
Lipolysis, 251

Lipoprotein, 95, 98, 152
5-lipoxygenase, 23
Liquid scintillation counter, 105
Listeria monocytogenes, 92
Little gene (mouse), 56, 175, 179
Liver, 73, 88, 134, 145, 261
  GH receptor, 244, 248, 301-302
  transplant, 351-352
Lung
  carcinoma, 7, 262
  COPD, 195
Lupus erythematosus, 352
Luteinizing hormone (LH), 104, 193
Luteinizing hormone releasing hormone (LHRH), 117-118, 228
Lymphocytes, 104, 107, 110-111, 114, 120-121, 124
Lymphoid tissue, 95, 97-98, 104-105, 109, 116, 119-120, 264
Lymphoma, 40, 107, 261

Macrophages, 87, 90-92, 94-98, 109-110, 114, 168, 238
Magnetic resonance imaging (MRI), 225
Mammary cells, 88
Mammogram, 137
Mammotroph, 186-187
MCF-7 cell line, 260
MDA-231 cell line, 260
Meat animals, 317
Median eminence, 52-53, 69, 207, 217
Medical College of Wisconsin, 135
Mediterranean region, 279, 339, 343
Melatonin, 117-118
Men, aging, 133-147, 151, 154, 156, 167, 203-205, 223, 335, 339
Mental illness, 134
Messenger, second, 17-18, 20, 22-25, 28
Metabolism, 67, 69-70, 152, 155, 159-163, 175, 187-188, 252, 301, 342
Metallothionein growth hormone gene, 124-125
Metastasis, 262-263
^{35}S-methionine, 39, 76, 81, 305-306, 312
Methionine enkephalin amide, 177
α-methyl-ρ-tyrosine, 172-173
Methylscopolamine, 69
Milk production, 52, 88, 317, 325
Mitogens, 107, 109, 119-121, 124, 147, 261
MK 678 cells, 34, 37-42
Monkey, pituitary, 347
Monocytes, 88, 94
Monosodium glutamate, 50
Morphine, 69, 174
Mountain sickness, 195
Mouse
  aging, 95
  GH and, 243, 250
  GH receptor in, 282
  hypoxemic, 195
  IgG, 80
  leukemia prevention, 265
  leukocytes, 97
  little, 56, 175, 179
  mRNA, 245
  nude, 120
  Snell dwarf, 119-122
  SRIF receptor genes, 34
  3T3-Li cells, 313-314
  thymectomized, 119, 126
  thymus, 168, 265
  transgenic, 54-56, 124-125, 239
mRNA
  GH, 66, 74, 106-107, 109-111, 226-227, 328, 347-348
  GH receptor, 58, 245, 248, 302-304, 307, 314
  GHBP, 245, 302-303
  GHRH, 168, 185, 328-329, 334
  IGF-I, 95-97, 113-114, 260-261
  knockout studies, 38
  probes, 252
  prolactin receptor, 245
  SRIF receptors, 34, 36, 39, 41-42
Multiple pituitary hormone deficiencies (MPHD), 271-272
Muscarine, 69
Muscle, 27, 88, 145, 150
  atrophy, 342
  cells, 301-302
  mass, 70, 153, 159, 167
  strength, 338-340, 342-343
Mutagenesis, 39-40
  alanine-scanning, 232
  homolog-scanning, 232

Mutants. *See also* Gene for
of GH, 93–94, 232–233, 235, 238–239
of GH receptor gene, 282
of IGF-I receptors, 74–80
nonsense, 282
Mutational analysis, 232, 236–239
Mycobacterium avium, 94
Mycostatin, 111
Myeloid cells, 90, 92–93, 95, 97, 232, 235
Myocardial infarction, 143, 156, 160–162

N-terminal, 6, 8–9, 34, 36, 40, 108, 232, 237, 245, 301, 303–304
D-β-naphthyl-2-Ala (Nal), 6–8
Nalorphine, 69
Naloxone, 69
National Pituitary Agency (NIDDK), 105–106, 176
Natriuresis, 195
Natural killer cells, 109–110, 114, 168
Nb2 lymphoma cells, 107, 239
Negative feedback, 3, 66–67, 157, 184–185, 188, 210, 325, 328
Nensorb cartridges, 106, 111
Neomycin, 75, 312
Neoplasm, 126, 134, 259, 263–264
Nephropathy, 187
Nernst relation, 27
Nerves, 116, 136–137
Neuroendocrine-immune interactions, 87–99, 104, 106–114, 116–127, 155–156, 167–168
Neuro-GI peptides, 3–4, 9–10
Neurologic examination, 135
Neuromedin B, 8
Neurons, 27, 33, 40, 49, 52–54, 61
Neurotransmitters, 33–34, 38
Neutrophils, 90, 92–95, 109
NIDDK-anti-rGH-S-5 antiserum, 105–106
Nifedipine, 27
Nitrogen
balance, 161, 355
excretion, 283
metabolism, 159–162
retention, 151–154, 167, 231
Nitroglycerine, 135
NK cells, 109–110, 114, 168
Nonsense mutation, 282
Noonan's syndrome, 291
Norepinephrine, 69
Northern analysis, 34, 36, 41–42, 97, 107, 227, 302
Notre Dame Hospital, 160–161
Nuclear magnetic resonance, 7
Nucleotide sequences, 106–107, 301
Numbness, 135, 145
Nurse cell, 98
Nutrition, GH and, 66, 68, 83, 295

Obesity, 68–70, 157, 344, 281, 287-288, 292-293, 295–296
Octadecasilyl (ODS)-silica extraction, 105
Octapeptides, 5–7, 37
Octreotide (SMS 201-995), 6, 8
Oligodeoxynucleotide (ODN), 105–106, 110–112
Oligosaccharides, 39–40
Opiates, 3–4, 8, 69, 117–118, 211–212
Osmolality, 195
Osteoblasts, 150
Osteocalcin (OC), 151–152, 335, 339–343
Osteomyelitis, 134
Osteopenia, 281, 334, 338
Osteoporosis, 120, 146, 150, 153, 280, 338, 342
Osteosarcoma, 150
Ovary dysgenesis, 120
Oxandrolone, 274–275
Oxygen metabolites, 90
Oxytocin (OT), 52–53

PA-II tumor, 262–263
PACAP-38, 10
Pain, in arm or wrist, 135, 145
Pancreas
beta cells, 301
carcinoma, 7, 9
dysfunction, 183, 188
islets, 33
SRIF receptors, 39, 41
VIP receptor, 11

Papilloedema, 283
Paracrine action, 74, 96-97, 210, 213, 217, 261
Parathormone (PTH), 150-151, 339-343
Parathormone/osteocalcin ratio, 340, 342
Pasteurella multocida, 91
Penicillin, 105, 111
Penis, micro, 280-281
Pentobarbital, 171, 175-176, 211, 217
Peptides, 4-5, 7, 11. *See also* Growth hormone binding protein; Growth hormone releasing peptide
Perforated patch method, 21-22, 24-26
Pertussis toxin, 20, 22, 28, 33, 37-38
Phagocytosis, 90-92, 193
Phorbol myristate acetate (PMA), 92
Phosphatase, 33, 152
Phosphate, 283, 335
Phosphocalcic metabolism, 159
Phosphodiesterase, 17, 23
Phosphoinositide, 22
Phospholipase, 22-23
Phosphorylation, 18-20, 40-41, 76-78, 235
Phosphotyrosine, 76-77
Phytohemagglutinin (PHA), 109
Pig. *See also* Barrow; Gilt
  castrated, 317, 325. *See also* Barrow
  Duroc, 317
  GH, 91, 236-237
  GH implants, 318, 322-325
  GH injections, 318-322, 325
  GH receptor, 282
  hypothalamus, 206
  macrophages, 91-92
  Yorkshire, 317
Pineal gland, 118
Pirenzapine, 69
Pituitary. *See also* Somatostatin receptors; Somatotrophs
  adenoma, 56.
  anterior, 177, 228-229, 349
  atrophy of, 56, 119, 167, 184
  cell culture, 6-9, 12, 21-22, 38-40, 207-210, 226-228
  cells implanted in rats, 94-95, 124
  enlargement, 54, 181
  GH receptors, 5, 9
  GH secretion, 4, 6-7, 12, 56, 150, 333
  GH stores, 22, 49, 56, 69-70, 156, 175
  GHRH hyposensitivity, 157, 162, 168, 185, 328
  IGF-I in, 66
  membrane, 157, 169, 218
  multiple hormone deficiences (MPHD), 271-272
  neoplastic, 126
  pseudoposterior, 225
  rat. *See* Rat
  removal of, 88-92, 110
  stalk transection, 225, 227
  thymus interactions, 117-127
  tumor, 74, 182-183, 186
  weight, 181-183, 186, 188
Placebo treatments, 89-91, 151-152, 262-263, 273, 275, 339, 352-353
Placental lactogen, 238
Plaque formation, 228-229
Plasma. *See* Blood
Plasma membrane, 17-18, 20-24, 28, 38, 40, 91, 260, 314
Platelet-derived growth factor, 105
Poly-L-lysine, 228
Polyacrylamide gel, 306, 310, 312
Polyadenylation, 107, 302
Polyclonal antibodies, 105, 247, 288
Polymerase chain reaction (PCR), 97, 107, 282
Polymorphonuclear neutrophils (PMN), 92-95
Polymyalgia rheumatica, 352
Posttranslation processing, 39-40, 302, 314
Potassium, 18-28, 33, 38, 40-41, 136, 151
Pre-adipocytes, 249
Prednisone, 353-354
Pregnancy, GH and, 248
Prepro-IGF-I peptide, 97
Priming
  of macrophages, 90-92, 168
  of neutrophils, 92-94
Probenecid, 135
Procollagen, 152, 283
Progeria, 338

Prolactin (PRL)
acromegaly and, 121
binding proteins, 93, 231-233, 235, 238
deficiency, 119
diurnal variation, 291, 296
function of, 88, 91-94, 117-118, 124-125
GH3 implants, 124
increase in, 123, 155
injection, 88, 91-92
level, 145-146, 176, 182-184, 186, 188, 194, 288, 291
recombinant, 92
secretion, 126, 177, 186
stored, 182-183
synthesis, 97
Prolactin receptor
cloned, 231
GH and, 88, 93-94, 125-126, 231-233, 235, 238
GH receptor and, 231, 245, 282, 314
mRNA, 245
Prolactinemia, 239
Prolactinoma, 94, 123, 125
Promoter, GH gene, 106, 111
Propranolol, 287
Prostate
carcinoma, 143, 262-263
hypertrophy, 135
Protein-A, 228, 306-308
Protein
catabolism, 355
GTP binding, 17-19
kinases, 18-19, 23-24, 56, 92
synthesis, 92, 114, 155, 186
wasting, 355
Proteinase K, 106
Proteinuria, 134
Proteolysis, 245, 302, 314
Proto-oncogene c-*myc*, 261
Psoriasis, 135
Puberty, 67-68, 124, 167, 270-272, 276, 280-281, 286-288, 290, 294, 296, 325
precocious, 287-288, 292-293, 295
Pulmonary disease, 134-135, 195
PULSAR, 66
Pyridostigmine (PD), 169, 211, 239, 350-352, 355-356

Quality of life, 167

Rabbit
antibodies (As1615), 304-307, 309-310, 312
GH receptor, 244, 282, 302
GHBP, 244
Radiation, carcinogenic, 264
Radioimmunoassay (RIA), 58, 105, 109, 111-113, 146, 158, 171-172, 175-176, 194, 246-247, 260, 280-281, 283, 288, 318
Radiolabeling, 75-81, 243, 246, 269, 305, 307-310
Radiotherapy, 265, 287-288, 292-293, 295-296
Rapid-eye-movement (REM) sleep, 68
Rat
β-actin, 96
adenocarcinoma, 262-263
adipocytes, 302, 304-307, 314
adrenalectomized, 348
aging, 94-95, 124, 156-158, 168-188, 328
analog screening, 8-9, 12
anesthetized, 171-172, 175-176
antibody treatment, 225-226
bladder cancer, 261-262
brain, 34, 36, 37, 39, 41-42
Buffalo, 121, 124
cachexia, 262
cDNA library, 34, 42
cell culture, 104-105, 108, 110-111, 157, 175, 177, 207-209, 226, 228
conscious, 47-50, 53-54, 58-59, 174, 217, 225-226, 348
dexamethasone, 355
diabetic, 95
dwarf, 56-57, 59-60, 175, 177, 179-180, 248, 251
Fischer, 170, 182
GH and, 106, 156-157, 168-169, 171-188, 207-208, 212, 214-217, 242, 248, 250-252, 348
GH receptor, 248, 282, 302, 304-305 312, 314
GHBP in, 247-248, 250-251, 302

Rat (*continued*)
GHRH and, 52, 156–157, 168–188, 207–208, 212, 214–217, 226, 328
GHRH receptors in, 157, 169, 172–173, 178–179, 185–186
GHRPs and, 156, 168–188, 207–208, 214–217
glucocorticoid-treated, 348–349, 355
-human hybrid IGF-I receptors, 78–82
hyperadrenal, 348–349
hypophysectomized, 88–90, 91–92, 95, 121, 125, 248, 251, 261
hypoxemic, 195
lymphocytes, 105, 107
macrophages, 94, 96–97
mRNA, 245, 248
Nb2 cells, 239
necropsy, 176, 181
pentobarbital study, 217
pituitary, 6–9, 12, 22, 36, 39, 74–82, 156–157, 175, 177, 214, 328, 347
prolactin receptor, 245
S. typhimurium injection, 88–90
sex differences, 48–49, 51–52, 54, 57, 60, 248
somatotrophs, 18–19, 21, 25
Sprague-Dawley, 104, 111
SRIF and, 158, 209, 216–217, 226, 334
transfected pituitary cells, 74–82
tumorigenesis, 263
Recombinant proteins, 87, 92, 106, 153, 283
Renal. *See* Kidney
Restriction enzyme analysis, 107–108
Reverse hemolytic plaque assay (RHPA), 228
Reverse transcription, 107
Rhodamine labeling, 304–305
RIPA buffer, 310, 312
RNA. *See also* mRNA
polyadenylated, 107, 302
polymerase, 282
synthesis, 125
total, 96, 106, 282, 302
RNase protection assays, 96–97
RPMI-1640 medium, 105, 111–113
S49 lymphoma cells, 40
Salmonella typhimurium, 88–90
Salt load, 195
Scandinavia, GH trials, 339, 343
Scatchard analysis, 75–76, 211, 301, 307–308, 310, 313
Scintillation counter, 111
Sclera, blue, 280–281
Scleroderma, 352
SDS-polyacrylamide gel, 76
Second messenger, 17–18, 20, 22–25, 28, 175, 177, 179, 186
Secretagogue, 19, 23, 27–28, 55–56, 168, 170–171, 175, 177–179, 181, 183–186, 188, 229
Secretin, 9–10
Senescence, 167, 170, 174, 184, 187–188, 342–343. *See also* Aging
Sequences
amino acid, 5–10, 34–35, 39, 41, 244–245, 301
nucleotide, 106–107, 301
Serotinin, 69
Serum. *See* Blood
Sex differences, GH response, 68, 154, 160, 223, 248, 263, 290, 317–318, 320–322, 325
Sex hormone binding globulin (SHBG), 339–340, 342–343
Short stature, 196, 223–225, 227, 247, 259, 264, 269, 271–272, 279–281, 286, 292, 295–296
Signal, GHRH
neuronal, 52–54, 61
secretory, 47–52, 57–60
trophic, 54–57
Signal peptide, 281–282
Signal transduction, 21, 33, 56, 74, 77, 81–82, 178, 245–246, 252, 314, 329
Silver-Russell syndrome, 273
Skeletal maturation, 273, 275–276, 287
Skin thickness, 119, 152, 159
Skin-graft survival, 119
Sleep, GH and, 66, 68, 287–289, 292, 294–296, 334
Slot blot analysis, 113
SMS-201-995 octapeptide, 37, 41–42
Snell dwarf mice, 119

Sodium
channels, 18–21
excretion, 151–152, 283
Solid-phase synthesizer, 105
Somatomammotrophs, 186
Somatomedin C. *See* Insulin-like growth factor I
Somatopause, 334
Somatostatin
action, 17, 19–20, 22–28, 33, 47–48, 57–58, 60–61, 66–67, 70, 73, 108–109, 114, 207–210, 212, 216–217
aging and, 156, 158, 169
analogs, 3–8, 37, 228
antagonist, 170
antibodies to, 158, 169, 211, 216, 225–226, 349, 355
gene expression, 68
GH and, 156, 158, 169, 207–212, 216–219, 224–225, 228, 328–329, 347, 349
GHRH and, 187, 207–210, 212, 216–219, 225, 228
GHRPs and, 187, 207–213, 216–217, 225–226, 228
IGF-I and, 328
infusion of, 49–51, 67, 225
level, 218, 348
pulse, 67
pyridostigmine and, 350
secretion, 48, 51–54, 57, 68, 158–159, 162, 187, 216, 226, 334, 350
sequence, 5
species differences, 50–51
suppression of, 50, 52, 57, 171, 175, 207, 212–213, 334
synthesis, 158
tone, 48–49, 51, 54, 158–159, 224, 329, 334, 349, 351, 355–356
Somatostatin-14, 208–210
Somatostatin-28, 36–37, 41, 208–210
Somatostatin receptors
antagonists, 8–9
antibodies to, 38–39, 41–42
clones, 34–42
function, 19–20, 22, 28, 73, 114
genes, 22, 34
mRNA, 34, 36, 39, 41–42
subtypes, 4, 7, 9, 20, 22, 33–42
Somatotrophs, 17–22, 24–28, 49, 52, 54–56, 66, 73–74, 78, 108, 156, 158, 169–170, 186, 207, 210, 216, 227–229, 235, 328, 330, 334, 347–348, 350
Somatotropin. *See* Growth hormone
Southern analysis, 75, 107
Spleen, 90, 96, 104, 109, 125
hypoplasia, 120
involution of, 264
Splenocytes, 111–114, 119, 168, 264
Splice junction, 282
Splicing, 302–304
Spontaneous release studies, GH, 286–296
Sprague-Dawley rats, 104, 111
SRIF. *See* Somatostatin
SS. *See* Somatostatin
SSTR clones of SRIF receptor, 7, 9, 34–42
Stem cells, 110, 117
Steroids. *See also* Adrenocorticotropin; Glucocorticoid hormones
gonadal, 60, 67–68, 73, 117–118, 271–272, 295, 325, 342
oxandrolone, 274–275
therapy, 272, 348–356
STOP transfectants, 75–77, 80–82
Streptomycin, 105, 111
Sulfhydryl group inactivators, 245
Superoxide anion (O_2^-), 90–95, 99, 109–110
Superoxide catalase, 91
Superoxide dismutase, 90
SYBYL biopolymer modeling, 11
Synergism, 178, 185, 203–205, 207–208, 213–216, 218, 226, 228

T lymphocytes, 87–88, 93–95, 97–99, 109–110, 114, 117–118, 121, 124, 126, 167–168, 187
helper/killer ratio, 187
Tanner stages, 287, 290, 294, 296
Testis, 194
Testosterone, 68, 194, 288, 290, 294–295
12-0-tetradecanoylphorbol-13-acetate, 74

Tetradotoxin (TTX), 21
L-2,3,4,5,-tetrahydro-β-carboline-4-carboxylic acid (Tcc), 6–7
TF-5 thymic extract, 126
Theophylline, 156
3T3-F433A pre-adipocytes, 249
3T3-L1 cells, 311–313
Thymectomy, 116, 119, 126
3H-Thymidine, 105, 261, 264
Thymocytes, 104–105, 118, 125–126, 264
Thymopoietin, 117
Thymosins, 117
Thymosin α1, 117, 317–320
Thymosin β4
  barrows, 318, 320, 322–323
  concentrations, 117, 318–320, 322–325
  GH and, 317–319, 324
  gilts, 318–320, 322
Thymulin (ZnFTS), 117–118, 120–123, 125
Thymus. *See also* Thymectomy
  aging and, 94–95, 118, 124, 126–127, 167–168, 317
  enlarged, 121
  epithelial cells, 94, 98–99, 117, 124–125
  GH and, 124–126, 167–168
  hypotrophy, 119–121
  IGF-I production, 98–99
  involution of, 94–95, 117–118, 124, 167–168, 264–265, 317
  -pituitary interactions, 117–121
  regrowth, 118, 124
  weight, 122, 126, 181–182, 187–188, 317–319, 324–325
  zinc and, 117–119, 121, 123, 126
Thyroid, 73, 117–118, 120, 134, 152, 155, 340, 342–343
Thyroid stimulating hormone (TSH), 119, 193–194, 339, 343
Thyrotropin, 104–105
Thyrotropin releasing hormone (TRH), 108–109, 193–194, 228
Thyroxine, 118–119, 339, 343
Tinel's test, 135–137, 145
Tissue regeneration, 155–156, 160–163
Tooth decay, 281
Transcription, 36, 54, 66, 74, 97, 109, 186, 347
Transcripts, 96, 106–107, 302–304
  antisense, 96, 106, 110–112
Transduction, signal, 21, 33, 56, 74, 77, 81–82, 178, 245–246, 252, 314, 329
Transfection, 9, 74–75, 77–80, 82, 232, 235, 239, 311–314
Transforming growth factor, 105
Transgenic mice, 54–56, 124–125, 239
Translation, 92, 114, 155, 186
Transmembrane domain, GH receptor, 34–35, 41, 58, 74–75, 82, 231–232, 245, 282, 301–304, 314
Triiodothyronine, 74
tRNA, 96
Trunk/limb disproportion, 271, 280–281
Trypan blue exclusion, 105
Trypsin, 310–311
Tuberculosis, 134
Tumor necrosis factor α (TNF-α), 90, 109
Tumorigenesis, 259, 261, 263–266
  GH and, 259, 261–265
  IGF-I and, 259–262, 264–265
Tumors. *See also* Carcinoma
  brain, 266, 271, 287
  GH-producing, 125
  GHRH secreting, 68
  growth of, 259–262, 264
  pituitary, 74
  relapse rate, 265–266
Turner syndrome (TS), 269, 274–276, 287, 291–293, 295
Tyrosine^{950}, 74, 78–82
Tyrosine, phosphorylation of, 76–77
Tyrosine kinase domain, 75
Tyrosine phosphatase, 33

U-factor, 213, 216–218
United States
  GH trials, 339, 343
  Laron syndrome, 279, 283
  leukemia and GH, 265
Urea elution, 307–308
Uremia, children, 193–194
Urinalysis, 134, 141

Urinary excretion, 151–152, 195, 283
Urinary tract infection, 134
Uterus, hypoplasia, 120

Vasointestinal polypeptide (VIP), 9–11, 157
Vasopressin (AVP), 52–53, 195
Vitamin D, 151
Voltage-clamp studies, 21–22, 25
Voltage-sensitive Ca^{++} channels (VSCC), 18, 20–22, 24

Ward's triangle, 152
Weight gain, 127, 251, 325
Weight loss, 349
Weimaraner dogs, 121
Well-being, 262, 338–339, 343
Western ligand blotting (WLB), 281, 283, 307
Whittaker Mini-Mash Harvester, 105
Women, aging, 145–147, 151–152, 154, 203–204, 328–329, 335, 339
Wound healing, 97, 160

X-ray crystal structure, 236–238
X308 cell line, 261
XR-26 unit, 136

Yalow and Berson method, 176
Yeast, 87, 96

Zinc, 93, 117–118, 121, 123, 125–126, 233, 235
  deficiency of, 117–119
  thymus and, 117–119, 121, 123, 126
ZR-75-1 cell line, 260
Zymosan, 91–92

MIX
Papier aus verantwortungsvollen Quellen
Paper from responsible sources
FSC® C105338

If you have any concerns about our products,
you can contact us on
ProductSafety@springernature.com

In case Publisher is established outside the EU,
the EU authorized representative is:
Springer Nature Customer Service Center GmbH
Europaplatz 3, 69115 Heidelberg, Germany

Printed by Libri Plureos GmbH
in Hamburg, Germany